Advances in Meat Research – Volume 11

Production and Processing of Healthy Meat, Poultry and Fish Products

The *Advances in Meat Research* series reviews recent advances in meat science and technology. Each volume concentrates on one specific topic and discusses it in depth. The chapter authors are recognized as authorities in their fields and are drawn from around the world providing an international perspective.

The following volumes are also available:

Volume 6	Meat and Health
Volume 7	Growth Regulations in Farm Animals
Volume 8	Inedible Meat By-Products
Volume 9	Quality Attributes and their Measurement in Meat, Poultry and Fish Products
Volume 10	HACCP in Meat, Poultry and Fish Processing

Advances in Meat Research – Volume 11

Production and Processing of Healthy Meat, Poultry and Fish Products

Edited by
A.M. PEARSON
College of Agricultural Sciences
Oregon State University
Corvallis
Oregon
USA

and

T.R. DUTSON
College of Agricultural Sciences
Agricultural Experimental Station
Oregon State University
Corvallis
Oregon
USA

BLACKIE ACADEMIC & PROFESSIONAL
An Imprint of Chapman & Hall
London · Weinheim · New York · Tokyo · Melbourne · Madras

Published by Blackie Academic and Professional, an imprint of Chapman & Hall, 2–6 Boundary Row, London SE1 8HN, UK

Chapman & Hall, 2–6 Boundary Row, London SE1 8HN, UK

Chapman & Hall GmbH, Pappelallee 3, 69469 Weinheim, Germany

Chapman & Hall USA, 115 Fifth Avenue, New York, NY 10003, USA

Chapman & Hall Japan, ITP-Japan, Kyowa Building, 3F, 2-2-1 Hirakawacho, Chiyoda-ku, Tokyo 102, Japan

DA Book (Aust.) Pty Ltd, 648 Whitehorse Road, Mitcham 3132, Victoria, Australia

Chapman & Hall India, R. Seshadri, 32 Second Main Road, CIT East, Madras 600 035, India

First edition 1997

Typeset in 10/12 pt Times by Acorn Bookwork, Salisbury

Printed in Great Britain by T.J. Press, Padstow, Cornwall

ISBN 0 7514 0390 3

A catalogue record for this book is available from the British Library

Library of Congress Catalog Card Number: 97-66369

∞ Printed on acid-free text paper, manufactured in accordance with ANSI/NISO Z39.48-1992 (Permanence of Paper)

Contents

Contributors

T.F.T. Antonios Senior Registrar, Blood Pressure Unit, Department of Medicine, St George's Hospital Medical School, Cranmer Terrace, London SW17 0RE, UK

A.E. Bender Professor Emeritus, University of London, 2, Willow Vale, Fetcham, Leatherhead, Surrey KT22 9TE, UK

A.D. Clarke Department of Food Science and Human Nutrition, University of Missouri, Columbia, MO 65211, USA

J.N. Clarke Ruakura Agricultural Research Centre, Private Bag 3123, Hamilton, New Zealand

J.E. Collins Vice President, Scientific and Technical Affairs, American Meat Institute, 1700 N. Moore Street, Suite 1600, Arlington, VA 22209, USA

M.E. Dikeman Department of Animal Sciences and Industry, Kansas State University, Manhattan, KS 66506-0201, USA

S.J. Eilert Manager, Fresh/Processed Pork Research, Excel Product Development Center, 2901 N. Mead, Wichita, KS 67219, USA

D.L. Huffman Professor of Meat Science Emeritus, Food Technology Institute, 2 Comer Hall, Auburn University, Auburn AL 36849-5430, USA

Address for correspondence: 219 Deer Run Road, Auburn, AL 36832, USA

R.D. Huffman Manager, Meat Product Development, Koch Beef Co., P.O. Box 2256, Wichita, KS 67201, USA

A.H. Kirton Ruakura Agricultural Research Centre, Private Bag 3123, Hamilton, New Zealand

L. Leistner International Food Consultant, formerly Director and Professor of the Federal Centre of Meat Research, An den Weinbergen 20, D-95326 Kulmbach, Germany

G.A. MacGregor Professor of Cardiovascular Medicine, Blood Pressure Unit, Department of Medicine, St George's Hospital Medical School, Cranmer Terrace, London SW17 0RE, UK

R.W. Mandigo Professor, Department of Animal Science, University of Nebraska, P.O. Box 830908, Lincoln, NE 68583-0908, USA

F.J. Monahan Department of Food Science, Faculty of Agriculture, University College Dublin, Belfield, Dublin 4, Ireland

C.A. Morris Ruakura Agricultural Research Centre, Private Bag 3123, Hamilton, New Zealand

M.T. Morrissey Director, Oregon State University Seafood Laboratory, 250 36th Street, Astoria, OR 97103, USA

A.M. Pearson Professor Emeritus, Office of the Dean, College of Agricultural Sciences, Oregon State University, 126 Strand Agriculture Hall, Corvallis, OR 97331-2212, USA

J.F. Price Meat Laboratory, Department of Food Science and Human Nutrition, Michigan State University, East Lansing, MI 48824, USA

J. Quick Judith Quick and Associates, 1755 Edna Court, Tracy, CA 95376, USA

P.J. Shand Department of Applied Microbiology and Food Science, University of Saskatchewan, 51 Campus Drive, Saskatoon, Saskatchewan, Canada S7N 5A8

D.M. Smith Professor, Department of Food Science and Human Nutrition, Michigan State University, East Lansing, MI 48824-1224, USA
Formerly Head of Nutrition, Diet and Health Department, Institute of Food Research, Norwich, UK

D.A.T. Southgate 8 Penryn Close, Norwich NR4 7LY, UK

P.A. Speck Ruakura Agricultural Research Centre, Private Bag 3123, Hamilton, New Zealand

D.J. Troy Teagasc, The National Food Centre, Dunsinea, Castleknock, Dublin 15, Ireland

Preface

The central theme for this volume was chosen since consumers have great interest in purchasing low fat, low salt and reduced cholesterol meat, poultry and fish products. As in past volumes, experts in the field have been chosen to write chapters with emphasis on their breadth of knowledge in each specific area. Efforts were also made to obtain authors from different countries in order to give the book a worldwide perspective.

Chapter 1 stresses the nutritional and sensory properties that meat, poultry and fish products make to healthful diets and discusses consumer concerns about these products. Chapter 2 covers dietary recommendations in major consumer nations, along with data from food composition tables and the dietary contributions of meat, poultry and fish to meeting dietary needs. Chapter 3 discusses the labeling of low and reduced fat/salt products which, although written mainly from the US viewpoint, may serve as a model for labeling in other countries. Chapter 4 reviews the rationale for reducing fat-energy levels in muscle foods, problems encountered in their production and how these may be solved. Chapter 5 discusses the scientific basis for reducing the salt (sodium) content in food products and the health benefits derived from lowering salt intake. Methods of reducing the cholesterol content of these animal products is reviewed in Chapter 6.

Chapter 7 outlines procedures that can be used in producing animals and poultry with a reduced fat content by breeding and selection and also discusses the removal of excess fat from carcasses and cuts and its effect on fat content. Chapter 8 extends methods for reducing fat content to the influence of production practices, such as sex condition, stage of the growth curve, biological type and the effects of various growth promotants on composition. Chapter 9 reviews the possible use of various synthetic and mimetic fat replacers by the meat industry, explaining their structures, action and probable usefulness. Chapter 10 follows up by reviewing the plant and animal products that are currently being used or show promise for usage in producing low fat meat, poultry and fish products. Chapter 11 discusses production of low and reduced fat ground beef, drawing information from the literature and the authors' personal experience. Chapter 12 then follows up by covering production of low fat–low salt cured meat products with some examples of the calculations necessary to achieve the desired reduction.

Chapter 13 discusses overcoming flavor, color and texture problems that are often encountered in the production of low salt and low fat products. Chapter 14 reviews microbial problems that are inherent to low salt products and some procedures that can be used to reduce salt levels while still maintaining food safety and freedom from spoilage. Chapter 15 follows up by discussing production of low fat and low salt poultry products and some problems occurring in their production. Chapter 16 covers the fat content of various seafoods and the influence of processing on composition. This chapter also reviews the beneficial effects of fish lipids and concludes by discussing production of some special products, including surimi, fish protein concentrate, marine beef and fish protein hydrolysate. Finally, Chapter 17 reviews some microbial problems and the safety of reduced fat and low salt products. Emphasis is placed on the use of hurdle technology to improve the safety of these products.

A unique feature of the book is a section on research needs in each chapter, which provides readers with knowledge of the research needed in each area.

November 1996

A.M. Pearson
Corvallis, OR

1 Demand for healthful meat, poultry and fish products

D.A.T. SOUTHGATE

1.1 Introduction

Animal foods have been major components of the human diet for most of human evolutionary history. While it is not entirely clear whether man first scavenged meat or killed animals by hunting, the central role of meat in the diet is borne out by the large accumulations of animal bones with clear evidence of butchering marks on them. Leakey (1994) argues that the consumption of meat marked a key part in human evolutionary development since it provided a protein and energy-dense food and, thus, relieved the early humanoids of the continuous search for food in which the plant eater must engage. Estimates of the energy needs of primates show that the consumption of animal foods and the fats and proteins which they contain is critically important in determining the bulk of foods required in the diet (Southgate, 1991). The early human settlements, based around streams and lakes, would analogously have eaten fish as an important component of their diets. Thus, for early man the consumption of animal foods was a significant factor in the development of human society. The concept that these products constitute a health risk is, therefore, at odds with their role in the diet during development of human society.

This chapter first considers the role of these foods in the present diet and then examines the bases for health concerns regarding these foods as they impact on the consumer and consumers' attitudes to their food choices. Finally, the rationale for producing healthy animal foods and the response of the producers and processors to the consumer demand for healthy meat, poultry and fish products are considered. In this the term 'meat products' will be widened from the conventional definition to include carcass meats.

1.2 Role of meat, poultry and fish products in the diet

In the UK, the Reports of the National Food Survey, published annually by the Ministry of Agriculture, Fisheries and Food (MAAF), provide a series of records of food and nutrient consumption in the home over the past 50 years (MAAF, 1990). The report for 1994 (MAAF, 1995) shows that nearly 90% of households purchased meats and meat products during the week of the study. Of these, carcass meats made up 28% of purchases, with beef and veal making up 50% of the carcass meats, lamb 20.6% and pork 29.3%. The consumption pattern for meat products was more complex because the classification includes poultry (Table 1.1). Fish consumption was 145 g per head per week for 1994, distributed over a number of types.

1.2.1 Nutritional contributions

The nutritional role of these foods in the provision of animal protein is less prominent than previously but, nevertheless, the fact remains that meats and fish provide protein that has a high biological value. In many diets in the world, the proportion of animal protein is critical in determining the nutritional value of the diet for growing infants and children. The major nutritional focus has shifted to the provision of fat and other nutrients, especially iron, zinc and a range of trace elements. The details of the nutritional contribution of meat, poultry and fish in the diet are considered in Chapter 2 and here the focus is on the more general aspects of the role of these foods in the diet.

1.2.2 Social–cultural aspects

Focusing on the nutritional aspects of meat discounts the very important social–cultural role of these foods in the diet (Ebling, 1982; Fiddes, 1994; Murcott, 1995). Hunting animals of any size involves the development of tracking and stalking skills and, with the larger animals, a considerable element of danger to the hunter, both in killing the animal and protecting it from scavengers. Cave paintings provide evidence for the involvement of magical or early religious rites associated with animals and hunting. They also reflect the power of the animals and the awe which they attracted. Most present-day hunter–gatherers exhibit respect for the animals which they are fortunate enough to kill. There remains some residual evidence of this in the customs of some European hunters at the present time. Thus, to early man animal foods were associated with a prestigious activity, usually restricted to the adult males, and the food which it provided acquired a high status.

The socio-cultural significance of meat and other animal foods in the diet is probably the most important determinant of meat consumption and in many respects is very deeply rooted. Thus, if there are real health concerns about the consumption of meat, there is in a sense an obligation to develop strategies to respond to these concerns because it is probable

Table 1.1 Patterns of consumption of meat, poultry and fish in the UK for 1992, 1993 and 1994

	1992		1993		1994	
	Consumption (g/head/week)	Contribution to fat intake (%)	Consumption (g/head/week)	Contribution to fat intake (%)	Consumption (g/head/week)	Contribution to fat intake (%)
Meat:						
Beef and veal	141	na[b]	133	na	131	na
Mutton and lamb	71	na	166	na	54	na
Pork	72	na	80	na	77	na
Total carcase meats	284	9.3	279	9.4	262	6.6
Bacon and ham	77	na	77	na	77	na
Poultry, uncooked	216	na	222	na	208	na
Other meat products	373	na	379	na	395	na
Total meat and meat products	950	24.8	956	25.4	943	23.3
Fish:						
Fresh	29	na	30	na	29	na
Processed and shellfish	16	na	15	na	15	na
Prepared	54	na	56	na	56	na
Frozen and processed	43	na	44	na	44	na
Total fish	142	na	144	na	145	na

[a]Source: MAFF (1995).
[b]na indicates that information was not given in the publication.

that implementing dietary changes which displace these foods from their central role in the diet is likely to be very difficult indeed (Cuthbertson, 1980).

1.2.3 Religious aspects

There also developed, alongside this, religious concepts of which animals could properly be considered food and which were regarded as not appropriate and proscribed in religious canon. In both Judaism and Islam, the religious constraints also extended into prescribing how the animals should be slaughtered and the associated religious rites for the slaughter of animals for food. In some cultures the religious laws developed to proscribe the taking of life in any form and, thus, prohibited the killing of animals for food. This reverence for all living things and the refusal to accept that humans are uniquely separated from the rest of the animal kingdom could be seen to lie at the basis of the vegan dietary code and as the foundation of vegetarianism (Webster, 1994; Murcott, 1995).

In most western cultures, despite the growing influence of the development of multicultural societies, meat and, to a lesser extent, fish play a central role in the structure of the diet. Their consumption is identified with a 'proper' diet. In menu planning, the initial focus is on these foods with the selection of other items based around the meat or fish. Thus, on most restaurant menus, the first selection is the type of meat or fish, with vegetables virtually being regarded as accessories to add colour and other textures to make the plated meal more attractive. This, of course, represents what might be regarded as the dietary customs of the affluent, but in essence it also applies to many of the less affluent peasant types of food cultures where the animal food is an important, albeit quantitatively minor, part of the dish, and cheaper cereal foods in the form of pasta, rice, flour-based dumplings or potatoes serve to extend the animal component of the diet.

1.3 Health concerns

Despite clear evidence for meat and other animal foods as long-standing and central components of the human diet, in recent years a number of questions have been raised about the safety of unlimited consumption of these foods. The most significant of these has arisen regarding nutritional concerns about the consumption of these foods in relation to diet and health. In addition, there have been concerns raised by environmental groups about contamination and the use of growth promoters and additives.

1.3.1 Changing nutritional concerns

In the first half of the twentieth century, the major concern of the nutritional sciences in relation to diet and health of the population was ensuring the provision of an adequate supply of nutrients, especially protein and energy. These concerns still remain of paramount importance in many parts of the world, although one would now argue that the emphasis placed on protein is less prominent than that given to ensuring the adequacy of the energy supply.

In the 1930s, as knowledge of the vitamins and mineral nutrition expanded, the nutritional priorities for a healthy diet were enlarged to include the provision of 'protective foods', those that provided minerals and vitamins: dairy foods for vitamins A and D, calcium and riboflavin, cereals for thiamin and niacin and fruits and vegetables for vitamin C and the B vitamins. Meats and fish were seen as important for the provision of first-class proteins of high biological value, iron, zinc and the B vitamins, especially vitamin B_{12} which is the sole source in the diet. It is interesting to note even in the early 1930s that Sherman (1933), when discussing dietary patterns in the USA, argued that the consumption of meats should be reduced in order to allow increased expenditure on the protective foods.

The focus on diets to prevent nutritional deficiencies continued until well after the Second World War and even in the 1970s was evident in official nutritional publications (Department of Health, 1978). However, there was growing evidence at that time that the pattern of mortality in many developing countries was changing; improved nutrition and especially the reduction in the mortality from infective diseases due to the use of antibiotics had resulted in substantial increases in life expectancy at birth, whereas the incidence of deaths due to chronic degenerative diseases, such as coronary heart disease (CHD) and cancers, was rising. Furthermore, there was evidence that the mortality from CHD in men was occurring earlier.

There was evidence that environmental factors were involved in the changes in mortality and concerns that diet might be among these factors resulted in a shift of nutritional concerns in the diet and health context away from nutritional deficiency towards concerns about dietary patterns since there were profound differences in the incidence of these chronic degenerative diseases internationally which could not be attributed to genetic differences (Department of Health and Social Security, 1984; National Research Council, 1989).

One common observation in the developing countries from the early 1970s with growing affluence (in fact, the incidence of CHD was shown anecdotally to be related to many indicators of affluence, such as the number of television licences purchased in the UK). With the increase in the availability of private cars and the mechanization of industry, there was a rise in the incidence of obesity, probably the consequence of an abundant food supply and a reduction in energy expenditure in the workplace and in leisure activities, the implication being that the over-consumption of food energy *per se* was the primary dietary factor associated with the rise in the incidence of degenerative diseases. While obesity remains a risk factor for many degenerative diseases, epidemiological studies, however, began to indicate that the pattern of the diet, especially the proportion of the energy intake derived from fats, was probably more important than the imbalance in energy intake and expenditure.

1.3.2 Total fat intake

One of the seminal epidemiological studies was the so-called 'Seven Countries Study' undertaken in Europe and initiated by Ancel Keys (1970, 1980). This study measured dietary intake and serum cholesterol levels in seven countries in Europe which showed substantial differences in mortality from CHD. The study showed a highly significant relationship between dietary fat intake as a proportion of the energy intake and serum cholesterol levels, which from other studies seemed to be an important predictor of coronary heart diseases risk (Keys, 1957). This study provides the foundation for much of the current dietary advice about reducing fat intakes and the concerns about cholesterol levels, although many epidemiological and clinical observations since this initial study have been reported. It is interesting to note that in this study the southern Mediterranean countries, especially Crete, appear anomalous in having a high fat intake but low serum cholesterol levels, and this was the origin of the concept of the 'Mediterranean diet' as protective and the clear indication that the total fat intake is not the only factor which needs to be considered. The fatty acid composition of the fat was also shown to be critical.

Subsequent studies have substantially confirmed that the total fat intake is a risk factor for CHD, primarily because of its effects in raising serum cholesterol levels, although it is recognized that the saturated component of fat intake is the most potent factor (National Research Council, 1989; Department of Health, 1994).

There also are other concerns about the total fat intake which relate to obesity and to the incidence of some cancers. Some of these concerns will be addressed subsequently.

(a) Obesity and Fat. Obesity is a risk factor for many chronic diseases and it is now clear that fat intake is of major importance. Clearly, for obesity to develop the energy intake must exceed expenditure and, except for a very small number of individuals with genetic metabolic faults,

obesity in most individuals arises because of the over-consumption of food energy coupled with inadequate physical activity. Whereas in the past the over-consumption of carbohydrate energy was given undue focus, it is now clear that the body has mechanisms which compensate for the over-consumption of carbohydrate by increasing carbohydrate oxidation (and suppressing fat oxidation). The capacity to increase fat oxidation under comparable circumstances is very limited and excess fat is deposited as body fat (Jequier, 1995). The prevention of obesity, which appears to be the only public health strategy (National Research Council, 1989) that has a chance of success, therefore, depends on matching food energy intake to expenditure and reducing fat consumption and increasing the proportional intake of carbohydrates (Department of Health, 1994).

(b) Fat and cancer. Drasar and Irving (1973), in a meta-analysis of epidemiological studies of cancer, showed that the incidence of colonic and breast cancers was highly correlated, and the data showed a strong relationship with total fat intake. Despite some initial concerns about the validity of meta-analyses and the variable quality of the data, especially the data on fat intake, this relationship has been confirmed in many subsequent studies (National Research Council, 1989; World Cancer Research Fund, 1994), most strongly for colonic cancer and more weakly with pancreatic and breast cancer. Although such studies suggest that the saturated component is most strongly involved, there is experimental evidence from animal studies that excessive intake of polyunsaturated n-6 fatty acids may promote carcinogenesis, whereas n-3 fatty acids have an inhibitory effect.

In a recent expert review of the literature, a group convened by the World Cancer Research Fund (1994) was of the opinion that the intake of total dietary fat was strongly implicated as a risk factor for cancer of the colon, rectum, pancreas and endometrium. The group also cited excessive energy intake and obesity as significant risk factors, especially for pancreatic cancer, the hormone-related cancers of the endometrium and post-menopausal breast and prostatic cancer.

1.3.3 Composition of fat intake

(a) Effects of fatty acids on serum cholesterol. The evidence with regard to the effects of fat intake on serum cholesterol levels and the incidence of CHD show that the hypercholesterolaemic effects are strongly dependent on the composition of the fat, particularly the proportions of saturated (SFA), monounsaturated (MFA) and polyunsaturated fatty acids (PUFA) (National Research Council, 1989). Early dietary advice in this respect focused on the polyunsaturated to saturated (P/S) ratio. A formula proposed by Keys (1957) was designed to predict the change in serum choles-

terol concentration from the proportion of SFAs and PUFAs in the diet. This formula has undergone many modifications since it was first proposed. Current evidence suggests that the effects of fatty acid composition on serum cholesterol are much more complex (Ulbricht and Southgate, 1991). This is due to two main reasons: first, the form of cholesterol in the serum has a critical influence on the development of arteriosclerosis, thus low density lipoprotein cholesterol (LDLc) is the form involved in development of plaques in the arterial wall, and high density lipoprotein cholesterol (HDLc) appears to be protective. The second reason is that the effects of the saturated fatty acids on raising LDLc depends on chain length, thus C_{12} (lauric), C_{14} (myristic) and C_{16} (palmitic) are most active, with the lower and higher chain lengths having little or no effect on serum cholesterol levels. There is, in addition, growing evidence that the MFAs, which were previously discounted in their influence on cholesterol, now appear to have effects in raising HDLc levels. Hence MFAs should be regarded as being advantageous and included in any assessment of fat intake on CHD (Mattson and Grundy, 1985; Ulbricht and Southgate, 1991).

(b) Effects of trans-*fatty acids.* The dietary fats also contain minor amounts of *trans*-monounsaturated fatty acids derived from the hydrogenation of vegetable and fish oils in the production of margarines and cooking fats. The meat and milk fats from ruminant species also contain some *trans*-fatty acids formed by hydrogenation in the rumen. The *trans*-fatty acids appear to have a cholesterol-raising effect that is smaller than that of saturated fats. There also is some evidence that they may compete in the metabolic pathways with the essential fatty acids (Department of Health, 1994). An epidemiological study (Willett *et al.*, 1993) showed a relationship between the consumption of *trans*-fatty acids in margarines with CHD, but not with the *trans*-fatty acid isomers found in meats.

(c) Factors associated with coronary heart disease. It should be recognized that CHD has two elements in its aetiology: (1) atheriosclerosis, the development of arterial plaques and the narrowing of the arterial lumen, and (2) thrombosis, the formation of a thrombus which blocks the lumen, leading to the myocardial infarction. The influence of fatty acids on the thrombotic events depends considerably on the ratio of the n-6 and n-3 PUFAs. Ulbricht and Southgate (1991) suggested that fat intakes should be judged by two indices, an atherosclerotic index and a thromobotic index.

In any discussion of the relationship between diet and disease, it is important to recognize that there will be an interaction between diet and the genetic constitution of the individual, and that in any population a number of early deaths will be due to familial hyperlipidaemias. Furthermore, there is a range of risk factors that increase the risk of CHD, which include smoking and hypertension (Ashwell, 1993). The evidence suggests, however, that the increased risk due to fat is comparable to that arising from smoking or hypertension (which itself may have a dietary component in its aetiology).

1.3.4 Cholesterol and coronary heart disease

High serum cholesterol is a risk factor for CHD, which appears to be causal (Haq *et al.*, 1995). The relationship is so unequivocal that it is possible to estimate the rise in CHD incidence which is associated with the population change in serum cholesterol (Marmot, 1994) and, conversely to estimate the benefits that would follow from a reduction in serum cholesterol.

When the relationship began to emerge, the initial nutritional focus was on the amount of cholesterol in the diet and dietary advice reflected the desirability of reducing the intake of cholesterol, which is derived exclusively from foods of animal origin. It is now clear that the endogenous synthesis of cholesterol is of much greater magnitude than that derived from the diet and, moreover, that in most individuals there is a feedback regulation of cholesterol synthesis, which responds to the level of intake. Most population studies show no relationship between dietary cholesterol intakes and blood serum levels (National Research Council, 1989). Moreover, it is evident that dietary intake of fat is more significant in influencing the levels of the serum LDLc, which is involved in arterial plaque formation, in that oxidized LDLc seems to be most important. There is considerable individual variability in the response to dietary cholesterol and serum LDL levels. There is evidence that there may be some individuals who are responsive to dietary cholesterol. For this reason, most dietary advice recommends moderation in cholesterol intakes.

1.3.5 Salt – relationship to stroke and coronary heart disease

In epidemiological studies, there is a significant positive association between habitual sodium intake (which is derived principally from salt in foods, but there are also contributions from other sodium salts, such as carbonate, phosphates and monoglutamates) and blood pressure (Elliot *et al.*, 1996). Raised blood pressure is an important risk factor for stroke and CHD. Sodium intakes in most developed countries are greatly in excess of the physiological requirements, and it has been argued that health benefits would follow from a reduction to about 25% of the current intake of sodium. The situation is slightly more complex than might at first appear since epidemiologically there is evidence for an interaction between sodium and potassium, suggesting that it is the ratio between sodium and potassium intakes which is the critical determinant of blood pressure. Current dietary advice (Department of Health, 1994), therefore, includes the recommendation to increase potassium intake (the major sources of potassium in the diet are plant foods) alongside the recommendation for a reduction in salt intake.

1.3.6 Protein – meat consumption and cancer

The evidence for an association between protein intake and cancer is difficult to establish by epidemiological studies because protein intake is so closely linked with energy and fat intakes. Some of the earlier studies showed associations between the intake of animal protein and colonic and breast cancers (Drasar and Irving, 1973). Armstrong and Doll (1975) also reported an association between consumption of total protein, e.g. animal protein (meat and fish), with cancers at a range of sites. These associations have been confirmed in other studies. Nevertheless, it is difficult to establish an independent role of protein *per se* on the incidence of cancer (World Cancer Research Fund, 1994). Willett *et al.* (1990) found that the risk of colonic cancer was strongly associated with the consumption of red meats (beef and lamb), but not of low fat chicken or fish. These findings suggest an association with red meat rather than with animal protein. The mechanisms whereby protein may influence carcinogenesis are obscure, but may relate to its effects on bile secretion or to an increase in nitrogenous material entering the colon and raising ammonia levels. The carcinogenic properties of meat rather than protein may be due to the range of mutagenic compounds formed in the cooking of meat (Chen, Pearson and Gray, 1990) and to the preservatives present in preserved and cured meats (NAS, 1981).

Comparisons between the incidence of cancers in vegetarian and meat-eating groups are often confounded by the large number of differences in lifestyle, which characterize the groups. In addition, their dietary patterns may also tend to differ in many respects other than the consumption of animal foods (National Research Council, 1989).

1.3.7 Contaminants

The origin of these concerns lies not in any specific demonstration of risk but in more general environmental concerns about the contamination of the environment with industrial chemicals and pesticide and herbicide residues. With intensely reared animals, there have also been concerns about the persistence of veterinary medicines in meat and milk and antimicrobial and antifungal agents in fish which have been farmed.

(a) Polychlorinated biphenyls (PCBs). These are highly persistent chemicals and exposure may occur through food, water and air. The major

route for these into meats and poultry is probably from fish used in animal feed. In extreme exposures through industrial accidents these substances have been associated with a range of diseases, including liver cancer. It is accepted that there is evidence for limited cancer risk from PCBs and regular monitoring of them in the food chain is carried out in most developed countries.

(b) Polyvinyl chloride (PVC). PVC contaminates food primarily by contact with packaging materials. Levels in meats and meat products are generally low, except where fatty products have been packed or microwaved in unsuitable cooking films. There is a little evidence that PVC is carcinogenic in man.

(c) Polycyclic aromatic hydrocarbons (PAHs). A large number of PAHs have been identified as contaminants of foods. The major source of the contamination arises during the smoking and cooking of foods, especially grilling and roasting of meats and fish. Grilling meat or fish over an open flame in which the fat drips on to the flame and is pyrolysed contributes significantly to human dietary exposure. Benzopyrene levels as high as 50 $\mu g\ kg^{-1}$ have been detected in charcoal-grilled steaks (National Research Council, 1989). A very wide range of PAHs are formed by pyrolysis. Smoking of foods also exposes the food to a wide range of pyrolysis products. The risk can be avoided through alternative processing and cooking methods.

Many of the PAHs found in the environment are mutagenic (Ames, 1983) and many have been shown to be carcinogenic in animal studies, especially benzopyrene, dibenzanthracene and benzanthracene. These and the heterocyclic amines imidazole quinoline and phenylimidazole-pyrimidine formed from amines and amino acids during cooking may account for the association between colonic cancer and meat and fish consumption in epidemiological studies.

(d) Herbicide and pesticide residues. Many of these are highly persistent and remain in the soil, from where they contaminate crops and animal feeds. The major pesticide residues of concern are dioxin, phenoxyacetic acid herbicides and a range of organochlorine compounds, which tend to accumulate in animal fat depots. The cancer risk of many herbicides has not been fully evaluated, but dioxin and the phenoxyacetic acid herbicides have not always been associated with increased cancer mortality. However, it is considered that many of these compounds do pose a cancer risk to man and the use of many of the chlorinated compounds is no longer permitted. In the UK, regular monitoring of the food supply is carried out, especially of imported foods, because the residual stocks of many chlorinated materials are still considerable. There is, however, little

evidence to indicate that the levels in animal products, except from wilful or accidental misusage, pose risks to the health of the consumer. Dioxin can be formed from the incineration of plastic wastes and may be present in flue ash or contaminate pasture downwind from incinerators. This is a further reason for regular surveillance of levels in the food supply (MAFF, 1995).

(e) Antibiotic residues. Two issues of concern arise here: first, the persistence in foods of veterinary residues used to treat animals, and second, the use of antibiotics in feeds to promote growth. The concerns are primarily whether these antibiotic residues would encourage the development of resistance in bacteria and so limit the therapeutic use of antibiotics in human medicine. As a general principle the antibiotics used in veterinary practice are not those in human use, but the appearance of resistant intestinal organisms has led to the indiscriminate use of antibiotics in feeds being prohibited or restricted. There is little or no evidence of risk to human health being attributable to antibiotic residues in meats, poultry or fish products, but the levels in the food supply are regularly monitored in most developed countries to detect misuse.

1.3.8 Hormones

A number of hormones and synthetic analogues of natural sex hormones were widely used as growth promoters in beef, lamb and poultry. For example, diethylstilbestrol (DES) was widely used in the USA under defined conditions for beef and lamb. The hormone was usually given as an implanted depot, which was placed in a part of the animal not usually eaten as human food. There was no evidence of human cancer from the consumption of food, but the role of DES in cancer in women has led to its use being discontinued. About 20 growth hormones are still in use in the USA. Despite the lack of diseases attributable to the use of these hormones, the possibility that depot sites may appear in foods has led to the prohibition of the use of hormones as growth promoters in the European Union, although regular monitoring is still the rule in the UK and other countries to detect illegal usage.

1.3.9 Additives

Some substances are added intentionally to foods during processing. The most important in the context of animal foods are preservatives and antioxidants. In the European Union and the USA and most developed countries there are elaborate regulatory procedures which control the use of additives, and the use of an additive is conditional on the demonstration of a need for the additive for technological or food quality reasons coupled with evidence that the additive is safe based on extensive toxicological studies in animals.

(a) Preservation of cured meats with sodium chloride and sodium nitrite. There is growing evidence that the incidence of gastric cancer is linked with excessive consumption of preserved meats (and some vegetables). The development of gastric cancer is believed to follow as a sequence following superficial gastritis and infection with the organism *Helicobacter pylori.* This proceeds to an atrophic gastritis with a rise in pH, which permits colonization with other bacteria that produce mutagens from dietary constituents. The production of *N*-nitrosamines from nitrates in vegetables and nitrites in the preserving salts used in meats are believed to be the initiators. Nitrite can react with a range of amides, amines and other proteins to form *N*-nitroso compounds. These are highly mutagenic in animals, although this has not been shown unequivocally in humans.

Epidemiologically the associations are difficult to establish because of the complexity of the interactions; for example, the presence of vitamin C, a powerful reducing agent, may protect against nitrosamine formation. The high concentrations of salt used in preserved meats may act as an irritant and are capable of producing atrophic changes. In addition, they could lead to the induction of enzymes involved in DNA synthesis and cell replication, thus acting as a tumour promoter (World Cancer Research Fund, 1994).

(b) Phenolic antioxidants. Butylated hydroxyanisole (BHA) and butylated hydroxytoluene (BHT) are widely used in food products at low levels as antioxidants and preservatives to prevent lipid oxidation. There is no epidemiological evidence to suggest that their consumption is related to the development of cancers. BHA is not genotoxic, and there is some evidence from animal studies that exposure to BHA before exposure to some PAH carcinogens may confer protection. BHT is similarly not genotoxic and any antitumour properties may result from stimulation of liver detoxifying systems (National Research Council, 1989).

1.3.10 Microbiological safety

Most of the health risks discussed so far relate to the chronic effects of consuming foods, and in many people's perception, because they lie in the future, are discounted in relation to other risks (Conning, 1988). The effects of microbial contamination of foods are usually obvious and immediate. Their effects on consumer demand follow a similar time course, which may be very long-lasting and potentially damaging to the industry (Waites and Arbuthnott, 1990).

(a) Sources of contamination. The tissues of meat and fish are ideal media for bacterial growth and the maintenance of good manufacturing practice is essential to maintain hygienic standards. In the healthy, living animal the muscle tissue of animal foods is essentially free from contamination and most contamination occurs during transport, slaughter and subsequent processing. Contamination of the hides of animals during transport (Madden, 1994) leads to transfer to the flesh during slaughter and the initial butchering stages. Contamination of poultry with gastrointestinal organisms is effectively facilitated by the processing procedures (Baird-Parker, 1990). Fish and fish products tend to be less microbiologically stable and contamination from their environment during handling is an ever-present potential hazard. Contamination with heavy metals, bacteria and viruses of molluscs and crustacea from their environment represents a significant food safety hazard (Appleton, 1990).

Despite growing awareness by the industry of the need for strict microbiological safety, growth in the number of food poisoning outbreaks in the UK and other industrialized countries (Cooke, 1990; Todd, 1990) leads one to the assumption that food handling in the home, and possibly to a lesser extent in catering establishments, is inadequate. This indicates the need for improved consumer education about the safe handling of animal foods.

(b) Bovine spongiform encephalopathy. A further concern has arisen in the UK concerning the emergence of bovine spongiform encephalopathy (BSE) in cattle. This disease is believed to have arisen from the use of rendered tissues from sheep infected with scrapie in cattle feeds. The disease has the characteristics of a 'scrapie-like' prion disease from which the infective agent has resisted attempts at isolation (Collee, 1990). By analogy with scrapie, the risk of transfer to humans via an oral route from beef appears highly improbable, although transfer to several other ruminant and non-ruminant species appears to have taken place. As a precautionary move, infected animals have been slaughtered and removed from the food chain and the tissues most likely to be involved in transmission (brain, spleen, spinal cord, thymus, intestine and tonsils) are no longer permitted in foods. Mechanically recovered meat is also excluded. The very unusual and poorly understood nature of the infective agent has made unequivocal statements on the safety of beef difficult (Anon., 1995). Concern that a related human disease, Creuzfeldt–Jakob disease (CJD), may be linked with BSE has provoked the media and some of the scientific community to express concern about the safety of beef. However, a critical examination of the evidence, and, by analogy, with scrapie, which has been present in sheep for some 250 years, suggests that the human risks from eating beef are very small indeed (Tyler, 1995). Monitoring of

the incidence of CJD has, however, recently identified the emergence of a new form of the disease which may be associated with the consumption of beef products before the selected offals ban was introduced (Will *et al.*, 1996). The risk from BSE, however, is nevertheless still estimated to be very small. Nevertheless, consumer concerns may not be dispelled by rational arguments because of the complex ways in which risks are perceived and form the basis for action (Wheelock, 1988). Food fears can be exploited by the media or activists with other agendas (Gofton, 1990).

1.4 The impact of health concerns on the consumer

So far, the health concerns have been presented in the context of the scientific evidence drawn from epidemiological studies and related biochemical evidence for the mechanisms involved. It is now necessary to consider how this scientific evidence has been presented to the consumer.

1.4.1 Presentation of dietary and health information and advice

The first major attempt at presenting dietary advice in this context came with the McGovern Report on 'Dietary Goals for the United States' (Senate Select Committee on Nutrition and Human Needs, 1977). This report presented the evidence for a link between changes in the US diet and the changing incidence of chronic diseases and set out the dietary changes which were considered necessary to reverse the increasing incidence. Initially the report included cost–benefit analyses in relation to health costs. The Committee used population data for dietary intakes and recommended changes that would alter the average diet. The recommendations are summarized in Table 1.2. They were nutrient based but broad recommendations regarding the changes in food selection that were required. The major impact of these recommendations in relation to meats, poultry and fish derives from the recommendation to reduce the consumption of total fat and to reduce the proportion from saturated fat. This would require reduced consumption of red meats and dairy products, which were among the major sources of total fat and saturated fats in the USA, and increased consumption of poultry and fish. The recommendation with regard to salt was of importance for the consumption of meat products, especially preserved meats. The other dietary changes related to the replacement of the energy from fat with complex carbohydrate-containing foods. Subsequent comment drew attention to the fact that translation of the recommendations in terms of foods at the table would have implied substantial dietary changes (Peterkin, 1978; Pao, 1981). Similar reports followed in other countries (UK, Health Education Council, 1983; USA, Surgeon General, 1988).

Table 1.2 US dietary goals[a]

Nutrient	Recommended changes	Changes in food selection
Carbohydrate	Increase consumption to 55–60% of energy intake	Increase consumption of fruits, vegetables and whole grains
Fat	Reduce total from about 40% to 30% of energy intake	Decrease meat consumption, increase consumption of poultry and fish
Composition of dietary fat	Reduce saturated fat to about 10% of energy intake. Balance intake of polyunsaturated and monounsaturated fatty acids to about 10% of energy intake each	Decrease consumption of fat-rich foods; partially substitute polyunsaturated fat for saturated fat. Substitute non-fat milk for whole milk
Cholesterol	Reduce intake to about 300 mg per day	Decrease consumption of butterfat, eggs and other high-cholesterol foods
Sugar	Reduce consumption to about 15% of energy intake	Decrease consumption of sugar and sugar-rich foods
Salt	Reduce to approximately 3 g per day	Decrease consumption of salt and salt-rich foods

[a]Source: Senate Select Committee on Nutrition and Human Needs (1977).

The Health Education Council's (1983) recommendations were, however, expressed in more quantitative terms of foods, and it is interesting to note that they concluded that the reduction in meat fats to achieve the target was only 3 g per head per day.

All the reports tended to be nutrient focused and the suggestions regarding the actual changes required at the individual food choice level were in many respects naive. In part, this was due to the use of population averages to determine risk (which was to be expected from the epidemiological evidence base) that required the individual to establish his or her dietary patterns in relation to the national average. The Department of Health and Social Security's (1984) report on diet and cardiovascular disease did recognize that dietary changes would have to be made at the individual level, but this posed problems in the quantitative interpretation of the changes that were being advised (Cottrell, 1985). One of the major difficulties for the consumer, when interpreting the recommendations, lay in the expression of the fat recommendations in terms of percentages of the total energy intake (Leverkus *et al.*, 1988). For this to act as a guide, the consumer requires a sophisticated calculation based on a knowledge of the fat content of the portions consumed together with some running forecast of the day's energy intake.

(a) Recommendations on diet and health. The National Research Council (1989), in making its recommendations for diet and health, was

Table 1.3 Recommendations of the National Research Council[a]

Nutrient	Recommendation	Implications for food choice
Energy	Balance intake of food energy and energy expenditure to maintain appropriate body weight	
Fat	Reduce intake to 30% of energy intake	Replace fatty meats and whole milk dairy products with fish, poultry, lean meat and low fat dairy foods
Composition of dietary fats	Reduce saturated fatty acids to 10% of energy intake	Choose more vegetables, fruits, cereals and legumes
Cholesterol	Reduce intake to less than 300 mg per day	Limit consumption of fats, egg yolks and fatty foods
Carbohydrates	Increase to 55% of energy intake primarily as complex carbohydrates	Eat every day five or more servings of a combination of vegetables and fruits. Increase consumption of starches and complex carbohydrates by eating six servings of breads, cereals and legumes per day
Protein	Maintain at moderate intake	
Salt	Limit to 6 g per day or less	

[a]Source: National Research Council (1989).

more explicit about the amounts of plant foods which should be eaten (Table 1.3). However, the recommendations regarding meats were still couched in rather general terms and the fat recommendation was still based on the percentage of the energy intake. The overall composition of the macronutrients in the diet was built into the dietary recommendations for nutrients in the Department of Health (1991) report on Dietary Reference Values. Here again, the percentage of the total energy intake was used as the base.

In the report on Diet and Cardiovascular Disease by the UK Department of Health (1994) these recommendations were reinforced, but the report did provide indications of the extent of the changes which would be required in the average food intake patterns. The recommendations regarding meat were to maintain the consumption of beef and lamb but to choose lean cuts and, where there was a consumer preference for a fatty cut, to remove the fat on the plate and discard it. Pork and poultry were considered together and the advice required no change. The recommendation was to increase fish consumption to one portion of white fish and one of oily fish per week. The major change of importance to the meat industry was to recommend a halving of intake of processed meat products, and this carried an assumption that the salt content of meat products could be reduced by 30%. The diet and health messages to the consumer, therefore, have gathered momentum and some consistency in as much as they relate to meat, poultry and fish:

1. to consume a variety of foods and to try to balance energy intake and expenditure to maintain an appropriate body weight, usually assessed as a body mass index (height in metres divided by the body weight in kg squared) within the range 20.0–25.0 (Garrow, 1981);
2. to reduce total fat intake to provide between 30 and 35% of total energy intake; the lower value is favoured by those concerned with reducing cancer risk;
3. to limit the percentage of energy from saturated fatty acids to around 10% of energy;
4. to ensure that the diet contains long-chain n-3 polyunsaturated fatty acids;
5. to limit the consumption of preserved (salted and smoked) meat products; this primarily relates to reducing cancer risk;
6. to limit sodium consumption and to increase potassium intake;
7. the recommendations for plant foods relate, first, to their role in providing carbohydrate to replace the reduction in the energy from fat, and second, to their role in the provision of protective factors for both CHD and cancers (vitamins, dietary fibre and non-nutrient antioxidant and other biologically active constituents).

Although these recommendations have been primarily focused on the prevention of CHD, the growing evidence regarding the role of diet in cancer (World Cancer Research Fund, 1994) indicates that these dietary recommendations are also highly appropriate for reducing cancer risk.

(b) Pros and cons of recommendations. The advice has not been without its critics. This criticism has been argued from a number of premises: first, that life expectancy has increased substantially in the latter half of this century, and it is therefore difficult to say that the diet which was eaten over that period was unhealthy (Harper, 1980); second, that the dietary intake data used in many of the epidemiological studies are of low quality, or are based on food disappearance data, which are a poor indicator of actual food consumption; this is a valid point because the data are measured at the food commodity level, and therefore overestimate food consumption very substantially; and third, there is concern about the role of dietary fats in the aetiology of both CHD and cancers. It is true that the aetiological mechanisms involved are highly complex and in many cases not fully understood (National Research Council, 1989), and there is a major difficulty in developing ways of understanding how the various factors in the diet are integrated into the disease response (Southgate, 1991). In all the responses of an individual to diet, there are diet and gene interactions and with many chronic diseases the time-scale required for a

dietary pattern to exert its effects is often unknown. One significant criticism is that intervention studies, for example the reduction of fat intake, have usually failed to produce the expected changes in cholesterol levels (Haq *et al.*, 1995). Despite these valid criticisms the evidence for a relationship between diet and disease, however, is consistent in supporting the current advice summarized above.

1.4.2 Impact on consumer choices

Food choice is measured by the purchase of foods rather than the stated intention to purchase, and is governed by a number of factors, of which concern about diet and health is only one. This assumes that the consumers see the advice as relevant to themselves and accept or reject the risk that discounting the advice implies and can interpret the advice correctly.

Shepherd (1990) has summarized some of the more identifiable factors involved in a flow diagram (Figure 1.1). Many of the factors involved in the consumption of meat are significant in this diagram.

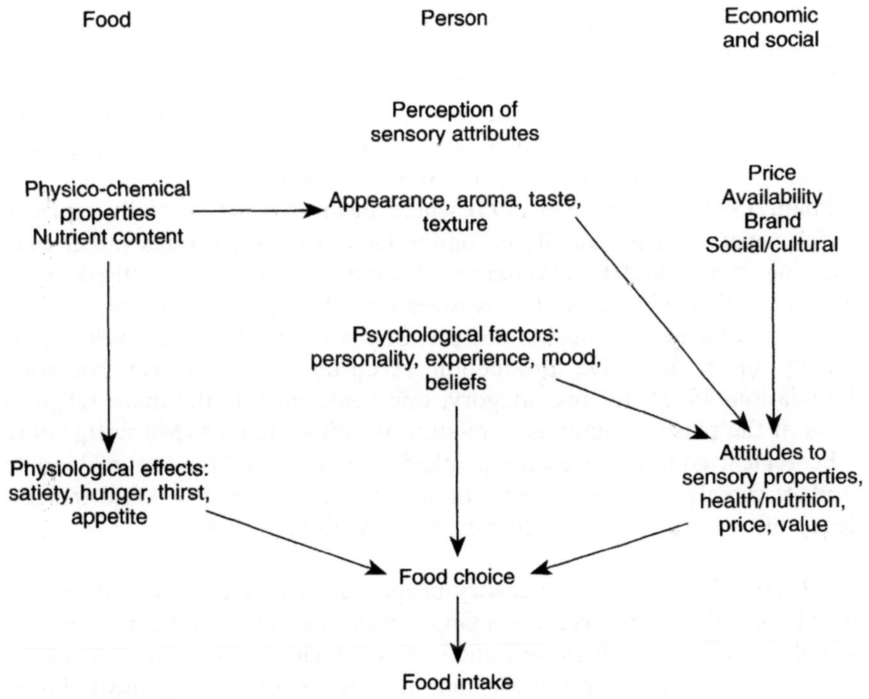

Figure 1.1 Diagram showing the interrelationships between the factors affecting food choice: categorized into those relating to foods, the person and socio-economic factors. (Source: modified from Shepherd, 1990.)

(a) Economic and social factors. The special role that meat has in most human dietary cultures was noted earlier and this, coupled with economic factors, is very important. Thus, the price elasticity of beef, for example, is high, whereas that of other less prestigious meats is less so. The decline in the consumption of beef in the UK over the past few years is probably explicable in terms of increases in price, and the same is true to a lesser extent for lamb. In contrast, the price of poultry has fallen, so what in the 1940–50s were prestige foods reserved for special occasions have become cheap foods with increased consumption. Although dietary advice may have reinforced this trend, it was largely of economic origin (MAFF, 1990).

Social and cultural factors regarding meat consumption are very important in this respect and the perceived importance of meat in the meal structures has been mentioned earlier. There are a number of factors which are acting as negative influences on meat consumption. Among these are cultural changes in lifestyle, the increasing numbers of women working outside the home, changes in family size and the radical changes in family meal patterns (MAFF, 1990), which reduce the number of occasions on which the family eat together. This has led to the increasing use of convenience products and light snack-type meals being eaten at an increased frequency.

Vegetarianism appears to be increasing and becoming more socially acceptable, as judged by the increasing appearance of vegetarian options on menus. In this respect, it appears that many people are in a transitional stage from 'demi-vegetarianism' (Richardson, Shepherd and Elliman, 1993; Richardson, 1994) where meat is being partially replaced by full vegetarianism, usually becoming lacto–ovo-vegetarians rather than adopting the complete avoidance of animal products practised by a limited number of vegans. The reasons for adoption of vegetarianism are complex but ethical concerns are predominant about animal welfare, especially under intensive production (Webster, 1994; British Nutrition Foundation, 1995). In this category, one could include the quasi-religious views of the place of humans in relation to other animals (Murcott, 1995).

Ecological concerns are also invoked in relation to the use of land to produce cereals for animal feeds (Blythe, 1976; Wardle, 1977) and the efficiency of meat production (Holmes, 1977; Webster, 1994).

(b) Personal factors. The sensory properties of cooked meats are highly prized and these are seen by many within the meat industry, and especially amongst butchers and chefs, as obstacles to reducing fat intake. The fat in the meat and the common practice of basting meat during cooking are closely connected with the succulence and flavour of the cooked meat. Current understanding of the flavour components in cooked meats suggest that the triacylglycerols are not major factors in flavour

Table 1.4 Fat content of some retail joints of meat

Meat species	Cut or joint	Fat content	(g per 100 g raw meat)
		1970[a]	1990[b]
Beef	Forerib	25.1	19.8
	Rump steak	13.5	10.1
	Sirloin steak	22.8	12.7
	Stewing steak	10.6	6.4
	Topside	11.2	12.9
Lamb	Breast	34.6	24.7
	Leg	18.7	12.3
	Loin chop	35.4	23.0
	Shoulder	28.0	18.3
Pork	Belly	35.5	20.2
	Leg	22.5	15.2
	Loin chop	29.5	21.7

[a]Paul and Southgate (1978).
[b]Chan *et al.* (1995).

generation, but that the intramuscular phospholipids with their polyunsaturated fats are the major factors with the amino acids in flavour generation (Farmer, 1994).

(c) Food choices. Shepherd (1990) included nutritional properties within this group of factors. It is significant that in a survey of attitudes to foods that were rich sources of fat in the diet, the concept of meat as healthy was one of the most strongly held beliefs about meat, with taste being the most strongly valued item. Of the nutritional attributes, meat as a source of protein and vitamins was much stronger than that of fat (Shepherd, 1990). Concerns about additives emerged as significant influences on meat consumption in the study by Richardson, Shepherd and Elliman (1993). The dietary advice does seem to have played a major role in the UK with respect to milks and other dairy foods, where there has been a substantial growth in the consumption of lower fat products (MAFF, 1990). Table 1.4 presents some comparisons of the fat contents of retail cuts of meat sampled in the 1990s (Chan *et al.*, 1995) with those sampled in the late 1960s (Paul and Southgate, 1978) and shows that there has been a substantial reduction in fat content over this period.

(d) Using the knowledge of food choices in developing healthy products. It is clear that in the interpretation of changes in food consumption and in analysing the factors which determine food choices and especially the consumption of meats, poultry and fish, simplistic assumptions are likely to be incorrect (Mela, 1995). In responding to the pressure of consumer demands for the production of more healthy meat products, it is also

important by analogy not to imagine that changes in composition alone will bring about the dietary changes which seem desirable for health.

1.5 Rationale for producing healthy muscle foods

There is sufficient consensus amongst the nutritional bodies concerned with diet and health and the advice generated by them to the consumer, and this has led to changes in consumer demand to which the meat industry must respond. Many of the official recommendations specifically call for responses from the food producer. This section first summarizes the rationale for the recommendations from the nutritional viewpoint and then considers the rationale for the response from the industry.

1.5.1 Nutritional rationale for change

Records of food and nutrient intakes derived from them in most developed countries support the views, first, that meat and meat products are a major source of dietary fat, and second, meat as a whole is an important source of saturated fatty acids in the diet. In the UK, for example, the records from the National Food Survey show that meat and meat products provided about 23% of the total fat intake (MAFF, 1995) and in The Netherlands about 18.7% (Voorlichtingsbureau voor de Voeding, 1993). Furthermore, meats were also an important source of saturated fats, providing 23.1% and 19.8% of the total intake of saturated fatty acids in the UK and The Netherlands, respectively. In the UK, where the data are presented in greater detail, it is evident that meat products provide the major proportion of the total meat fats. This is why the Department of Health's (1994) report specifically focused attention on reducing the consumption of meat products. It is therefore clear that any dietary campaign to reduce the consumption of total fat and saturated fats will focus on reducing the fat content of meats and meat products. The advice to shift consumption to poultry and fish rests, first, on the lower fat content of poultry meat when consumed without the skin and its associated fats, and second, on the beneficial effects of fish consumption observed in epidemiological studies (Kromhout, Bosscheiter and de Lezenne Coulander, 1985; Burr *et al.*, 1989), presumably derived from their content of long-chain n-3 fatty acids, particularly from fatty fish. The recommendation to reduce the consumption of meat products also includes the proviso that the salt content of these foods could and would be reduced as part of the strategy to reduce salt intake.

From the consumer's viewpoint, the visible fat on meat cuts represents an obvious source of fat and provides a clear means of reducing fat intake. The fat in meat products is so-called 'hidden' fat, and nutritional labelling is aimed at providing the consumer with information on which to base food choice (Shepherd and Towler, 1992; Towler and Shepherd, 1992). All the dietary advice has a quantitative foundation and reducing fat intake to a certain level requires judgements that demand knowledge and numeracy. These could act to restrict consumer behaviour to adopting the extreme response of avoiding all cuts which have visible fat or else to seeking a vegetarian diet.

1.5.2 Rationale for the producer and retailer

The response to the demand for lower fat cuts of carcase meats and the development of lower fat and lower salt meat products must be made on a firmer scientific and compositional basis than is implied by a simple reaction to reduce the fat content. It must also be accompanied by the development of information for the consumer to make them aware of, first, the essential role of fat for the physiology of the living animal and establish clearly the differences between the composition of the fat of different animal species. Second, it is also desirable to show how modern concepts of the interactions between the different saturated fatty acids, the new concepts relating to the role of the monounsaturated fatty acids and the growing recognition of the importance of the ratio of n-6 to n-3 fatty acids in thrombogenesis (Ulbricht and Southgate, 1991). These considerations mean that nutritionists, and especially those who are responsible for nutritional advice, need to avoid generalizations that appear to imply that all meat fats (and all poultry fats) are homogeneous in composition or in their nutritional effects. This element in the response of the meat producers will require considerable subtlety because of concerns about the objectivity of the providers of information (Richardson, Shepherd and Elliman, 1994).

(a) Composition of the fat of different animals. The fat of beef and lamb is substantially richer in saturated fatty acids than pork, although the composition of pork fats is highly dependent on the dietary fats used in pig feeding. However, the more highly saturated lamb fats are conversely richer in n-3 fatty acids (Ulbricht, 1995). Furthermore, meat fats are a rich source of monounsaturated fatty acids with meat and meat products providing 31% of the intake of monounsaturated fats in the diet of British adults (Gregory *et al.*, 1990). Of the polyunsaturated fatty acids, meats provided 19% of the intake, slightly more than fish, and 17% of the n-6 fatty acid intake.

The rationale for reducing the fat content of the carcase meats should thus focus on the ruminant species even though the pig is more amenable to reductions in carcase fats by selective breeding. However, any opportunity for cost-effective modification of pig fat should be explored. The need to reduce the fat content of meat products will possibly have a greater impact on pork, since pork fat is more widely used in meat products (Ulbricht, 1995).

It is interesting to note in the Department of Health's (1994) report that the recommendation for dietary changes considered poultry and pork together, a recognition of the need to consider beef and lamb separately. There are some implications here for poultry production in that the advice regarding poultry is based on the present recorded composition of poultry fat, and on the ease with which the flesh of poultry can be separated from the fatty layers associated with the skin. The advice would be negated if poultry rations were modified to include more saturated fat with consequential changes in the fat deposited or if the skin was normally consumed because this raises the fat content by about 10 g per 100 g (Chan *et al.*, 1995).

(b) Changes in taste. In studies of the attitudes to meat consumption, it is clear that the taste and value for money are most important for the consumer, and any movement towards reducing fat content must not be made at the expense of sensory characteristics. Anecdotally, most retail butchers believe that some fat is absolutely essential for the best eating quality, and the rationale for producing healthy meat must not reduce fat contents below levels where taste and eating quality decline. In addition, the study of Richardson, Shepherd and Elliman (1993) suggested that the availability of lower fat meats would not necessarily result in increased consumption.

1.6 Responses by the industry

The meat industry, despite reservations about some of the quality of the data used in deriving the nutritional advice and the financial implications of changing their practices, has already made some attempts to develop lower fat products. These will be discussed in depth in later chapters and will only be summarized here. The possible time-scales for the different options for changes are very different and it is possible to consider these under two headings: the reduction of fat in meat cuts by trimming-off fat and the reformulation of meat products for which the time-scale is immediate to moderate, and the development of carcases with a higher lean/fat ratio for which the time-scales are longer.

It is worth making the observation that the consumer can respond immediately to reducing his or her fat intake without making dietary changes or foregoing the pleasure of eating meat by discarding the fat on the plate or during food preparation (Snyder *et al.*, 1994).

1.6.1 Reduction in fat by trimming and reformulation

(a) Trimming and cutting. Trimming of the fat from meat can be made at the wholesale stage on the primal cuts during the division of the carcase or by the retailers when preparing retail cuts for sale. The latter option provides the opportunity of giving consumers additional pieces of fat for use in cooking, which would not be consumed *per se*.

Trimming would have the effect of increasing the price of meat because it would reduce the weight of saleable meat in a carcase or primal cut. It is probable that some consumers use the proportion of lean in a cut when estimating value for money so the trimming may not have adverse effects on retail income. Such practices would also exert pressure on the primary producer to produce leaner carcases. Ulbricht (1995) estimates that the extent of trimming has increased substantially in the UK over the period 1982–92, and this is borne out by the changes in the fat content of retail cuts (Table 1.4).

Butchering practices could also be changed to move to the 'muscle seaming' methods widely used in Europe, which results in very lean retail cuts even from fatty meats, such as lamb. Experience in Australia has led to the development of novel cutting practices to produce cuts with appreciably less visible fat (Hutchinson, Greenfield and Wills, 1987). Similar low fat strategies for retail cuts have been developed in the UK by the Meat and Livestock Commission (Chan *et al.*, 1995).

(b) Reformulation of meat products. Here there are two major 'health' objectives: a reduction in fat content and a reduction in salt content. For many products, fat is a major contributor to the texture of the product and the lower fat versions have proved less acceptable. The use of emulsifying agents to retain textural properties has some possibilities, but in some cases this results in a product which is contrary to existing compositional regulations. Replacement of meat protein with soya protein or isolated milk proteins has been used in sausages and found limited success. One is left with the impression that in order to retain their essential identity some conventional meat products must remain high fat products and be consumed infrequently, as they largely are at present. The records from the National Food Survey (MAFF, 1995) suggest that the consumption of some meat products is so low that reduction to the point of eliminating their fat content would have little real effect on overall fat consumption.

Reduction in salt content impinges on two aspects of the formulation of meat products. The first is the sensory properties, where there is good evidence that progressive reductions in salt content over time are not detectable by the consumer. The second aspect is much more difficult, where salt is used as a preservative in conjunction with sodium nitrite in preserved meats. Here protection against *Clostridium botulinum* poisoning is of much greater public health significance than the longer term risks from chronic disease. Thus, adequate levels of salt to ensure microbiological safety are essential and the industry must make the reasons for the use of these preservatives absolutely clear to consumers. These products must therefore be considered as products whose consumption should be limited to occasional use, as they largely are at present.

(c) Low salt and low fat products. Meeting the demand for products which are both low in salt and low in fat presents a double challenge to the food technologist. The muscle tissues of meat animals and fish are intrinsically low salt foods and trimming the fatty tissues provides a product which meets the requirements of the consumer in this respect. Processed meat and fish products are a major source of both salt (Shepherd, 1985) and fat (MAFF, 1995) in the diet of the UK, and the same is true for most developed countries. As discussed above (section 1.6.1(b)), where salt has been used as part of the preservative strategy in cured meats and fish products, any reduction in salt content carries with it the risk of compromising microbiological safety. Where the salt is primarily added for flavour, progressive reductions can rarely be detected by the consumer. Some producers of bread, for example, have been able to reduce salt levels without any loss of sales. It is important to note that the reductions in salt levels will increase water activity in the food and additions of preservatives and changes in recommended storage conditions will be necessary. Replacement of fat by, say, polysaccharides will have a similar effect on the microbiological safety of low salt and low fat products, and consumer education in the handling of these modified products will be essential.

1.6.2 Production of leaner carcases

In some respects, this requires a shift in conventional views of what a high-quality 'finished animal' should look like. It is true that where stock are reared under adverse weather conditions, a certain covering of fat is essential for the animal to survive and breed. In general, however, there is a need to use fat less in assessments of carcase quality and to focus on the proportion of lean. In many respects carcase quality requirements for grading are moving in this direction and pressures from retailers for leaner carcases will provide the 'market drive' for this to continue.

There are a number of options open in the production of leaner animals, starting with choice of breeds for meat production and the choice of optimum time for slaughter. With pigs especially, and to a lesser extent for beef and lamb, feeding regimes and their timing can be manipulated to give leaner carcases.

The longer term options involve breeding programmes to select for higher lean to fat ratios very specifically. With pigs a substantial move in this direction has already taken place, facilitated by their shorter breeding cycle. Indeed, in some cases, the content of intermuscular fat is now so low that the eating quality of the cooked meat is beginning to deteriorate. Clearly there is a need for reasonable action. All the nutritional advice calls for a reduction in fat in the diet, not its elimination. In his analysis of the fat in the UK food supply, Ulbricht (1995) points out that fats and oils are the major contributors to fat consumption and offer the most feasible opportunities for reducing fat consumption, while avoiding reduction of meat fat further as meats are a major source of n-3 fatty acids in the UK diet.

1.7 Research needs

These are considered under a number of headings in outline only, because more detailed proposals will be given in the chapters which follow.

1.7.1 Epidemiological and clinical studies

Much of the epidemiological evidence for an association between diet and CHD and cancers relies on data for food and nutrient intakes that are of variable quality, which can lead to difficulties when comparing the findings from different studies. In future epidemiological studies, it is important that due attention be given to the problems of measuring food and nutrient intakes.

Despite this, epidemiological data only can provide evidence for associations and must be supported by a better understanding of the mechanisms whereby dietary components influence the disease process. This is particularly needed where there is evidence of protective components so that the balance between components which promote and inhibit disease are better understood. Further evidence must also be sought from properly controlled intervention studies.

1.7.2 Mechanisms

The relative importance of the effects of the different saturated fatty acids on cholesterol metabolism needs to be further explored and extended to establish the significance of positional effects in the triacylglycerols. The significance of the ratio of n-3 to n-6 fatty acids in relation to thrombogenic effects needs to be understood. These two types of studies together with a better understanding of the role of dietary protective factors in atherogenesis should provide a firmer basis for comparison between diets than the P/S or other ratios. Further studies are needed on the effects of the ratio of sodium to potassium intakes in relation to hypertension and stroke.

1.7.3 Development of leaner products

The major thrusts should lie in the potential for modifying carcase fat content by altered feeding regimes coupled with the establishment of breeding programmes directed at selecting leaner carcases, but not at the expense of eating quality. Trimming of excess fat remains the quickest and most effective way of reducing the fat content of meat products. Fat reductions in cuts and carcases should be measured and aimed at reducing fat levels by, say, 25%.

1.7.4 Consumer attitudes and beliefs

The factors which affect food choices are very poorly understood, although it is clear that the attitudes and beliefs about consuming foods are important. Little is known about the ways in which these attitudes and beliefs are formed or the factors which influence them. Such understanding would be valuable in counteracting the development of beliefs which are based on irrational information.

1.7.5 Ethical and ecological concerns

The industry must address these concerns openly and work to mitigate genuine concerns about animal welfare or ecological issues by changing practices, and possibly by developing more open styles of communication.

1.8 Summary

Meat, poultry and fish make an important contribution to the nutritional and sensory properties of the diet and play a role in contributing to the healthy properties of the diet. Epidemiological, clinical and biochemical evidence indicates that excessive intakes of fats, especially saturated fats, constitute important risk factors for chronic diseases including obesity, CHD and some cancers.

Excessive sodium intakes, mainly from salt, almost certainly associated with inadequate potassium intakes, appear to be risk factors in the development of hypertension (itself a risk factor for CHD) and stroke. Some methods of cooking meats, notably grilling, especially over an open flame, lead to the formation of high levels of mutagens and potential carcinogens in the meat.

The meat industry has an ethical duty to respond to the concerns about health and a strong commercial incentive to provide the consumer with leaner meats and products with lower salt contents, provided that it is consistent with maintaining food safety. Consumer attitudes and beliefs about the consumption of meats are positively influenced by concepts of healthfulness, taste and value for money, and negatively by concerns about additives and ethical concerns about animal welfare and ecological concerns about pollution and land use. Where the consumer concerns about animal welfare are sound, it is in the interests of the industry to modify its practices to ensure that meat and meat products continue as the central element in the diet of most western cultures.

References

Ames, B.N. (1983) Dietary carcinogens and anticarcinogens. *Science*, **221**, 1256.

Anon. (1995) Creutzfeld–Jakob disease and bovine spongiform encephalopathy: any connection? *Br. Med. J.*, **311**, 1415.

Appleton, H. (1990) Foodborne viruses. *Lancet*, **336**, 1362.

Armstrong, B. and Doll, R. (1975) Environmental factors and cancer incidence and mortality in different countries, with special reference to dietary practices. *Int. J. Cancer*, **50**, 223.

Ashwell, M. (ed.) (1993) *Diet and Heart Disease: a Round Table of Factors*, British Nutrition Foundation, London.

Baird-Parker, A.C. (1990) Foodborne salmonellosis. *Lancet*, **366**, 1231.

British Nutrition Foundation (1995) *Vegetarianism*, BNF Briefing Paper, British Nutrition Foundation, London.

Blythe, C. (1976) Problems of diet and affluence. *Food Policy*, May, 91.

Burr, M.L., Fehilly, A.M., Gilbert, J.F. *et al.* (1989) Effects of changes in fat, fish and fibre intakes on death and myocardial reinfarction: diet and reinfarction trial (DART). *Lancet*, **ii**, 757.

Chan, W., Brown, J., Lee, S.M. and Buss, D.H. (1995) *Meat, Poultry and Game. Supplement to McCance and Widdowson's The Composition of Foods*, Royal Society of Chemistry, Cambridge.

Chen, C., Pearson, A.M. and Gray, J.I. (1990) Meat mutagens. *Adv. Food Nutr. Res.*, **34**, 387.

Collee, J.G. (1990) Bovine spongiform encephalopathy. *Lancet*, **366**, 1300.

Conning, D.M. (1988) The public perception of food safety. *J. R. Soc. Health*, **4**, 134.

Cooke, E.M. (1990) Epidemiology of foodborne illness: UK. *Lancet*, **366**, 790.

Cottrell, R.C. (1985) The 1984 COMA report – a new approach? *Hum. Nutr. Appl. Nutr.*, **39A**, 195.

Cuthbertson, A. (1980) The future of meat in the western diet. *Chem. Ind. (London)*, May, 397.

Department of Health (1978) *Eating for Health*. HM Stationery Office, London.

Department of Health (1991) *Dietary Reference Values for Food Energy and Nutrients for the United Kingdom. Reports on Health and Social Subjects 41*, HM Stationery Office, London.

Department of Health (1994) *Nutritional Aspects of Cardiovascular Disease. Reports on Health and Social Subjects 46*, HM Stationery Office, London.

Department of Health and Social Security (1984) *Diet and Cardiovascular Disease. Reports on Health and Social Subjects 28*, HM Stationery Office, London.

Drasar, B.S. and Irving, D. (1973) Environmental factors and cancer of the colon and breast. *Br. Cancer J.*, **27**, 167.

Ebling, J. (1982) Man the consumer. *Chem. Ind. (London)*, May 315.

Elliot, P., Stamler, J., Nichols, R. *et al.* (1996) Intersalt revisited, further analyses of 24-hour sodium excretion and blood pressure, within and across populations. *Br. Med. J.*, **312**, 1249.

Farmer, L.J. (1994) The role of nutrients in meat flavour formation. *Proc. Nutr. Soc.*, **53**, 327.

Fiddes, N. (1994) Social aspects of meat eating. *Proc. Nutr. Soc.*, **53**, 271.

Garrow, J.S. (1981) *Treat Obesity Seriously*. Churchill Livingstone, Edinburgh.

Gofton, L. (1990) Food fears and time famines, *BNF Nutr. Bull.*, **15**, Suppl. 1, 78.

Gregory, J.R., Foster, K., Tyler, H. and Wiseman, M. (1990) *The Dietary and Nutritional Survey of British Adults*. HM Stationery Office, London.

Haq, I.U., Yeo, W.W., Jackson, P.R. and Ramsay, L.E. (1995) The effect of dietary change on serum cholesterol. *Proc. Nutr. Soc.*, **54**, 601.

Harper, A.E. (1980) Healthy people: critique of the nutrition segments of the Surgeon General's report on health promotion and disease prevention. *Am. J. Clin. Nutr.* **33**, 1703.

Health Education Council (1983) *Proposals for Nutritional Guidelines for Health Education in Britain*, National Advisory Committee for Nutrition Education, London.

Holmes, W. (1977) Choosing between animals. *Philos. Trans. R. Soc. London, Ser. B*, **281**, 121.

Hutchinson, G.I., Greenfield, H. and Wills, R.B.H. (1987) Composition of Australian foods. 35. Pork. *Food Technol. Aust.*, **39**, 216.

Jequier, E. (1995) Nutrient effects: post absorptive interactions. *Proc. Nutr. Soc.*, **54**, 253.

Keys, A. (1957) Diet and the epidemiology of coronary heart disease. *J. Am. Med. Assoc.*, **164**, 1912.

Keys, A. (1970) Coronary heart disease in seven countries. *Circulation*, Suppl. 41, 11.

Keys, A. (1980) *Seven Countries: a Multivariate Analysis of Death and Coronary Heart Disease*. Harvard University Press, Cambridge, MA.

Kromhout, D., Bosscheiter, E.B. and de Lezenne Coulander, C. (1985) Inverse relation between fish consumption and 20-year mortality from coronary heart disease. *N. Engl. J. Med.*, **312**, 1205.

Leakey, R. (1994) *The Origin of Mankind*, Phoenix Books, London.

Leverkus, C., Cole-Hamilton, I., Gunner, K. *et al.* (1988) *The Great British Diet*, Century Publishing, London.

Madden, R.H. (1994) Microbial hazards in animal products. *Proc. Nutr. Soc.*, **53**, 309.

Marmot, M. (1994) The cholesterol papers. *Br. Med. J*, **308**, 351.

MAFF (1990) *Household Food Consumption and Expenditure 1990. Annual Report of the National Food Survey Committee*, HM Stationery Office, London.

MAFF (1995) *National Food Survey 1994*, HM Stationery Office, London.

Mattson, F.H. and Grundy, S.M. (1985) Comparison of the effects of saturated, mono-unsaturated and polyunsaturated fats on plasma lipids and lipoproteins. *J. Lipid Res.*, **26**, 194.

Mela, D.J. (1995) Understanding fat preference and consumption: applications of the behavioural sciences to a nutritional problem. *Proc. Nutr. Soc.*, **54**, 453.

Murcott, A. (1995) Social influences on food choice and dietary change: a sociological attitude. *Proc. Nutr. Soc.*, **54**, 729.

NAS (1981) The health effects of nitrate, nitrite and *N*-nitrose compounds. Part 1. National Academy of Sciences Committee on Nitrite and Alternative Curing Agents in Foods, National Academy Press, Washington, DC.

National Research Council (1989) *Diet and Health: Implications for Reducing Chronic Disease Risk*, National Academy Press, Washington, DC.

Pao, E.M. (1981) Changes in American food consumption patterns and their nutritional significance. *Food Technol.*, **2**, 43.

Paul, A.A. and Southgate, D.A.T. (1978) *McCance and Widdowson's The Composition of Foods*, 4th edn, HM Stationery Office, London.

Peterkin, B.B. (1978) The dietary goals and food on the table. *Food Technol.*, **2**, 34.

Richardson, N.J. (1994) UK consumer perceptions of meat. *Proc. Nutr. Soc.*, **53**, 281.

Richardson, N.J., Shepherd, R. and Elliman, N.A. (1993) Current attitudes and future influences on meat consumption in the U.K. *Appetite*, **21**, 41.

Richardson, N.J., Shepherd, R. and Elliman, N.A. (1994) Meat consumption, definition of meat and trust in information sources in the UK population and members of the Vegetarian Society. *Ecol. Food Nutr.*, **33**, 1.

Senate Select Committee on Nutrition and Human Needs (1977) *Dietary Goals for the United States*. US Senate, US Government Printing Office, Washington, DC.

Shepherd, R. (1985) Dietary salt intake. *Nutr. Food Sci.*, **96**, 10.

Shepherd, R. (1990) Overview of factors affecting food choice. *BNF Nutr. Bull.*, **15**, Suppl. 1, 12.

Shepherd, R. and Towler, G. (1992) Nutrition knowledge, attitudes and fat intake: application of the theory of reasoned action. *J. Hum. Nutr. Diet*, **5**, 387.

Sherman, H.C. (1933) *Chemistry of Food and Nutrition*, 4th edn, Macmillan, New York, pp. 518–538.

Snyder, M.P., Obarzanek, E., Montgomery, D.H. *et al.* (1994) *J. Am. Diet. Assoc.*, **94**, 1135.

Southgate, D.A.T. (1991) Nature and variability of human food consumption. *Philos. Trans. R. Soc. London Ser.* **B**, **334**, 281.

Surgeon General (1988) *Report on Nutrition and Health*, US Department of Health and Human Services, Washington, DC.

Todd, E. (1990) Epidemiology of food borne illness: North America. *Lancet*, **366**, 788.

Towler, G. and Shepherd, R. (1992) Application of Fishbein and Ajzen's expectancy-value model to understanding fat intake. *Appetite*, **18**, 15.

Tyler, K.L. (1995) Risk of human exposure to bovine spongiform encephalopathy. *Br. Med. J.*, **311**, 1420.

Ulbricht, T.L.V. (1995) *Fat in the Food Chain*. Ministry of Agriculture Fisheries and Food, London.

Ulbricht, T.L.V. and Southgate, D.A.T. (1991) Coronary heart disease: seven dietary factors. *Lancet*, **388**, 985.

Voorlichtingsbureau voor de Voeding (1993) *Zo eet Nederland 1992*. Ministerie van Welzijn, Volksgezondheit en Cultuur, Minsterie van Landbouw, Natuurbeheer, en Visserij, Den Haag.

Waites, W.M. and Arbuthnott, J.P. (1990) Food borne illness: an overview. *Lancet*, **366**, 722.

Wardle, C. (1977) *Changing Food Habits in the UK*, Earth Resources Research, London.

Webster, A.J.F. (1994) Meat and right: the ethical dilemma. *Proc. Nutr. Soci.*, **53**, 263.

Wheelock, V. (1988) Public perception of food safety. *J. R. Soc. Health*, **4**, 130.

Will, R.G., Ironside, J.W., Zeidler, M.Z. *et al.* (1996) A new variant of Creutzfeldt–Jakob disease in the UK. *Lancet*, **347**, 921.

Willett, W.C., Stampfer, M.J., Colditz, G.A. *et al.* (1990) Relation of meat, fat and fiber intake to the risk of colonic cancer in a prospective study among women. *N. Eng. J. Med.*, **323**, 1664.

Willett, W.C., Stampfer, M.J., Manson, J.E. *et al.* (1993) Intake of *trans* fatty acids and risk of coronary heart disease among women. *Lancet*, **343**, 581.

World Cancer Research Fund (1994) *Diet and Cancers. A Review of the Literature on Genetic, Cellular and Physiological Mechanisms*. WCRF, London.

2 Contribution of meat, fish and poultry to the human diet

A.E. BENDER

2.1 Introduction

There are very many diets in different regions of the world that supply adequate amounts of all the nutrients so long as the quantities of food eaten are sufficient. In the western world, diets are largely based on wheat, meat and milk, in other parts on rice and fish, with some communities relying heavily on single staples such as cassava, maize or potatoes – all supplemented to greatly varying extents with fruit and vegetables. Any discussion of the role of one type of food in the diet is confounded by the amounts consumed. Briggs and Schweigert (1990) quote a fivefold range for the consumption of red meat in the USA between light and heavy eaters (apart, of course, from vegetarians).

Generalizations can be made along the lines suggested by Bender and Bender (1995). Here foods are described as 'sources' of a nutrient if a (stated) portion (for meat products 150 g) supplies 10% of the recommended daily amount (RDA), a 'good source' with 20% and a 'rich source' with 30%. On this basis, 'meat' is a rich source of protein, niacin, riboflavin and iron and a source or good source of thiamine. Venison, horse meat, goose and game birds are exceptionally rich sources of iron (i.e. far more than 30% of the RDA) and pork is an exceptionally rich source of thiamine. This method of expressing nutrient content is probably a closer approximation to a correct description than the apparently precise, but somewhat misleading, figures quoted in food composition tables.

It is not possible to generalize about the role of fish in the diet since individual consumption varies enormously, not only with the preferences of the consumer but to a greater extent on availability. In some regions, especially where other concentrated sources of protein are limited, fish is an important source of protein. In regions where a wide variety of food is freely available, fish usually makes only a small contribution to the intake of most nutrients. Readers are also referred to Pearson and Dutson (1990).

2.2 Dietary recommendations

We know what is needed for growth, development and maintenance of health – 13 vitamins, some 20 minerals, certain fatty acids and an adequate supply of energy – but not how much of each. Differences between dietary recommendations of different authorities are due to different standards and to the paucity of evidence for many nutrients, which leaves the figures subject to opinion that changes from time to time. Tables 2.1–2.3 show the differences between RDA, RNI and PRI in the USA, UK and EU, respectively.

Such tables are valuable in assessing the adequacy of the diets of population groups, especially if surveys are carried out at intervals to reveal trends. They are based on intakes necessary to prevent clinical disorders (plus some overage). In recent times, additional possible criteria have arisen since intakes of antioxidant nutrients in greater amounts are thought to be beneficial in 'quenching' free radicals.

Until recently, nutrient intakes were considered to be adequate in a population group if the average intake reached the RDA [requirements plus 2 SD (standard deviations)], since then it is likely that all diets would reach the average requirements. In 1991 the UK Authorities (Department of Health, 1991) introduced three standards: estimated average requirement (EAR – as measured directly and indirectly), reference nutrient intake (RNI – equivalent to RDA) and lower reference nutrient intake (LRNI, which is 2 SD lower than the EAR). If the average nutrient intake reaches the RNI, the intakes would (in theory, assuming a normal distribution) satisfy the needs of 97% of the population. Individual intakes below LNRI would be of concern since they would almost certainly be inadequate.

2.3 Food composition tables

In nutrition surveys, individual foods and total diets are rarely analysed but are calculated from food composition tables. Many countries have their own tables and those of the United Nations Food and Agriculture Organization are often used in other countries. However, only three of these provide a broad coverage of the many types of foods available and full coverage of the nutrients. These are the US Department of Agriculture *Handbook No. 8* (USDA, 1979–86), the UK *McCance and Widdowson's Composition of Foods* (Holland *et al.*, 1991, Chan *et al.*, 1995) and the German *Food Composition and Nutrition Tables* (Souci, Fachman and Kraut, 1989).

For meat it is difficult to make direct comparisons between these three sets of tables, except in a few instances. Since the German tables deal only with raw meats, the methods of presentation differ in the tables and because of differences between local breeds, feed and management of animals, together with the availability of imported meat, which may well be seasonal, the data are hard to compare. In addition, there are differ-

Table 2.1 US Recommended Daily Amounts (RDA) of nutrients[a]

Age	Protein (g)	Vit. A (μg)	Vit. D (μg)	Vit. E (mg)	Vit. K (mg)	Vit. C (mg)	Vit. B_1 (mg)	Vit. B_2 (mg)	Niacin (mg)	Vit. B_6 (mg)	Folate (μg)	Vit. B_{12} (μg)	Ca (mg)	P (mg)	Mg (mg)	Fe (mg)	Zn (mg)	I (μg)	Se (μg)
0–6 m	13	375	7.5	3	5	30	0.3	0.4	5	0.3	25	0.3	400	300	40	6	5	40	10
6–12 m	14	375	10	4	10	35	0.4	0.5	6	0.6	35	0.5	600	500	60	10	5	50	15
1–3 y	16	400	10	6	15	40	0.7	0.8	9	1.0	50	0.7	800	800	80	10	10	70	20
4–6 y	24	500	10	7	20	45	0.9	1.1	12	1.1	75	1.0	800	800	120	10	10	90	20
7–10 y	28	700	10	7	30	45	1.0	1.2	13	1.4	100	1.4	800	800	170	10	10	120	30
Males:																			
11–14 y	45	1000	10	10	45	50	1.3	1.5	17	1.7	150	2.0	1200	1200	270	12	15	150	40
15–18 y	59	1000	10	10	65	60	1.5	1.8	20	2.0	200	2.0	1200	1200	400	12	15	150	50
19–24 y	58	1000	10	10	70	60	1.5	1.7	19	2.0	200	2.0	1200	1200	350	10	15	150	70
25–50 y	63	1000	5	10	80	60	1.5	1.7	19	2.0	200	2.0	800	800	350	10	15	150	70
51 + y	63	1000	5	10	80	60	1.2	1.4	15	2.0	200	2.0	800	800	350	10	15	150	70
Females:																			
11–14 y	46	800	10	8	45	50	1.1	1.3	15	1.4	150	2.0	1200	1200	280	15	12	150	45
15–18 y	44	800	10	8	55	60	1.1	1.3	15	1.5	180	2.0	1200	1200	300	15	12	150	50
19–24 y	46	800	10	8	60	60	1.1	1.3	15	1.6	180	2.0	1200	1200	280	15	12	150	55
25–50 y	50	800	5	8	65	60	1.1	1.3	15	1.6	180	2.0	800	800	280	15	12	150	55
51 + y	50	800	5	8	65	60	1.0	1.2	13	1.6	180	2.0	800	800	280	10	12	150	55
Pregnant	60	800	10	10	65	70	1.5	1.6	17	2.2	400	2.2	1200	1200	320	30	15	175	65
Lactating	65	1300	10	10	65	95	1.6	1.8	20	2.1	280	2.6	1200	1200	355	15	19	200	75

[a]Source: National Research Council (1989).

Table 2.2 UK Reference Nutrient Intakes (RNI)[a]

Age	Vit. B_1 (mg)	Vit. B_2 (mg)	Niacin (mg)	Vit. B_6 (mg)	Vit. B_{12} (μg)	Folate (μg)	Vit. C (mg)	Vit. A (μg)	Vit. D (μg)	Ca (mg)	P (mg)	Mg (mg)	Na (mg)	Fe (mg)	Zn (mg)	Cu (mg)	Se (μg)	I (μg)
0–3 m	0.2	0.4	3	0.2	0.3	50	25	350	8.5	525	400	55	210	1.7	4.0	0.2	10	50
4–6 m	0.2	0.4	3	0.2	0.3	50	25	350	8.5	525	400	60	280	4.3	4.0	0.3	13	60
7–9 m	0.2	0.4	4	0.3	0.4	50	25	350	7	525	400	75	320	7.8	5.0	0.3	10	60
10–12 m	0.3	0.4	5	0.4	0.4	50	25	350	7	525	400	80	350	7.8	5.0	0.3	10	60
1–3 y	0.5	0.6	8	0.7	0.5	70	30	400	7	350	270	85	500	6.9	5.0	0.4	15	70
4–6 y	0.7	0.8	11	0.9	0.8	100	30	500		450	350	120	700	6.1	6.5	0.6	20	100
7–10 y	0.7	1.0	12	1.0	1.0	150	30	500		550	450	200	1200	8.7	7.0	0.7	30	110
Males:																		
11–14 y	0.9	1.2	15	1.2	1.2	200	35	600		1000	775	280	1600	11.3	9.0	0.8	45	130
15–18 y	1.1	1.3	18	1.5	1.5	200	40	700		1000	775	300	1600	11.3	9.5	1.0	70	140
19–50 y	1.0	1.3	17	1.4	1.5	200	40	700		700	550	300	1600	8.7	9.5	1.2	75	140
50+ y	0.9	1.3	16	1.4	1.5	200	40	700	10	700	550	300	1600	8.7	9.5	1.2	75	140
Females:																		
11–14 y	0.7	1.1	12	1.0	1.2	200	35	600		800	625	280	1600	14.8	9.0	0.8	45	130
15–18 y	0.8	1.1	14	1.2	1.5	200	40	600		800	625	300	1600	14.8	7.0	1.0	60	140
19–50 y	0.8	1.1	13	1.2	1.5	200	40	600		700	550	270	1600	14.8	7.0	1.2	60	140
50+ y	0.8	1.1	12	1.2	1.5	200	40	600	10	700	550	270	1600	8.7	7.0	1.2	60	140
Pregnant	+0.1	+0.3				+100	+10	+100	10									
Lactating	+0.1	+0.5	+2		+0.5	+60	+30	+350	10	+550	+440	+50			+6.0	+0.3	+15	

[a]Source: Department of Health (1991).

Table 2.3 EU Population Reference Intakes (PRI) of nutrients[a]

Age	Protein (g)	Vit. A (μg)	Vit. B_1 (mg)	Vit. B_2 (mg)	Niacin (mg)	Vit. B_6 (mg)	Folate (μg)	Vit. B_{12} (μg)	Vit. C (mg)	Ca (mg)	P (mg)	Fe (mg)	Zn (mg)	Cu (mg)	Se (μg)	I (μg)
6–12 m	15	350	0.3	0.4	5	0.4	50	0.5	20	400	300	6	4	0.3	8	50
1–3 y	15	400	0.5	0.8	9	0.7	100	0.7	25	400	300	4	4	0.4	10	70
4–6 y	20	400	0.7	1.0	11	0.9	130	0.9	25	450	350	4	6	0.6	15	90
7–10 y	29	500	0.8	1.2	13	1.1	150	1.0	30	550	450	6	7	0.7	25	100
Males:																
11–14 y	44	600	1.0	1.4	15	1.3	180	1.3	35	1000	775	10	9	0.8	35	120
15–17 y	55	700	1.2	1.6	18	1.5	200	1.4	40	1000	775	13	9	1.0	45	130
18+ y	56	700	1.1	1.6	18	1.5	200	1.4	45	700	550	9	9.5	1.1	55	130
Females:																
11–14 y	42	600	0.9	1.2	14	1.1	180	1.3	35	800	625	18	9	0.8	35	120
15–17 y	46	600	0.9	1.3	14	1.1	200	1.4	40	800	625	17	7	1.0	45	130
18+ y	47	600	0.9	1.3	14	1.1	200	1.4	45	700	550	16[b]	7	1.1	55	130
Pregnant	57	700	1.0	1.6	14	1.3	400	1.6	55	700	550	16[b]	7	1.1	55	130
Lactating	63	950	1.1	1.7	16	1.4	350	1.9	70	1200	950	16	12	1.4	70	160

[a]Source: Scientific Committee for Food (1993).

Table 2.4 Effect of cooking on the energy and B vitamins in some meat and poultry products[a]

	Nutrients per 100 g (lean and fat)				
Product	*E* (kJ)	B_1 (mg)	B_2 (mg)	Nic. Eq. (mg)	B_6 (mg)
Beef:					
Braising steak, raw	670	0.07	0.27	8	0.42
braised	1030	0.05	0.26	12	0.33
Brisket, raw	900	0.08	0.21	8	0.43
boiled	1120	0.04	0.2	10	0.19
Fillet steak, raw	650	0.13	0.28	9	0.46
fried	800	0.12	0.25	12	0.58
grilled	790	0.12	0.26	12	0.60
Veal:					
Escalope, raw	450	0.12	0.23	12.6	0.65
fried	825	0.08	0.25	15	0.70
Lamb:					
Loin chop, raw	1150	0.13	0.22	8.6	0.23
grilled	1270	0.16	0.25	12.5	0.44
microwaved	1460	0.14	0.20	11	0.27
roasted	1490	0.14	0.31	12	0.29
Pork:					
Belly, raw	1070	0.43	0.22	8.4	0.39
roasted	1220	0.47	0.19	13	0.30
grilled	1330	0.60	0.18	12	0.38
Chicken:					
Leg, raw	800	0.12	0.27	8.3	0.23
casserolled	900	0.05	0.12	10	0.23
whole, meat and skin, raw	900	0.07	0.13	10.8	0.46
roasted	910	0.07	0.15	13.6	0.34
Duck:					
Meat, fat, skin, raw	1600	0.14	0.51	6	0.33
roasted	1750	0.18	0.51	8	0.31

[a]Source: Chan *et al.* (1995).

ences in cuts of meat, the amounts of fat preferred locally and methods of cooking, as well as in sampling and methodology in compiling tables. Table 2.4 illustrates the possible effects of cooking.

Table 2.5 exemplifies the problems by a comparison from Chan *et al.* (1995) listing the data from the same laboratories with the same methodology for lamb, New Zealand chilled and frozen available in the UK, and the effect of differing methods of cooking of fresh lamb, namely medium cooked and well done.

2.4 Contributions of meat and poultry

2.4.1 Protein

Meat is highly regarded as a concentrated source of high quality protein. In the USA, meat (excluding poultry) supplies between 15 and 75% of the

Table 2.5 Variations in the nutrients of roasted leg of lamb (lean and fat) differing in source, relative freshness and extent of cooking[a]

Component	New Zealand		UK (fresh)	
	Chilled	Frozen	Medium	Well done
Water (%)	58	57	57	56
Protein (%)	28	29	28	30
Fat (%)	13	14	14	14
Energy (cal)	230	240	240	240
Fe (mg)	2	2.4	2.9	2.2
Zn (mg)	4.3	4.8	4.3	4.5
Se (μg)	4	4	4	4
B_1 (mg)	0.15	0.10	0.12	0.13
B_2 (mg)	0.31	0.26	0.28	0.26
Nic. Eq.[b] (mg)	10.7	10.5	11.3	12.7
B_{12} (μg)	3	3	2	3

[a]Source: Chan *et al.* (1995).
[b]Niacin equivalent.

RDA protein, depending on whether the consumer is a light or heavy meat eater (Briggs and Schweigert, 1990).

In developing regions, where the greater part of the protein comes from plant sources, even a small amount of meat provides a significant proportion of the protein. Since cereal and most plant proteins are limited by lysine with a relative surplus of the sulphur amino acids, while meat is limited by the sulphur amino acids with a relative surplus of lysine, meat proteins complement the plant proteins. While there is little difference between the overall quality of dietary proteins in developing and industrialized countries (WHO, 1985), protein quality is important in low-protein diets.

(a) Species and tissue effects. The amino acid composition of meat protein is relatively constant irrespective of the source of the flesh. Crawford (1968) showed that despite differences in species and diet, ranging from grass to acacia bush for herbivores and from zebra to warthog for carnivores, the amino acid composition of protein tissues of leopard, hyena, man, warthog, buffalo, hartebeest, eland, giraffe, elephant and domestic food animals is remarkably constant. Even bulls fattened on a diet in which urea replaced 50% of the conventional protein showed no change in chemical composition. Different tissues have a similar amino acid composition. Mitchell, Beadles and Kruger (1927) suggested that different cuts of meat would vary in the value of their protein because of differences in their digestibility due to different proportions of connective tissue. More recently, Bender and Zia (1976) measured protein quality and connective tissue content on the same samples of meat. A sample of low

(eating) quality stewing steak, with 23.6% collagen, had an NPU of 69 while fillet steak with 2.5% collagen had an NPU of 82. Both NPUs were increased by supplementation with methionine, so both were limited by the sulphur amino acids.

2.4.2 Fat

An excessive intake of dietary fat has been implicated as a cause of coronary heart disease (CHD), in particular the saturated fatty acids (SFA), especially lauric, myristic and palmitic. Dietary fat has also been associated with cancer of the colon, breast and prostate (Chapter 4). The enormous ranges of fat content in samples of the same cuts of meat are shown in Table 2.6.

(a) Saturated fatty acids. Recommendations to the public to reduce the intake of SFA have led to over-simplified explanations of what the terms mean. It is usually explained by equating SFA with animal fats and polyunsaturated fatty acids (PUFA) with vegetable oils. Table 2.7 shows that SFA comprise only half to one third of many common animal fats. While these figures are higher than in the vegetable oils (except for the harder vegetable fats), the main difference between the two groups is the content of PUFA.

(b) Monounsaturated fatty acids. Monounsaturated fatty acids (MUFA) have been shown in some trials to be beneficial in reducing plasma cholesterol levels, but meta-analysis of 27 recent trials designed to compare MUFA and PUFA showed such reductions to be small but not statistically significant (Truswell, 1995). In any case, the MUFA content of meat fats is greater than that of the vegetable oils except olive and rapeseed oils.

Table 2.6 Ranges of trimmable fat[a]

	% of meat samples[b]	
Meat	No. of samples	Trimmable fat
Lamb, neck cutlets, grilled	33	5–49
Beef fillet, raw	17	0–11
fried	16	0–12
Rump steak, raw	10	1–23
grilled	10	1–23
Lamb loin chops, raw		
'extra lean'	15	19–41
Pork cuts	-c-	Up to 62%

[a]Source: Chan *et al.* (1995).
[b]Ranges for cuts and different methods of preparation.
-c- several hundred.

Table 2.7 Composition of meat fats in comparison with some other animal and plant fats[a]

	% of total fat		
Product	SFA[b]	MUFA[b]	PUFA[b]
Beef raw (1)	49	48	3
Lamb	55	41	5
Pork	38	44	18
Chicken skin (48% fat)	30	52	18
leg, meat, fat and skin	29	51	20
Duck, meat, fat and skin	30	54	16
Corn (2)	13	25	58
Cottonseed	26	21	48
Olive	14	70	11
Peanut	19	48	29
Soybean	15	23	57
Coconut (2)	85	7	2
Palm	45	42	8
Butter	66	24	3

[a]Sources: (1) figures rounded off from Chan *et al.* (1995); (2) figures rounded off from Holland *et al.* (1991).
[b]SFA = saturated; MUFA = monounsaturated; PUFA = polyunsaturated fatty acids.

(c) trans-*Fatty acids (TFA).* *trans*-Isomers of fatty acids are found in beef and mutton fat and in milk fat – 2–9% as 11-octadienoic acid – formed by the hydrogenation of the *cis* isomers in the rumen. Small amounts are found in the meat of pigs and poultry, which are derived from the feed. *trans*-Fatty acids are also formed in vegetable oils during hardening by hydrogenation – between 0.4 and 18%.

Animal products account for only a third of the total (TFA) intake. However, it is hydrogenation of vegetable oils that has given rise to the most concern. There is some uncertainty about the amounts of TFA consumed because of analytical complications, but it is estimated to be 8–13 g per day in the USA, 4–6 g in the UK and 8–10 g across Europe (BNF, 1995).

Conflicting conclusions regarding the relation between TFA intake and CHD have been drawn from various studies. Attempts have been made to correlate the incidence of CHD with the development of hydrogenation over the years. A high dosage of experimental animals with TFA has been shown to cause an increase in low-density lipoprotein (LDL) and a small decrease in high-density lipoprotein (HDL), but it is concluded that any risk from consumption of TFA at current average levels of intake appears to be small. Fatty acids are digested, absorbed and metabolized for the most part in the same way as *cis* isomers. It is concluded that these amounts of TFA have no greater effect as CHD risk factors than SFA, and that in dietary assessment should be included with them but that average intakes should not rise above current levels (BNF, 1995). There is evidence, in general, that intakes of TFA are falling and specific steps are being taken in some countries to reduce the amount of the TFA present in spreads for bread.

2.4.3 *Mineral salts*

The essential minerals supplied in significant amounts by meat are iron, copper, zinc and selenium. Attempts to relate the amounts to RDA are clouded by differing opinions of the RDA, so for practical purposes selected average figures have been taken from the three major tables of RDA – namely US tables, UK Reference Nutrient Intakes, and those of the European Union (Tables 2.8 and 2.9).

Table 2.8 Micronutrients in samples of meat (per 100 g, juices discarded)[a]

Meat	Fe (mg)	Cu (mg)	Zn (mg)	Se (μg)	B_1 (mg)	B_2 (mg)	Nic. Eq. (mg)	B_{12} (μg)
Beef, fried rump steak	2.7	0.02	4.7	9	0.13	0.27	11.3	2
Pork, roast leg	1.0	0.08	2.6	19	1	0.25	15	1
Chicken, meat, average roast	0.7	0.1	1.5	16	0.07	0.16	14.5	Trace
Duck, roast, meat	2.7	0.3	2.6	22	0.26	0.47	10.5	3
Turkey, roast	0.8	0.09	2.5	17	0.06	0.19	16.5	1
Pheasant, roast	2.2	0.1	1.3	14	0.02	0.29	15	3
Pigeon, roast	7.2	0.33	1.7	–	0.27	1.2	12	8

[a]Source: Chan *et al.* (1995).

Table 2.9 Micronutrients in meat

	Expressed as % of RDA, 150 g portion							
	Fe (mg)	Cu (mg)	Zn (mg)	Se (μg)	B_1 (mg)	B_2 (mg)	Nic. Eq. (mg)	B_{12} (μg)
RDA[b]	10	1.5	10	70	1.2	1.5	18	1.7
Beef	40	2	70	20	20	30	60	180
Pork	15	8	40	40	125	25	120	90
Chicken	10	10	25	35	10	15	120	–
Duck	40	30	40	50	35	45	90	260
Turkey	10	10	40	35	10	20	140	90
Pheasant	35	10	20	30	3	30	120	260
Pigeon	10	35	25	–	35	120	100	700
RDA[c]	USA	UK	EU	Figure used				
B_1 (mg)	1.5	0.9	1.1	1.2				
B_2 (mg)	1.7	1.3	1.6	1.5				
Nic. Eq. (mg)	19	17	18	18				
B_{12} (μg)	2	1.5	1.4	1.7				

[a]Source: Chan *et al.* (1995).

(a) Iron. Iron presents a particular problem since the distribution curve for the requirements of females is skewed to the right. In an attempt to standardize the figures for the nutritional value of meat, the RDA for adult males has been used in Table 2.9. The figures for the USA, UK and EU are 10, 8.7 and 9 mg, respectively, from which a reasonable figure of 10 mg has been selected.

The amount of iron absorbed from the diet depends on its chemical form, the simultaneous presence of dietary factors that enhance or hinder its absorption and various physiological factors including the iron status of the individual. Half the iron in meat is present as haem iron, which is well absorbed (15–20%) compared with only 1–10% of the iron from plant foods. Furthermore, meat enhances the absorption of iron from other foods. For example, the addition of meat to a legume–cereal diet can double the proportion absorbed. Hence meat plays a special role in iron absorption and thus in the prevention of anaemia.

(b) Zinc. While dietary deficiency of zinc is uncommon, meat is the richest source and among meat eaters accounts for one third to half of the intake. RDA values recommended in the USA, UK and EU are 15, 9.5 and 9.5 mg per day, respectively. The figure of 10 mg is used in Table 2.9.

(c) Copper. The UK Committee concluded that there are no adequate data on which to base EAR and RNLI but that there is sufficient biochemical evidence to arrive at a figure for RNI (equivalent to RDA), namely 1.2 mg. The US and EU figures are 1.5–3 and 1.1 mg, respectively. The figure of 1.5 mg is used in Table 2.9.

(d) Selenium. The US, UK and EU figures are 70, 75 and 55 μg per day, respectively. The figure of 70 μg is used in Table 2.9.

2.4.4 *Effects of cooking*

There is a highly variable loss of weight during cooking, mostly as water but also some as fat. Thus, in 20 samples of grilled fillet steak, the weight losses ranged between 15 and 41%, with a mean of 26%. In 16 samples of roasted pork loin chops, losses ranged between 24 and 57%, mean 38% (Chan *et al.*, 1995).

The loss of water and fat results in a proportionate increase in protein. So far as fat is concerned, there is a proportionate increase with the loss of water, but some loss occurs, which depends largely on the method of cooking.

There are considerable differences between the mineral contents of different cuts. The figures shown in Table 2.8 refer to specific cuts. The meat included the fat, but not the skin, and was weighed without the bone.

With the enormous differences in types of meat, cuts and methods of cooking, it is probably more useful to describe these products as 'sources', 'good sources' and 'rich sources' as described in the Introduction. Table 2.9 lists the same figures as Table 2.6 for the same cuts of meat but expressed as a percentage of the RDA per 150 g portion.

The data for iron range from a 'source' (10% RDA per 150 g portion) to rich source (30% or more). The figures for copper cover a much wider range, from 2% of the RDA in grilled beef rump steak to 30% in duck and pigeon. Similarly, zinc varies over a broad range. Selenium in all samples other than beef is present in similar amounts, namely 30–40% of the RDA.

2.4.5 Vitamins

Literature values for vitamins in foods differ considerably. Apart from true differences between samples, these may be due to variations in methodology and analytical problems in the determination of vitamins. It might be expected that determinations carried out by the same method in the same laboratory would yield acceptable, or at least comparable, results, but this was not so in the work on vitamin retention in frozen meat products by Mikkelsen, Rasmussen and Zinck (1984). These authors measured the retention of thiamine, riboflavin and vitamin B_6 in various samples of meat stored at –12 and –24 °C over a period of 1 year and analysed every 2 months. Some of the results showed wide fluctuations, rising as well as falling in successive periods, which the authors were unable to explain. This made it impossible, in some instances, to draw any conclusions.

The figures in Tables 2.8 and 2.9 were determined on several samples of the same type of meat. However, being the most recent at the time of writing, they may be considered as the best currently available (Chan *et al.*, 1995). As with the minerals the vitamins are quoted in Table 2.8 per 100 g of cooked food with juices discarded, and in Table 2.9 per 150 g portion, expressed as a percentage of the RDA of USA, UK and EU tables.

Although water and some fat are lost in cooking, the vitamin content expressed per 1000 kcal (1 kcal = 4.184/kJ) falls in all instances. Since the juices that were discarded for the analyses may be consumed along with the meat, there will be considerable recovery of the vitamins, except possibly for heat-labile thiamine.

2.5 Contributions of fish

Fish are commonly classified as white (low fat) and fatty.

2.5.1 *White versus fatty fish*

White fish (low fat) contain 1–2% fat and fatty fish between 5 and 20–30% fat. Since they are mainly captured from the wild, they are of different ages and from different environments, so possibly consuming different diets, and therefore have a variable composition. In general, white fish are a 'rich source' of protein, niacin, vitamins B_6 and B_{12}, iodine and selenium and a 'source' of iron, thiamine and riboflavin (see Introduction). They contain only a trace of vitamin D, while fatty fish, with the exception referred to below, contain 10–25 μg per 100 g. In both types the ratio of potassium to sodium is 3:1–5:1.

The fat content of fatty fish fluctuates with the seasons. Herring and mackerel, for example, can vary between 1 and 30% fat, minimum levels being present just after spawning. Atlantic salmon contains only a trace of vitamin D, whereas Pacific salmon contains 5–20 μg per 100 g. Fatty fish are a rich source of protein, niacin, vitamin B_6, iodine and selenium. They are a source of iron, zinc, thiamine and riboflavin and an extremely rich source of vitamin B_{12}.

(a) Effects of cooking. Fish can be cooked in many ways. The commonly eaten 'fried fish' may be coated with flour, breadcrumbs or batter of varying thickness. Since the coatings soak up oil, there are wide variations in the type and amount of fat in breaded products.

2.5.2 *Omega fatty acids*

Oily fish are a rich source of ω-3 fatty acids – eicosapentaenoic acid (EPA) and docosohexaenoic acid (DHA). White fish contain only small amounts. These, as does the plant ω-3 fatty acid linolenic acid, reduce liver synthesis of triglycerides and reduce clotting of blood platelets by decreasing the levels of lipoprotein (a) in individuals with high levels, and so may have beneficial effects in coronary heart disease and autoimmune diseases.

Small amounts of fish, as little as an average of 30 g per day, appear to have these beneficial effects but larger amounts do not increase their effects. It is not certain whether the effects are due to the oil or to other products in the fish, nor is it clear whether the consumption of isolated fish oils is as safe as the consumption of fish. The American Public Health Association recommends the consumption of three fish servings per week while, on the contrary, the consumption of ω-3 oils in excess of 4 g per day inhibits blood clotting (Brown, 1990).

2.5.3 *Shellfish*

Crustaceans and molluscs tend to accumulate cations from their environment, and so their iron, copper and zinc contents vary considerably with

Table 2.10 Cholesterol content of shellfish per 100 g (raw)[a]

Species	Content (mg per 100 g)
Clams	40
Crab, Alaska king	50
Crab, Dungeness	70
Crayfish	160
Lobster, northern	110
Lobster, spiny	80
Mussels, blue	30
Oyster	60
Scallops	40
Shrimp	170
Squid	270

[a]Source: Brown (1990) (figures rounded off).

the source. For example, oysters can contain between 6 and 100 mg of zinc per 100 g, and some species contain 20–40 mg of vitamin C per 100 g (Paul and Southgate, 1978).

For some years shellfish were avoided since they were considered to be relatively rich in cholesterol, but more recent results have revised the figures downwards (Brown, 1990) and, since they have a low content of total fat and saturated fatty acids, it is considered that they are no more likely to raise blood cholesterol levels than lean meat or skinless poultry (Table 2.10).

2.6 Research needs

In assessing the role of specific foods in the diet, research is needed into the health aspects of the ingredients of those foods. In the case of meat, poultry and fish products, particularly the mechanisms of the deleterious and the beneficial effects of the various fatty acids need further clarification. It is generally accepted that it would be beneficial to reduce dietary total fat to about 30% of the total energy intake (including that from alcohol or 35% of the energy from food) and saturated fatty acids to 10%. There is evidence of benefits from fatty fish and from monounsaturated fatty acids, which call for additional elucidation.

The overall effects of meat, especially red meat, on health need investigating in detail along with the results of cooking and food processing. This is particularly true with reference to the potentially harmful effects of overheated fats and *trans*-fatty acids.

A growing area of nutrition research is that of bioavailability of nutrients. Since we already know that iron in meat is more available than that from other sources and that meat can increase the latter, it is clear that the bioavailability of nutrients present in meat is worthy of further research.

2.7 Summary

As is stated in other chapters, meat, poultry and fish products are generally regarded as being desirable parts of the diet by the majority of most population groups and nutritionists, but are shunned for a variety of reasons by a growing minority of industrialized communities. Clearly, the nutrients supplied by meat are beneficial to human health, especially nutrients such as protein and a range of vitamins and minerals. This is particularly true since some of these nutrients are more concentrated and better absorbed than those from other foods. There are unresolved problems regarding the fat in meat as compared with vegetable oils, and fish oils and fish products, which needs further elucidation.

Some fatty acids in meat products are currently considered to be risk factors in a number of diseases, whereas others, such as those in fish, appear to be beneficial. There is no one specifically 'good' or 'bad' food and it is always essential to strike a balance between the beneficial and potentially detrimental effects of any particular foodstuff.

References

Bender, A.E. and Bender, D.A. (1995) *A Dictionary of Food and Nutrition*, Oxford University Press, Oxford.

Bender, A.E. and Zia, M. (1976) Meat quality and protein quality. *J. Food Technol.*, **11**, 494.

BNF (1995) *Trans Fatty Acids, Report of the British Nutrition Foundation Task Force*, British Nutrition Foundation, London.

Briggs, G.M. and Schweigert, B.S. (1990) An overview of meat in the diet. *Adv. Meat Res.*, **6**, 1.

Brown, W.V. (1990) in *The Mount Sinai School of Medicine Complete Book of Nutrition* (eds V. Herbert and G.J. Subak-Sharpe), St Martin's Press, New York, pp. 80, 451.

Chan, W., Brown, J., Lee, S.M. and Buss, D.H. (1995) *Meat, Poultry and Game. 5th Supplement to McCance and Widdowson's, The Composition of Foods*, 5th edn, Royal Society of Chemistry and Ministry of Agriculture, London.

Crawford, M.A. (1968) Food selection under natural conditions and the possible relationship to heart disease in man. *Proc. Nutr. Soc.*, **27**, 163.

Department of Health (1991) *Dietary Reference Values for Food Energy and Nutrients for the United Kingdom. Report on Health and Social Subjects No. 41.* HM Stationery Office, London.

Holland, B., Welch, A.A., Unwin, I.D. *et al.* (1991) *McCance and Widdowson's The Composition of Foods*, 5th edn, Royal Society of Chemistry and Ministry of Agriculture, London.

Mikkelsen, K., Rasmussen, E.L. and Zinck, O. (1984) Retention of vitamins B1, B2 and B6 in frozen meats, in *Thermal Processing and Quality of Foods* (ed. P. Zeuthen), Elsevier Applied Science, London, pp. 777–781.

Mitchell, H.H., Beadles, J.R. and Kruger, J.H. (1927) The relation of the connective tissue of meat to its protein value in nutrition. *J. Biol. Chem.*, **69**, 767.

National Research Council (1989) *Recommended Dietary Allowances*, 10th edn, National Academy of Sciences, Washington, DC.

Paul, A.A. and Southgate, D.A.T. (1978) *McCance and Widdowson's The Composition of Foods*, 4th edn, HM Stationery Office, London.
Pearson, A.M. and Dutson, T.R. (eds) (1990) *Meat and Health. Advances in Meat Research*, Vol. 6, Elsevier Applied Science, London.
Scientific Committee for Food (1993) *Nutrient and Energy Intakes for the European Community*, Commission of the European Communities, Luxemburg.
Souci, S.W., Fachman, W. and Kraut, H. (1989) *Food Composition and Nutrition Tables*, Wissenschaftliche Verlagsgesellschaft, Stuttgart.
Truswell, A.S. (1995) *Dietary Fat – Some Aspects of Nutrition and Health and Product Development*, ILSI Europe, Brussels.
USDA (1979–86) *USDA Handbook No. 8: Poultry Products* (1979), No. 8-5; *Sausages and Luncheon Meats* (1980), No. 8-7; *Pork Products* (1983), No. 8-10; *Beef Products* (1986), No. 8-13; *Finfish and Shellfish Products* (1987), No. 8-15, US Government Printing Office, Washington, DC.
WHO (1985) *Energy and Protein Requirements. Joint Report FAO/WHO/UNU. Technical Report Series, No. 724.* WHO, Geneva.

3 Labeling of low and reduced fat/salt products

J. QUICK

3.1 Introduction

Although lower fat and/or sodium products can frequently compete on a palatability basis with other products, they are mainly intended for consumers who wish to lower the fat and/or sodium contents of their diets. Because these products are marketed primarily for their special nutritional attributes, it is essential that these attributes be readily apparent to consumers. The product label is the best vehicle for informing consumers about a food product because the label is readily available at the time the purchase decision is made.

Recently, labeling regulations for nutrition information and nutrient content claims were substantially changed in the USA to accommodate the changing needs of consumers and to make nutrition information easier for consumers to understand. Other countries are also examining their rules for such information on labels and may be making similar changes in the next few years. The information in this chapter reflects the new US system unless specifically stated as being used by another country. Any food product sold in the USA, regardless of whether it is a domestic or imported product, must comply with these new regulations.

Although it is possible for highly educated consumers to differentiate lower fat and/or sodium foods from the nutrition label alone, the use of nutrient claims makes it easier for consumers to identify these products. Nutrient claims are usually prominently placed on the front (or principal display panel) of the label while nutrition labels, because of their size, frequently appear on the back or side of the package (the information panel). There are specific type size and placement requirements for nutrition labels and nutrient claims, as well as definitions for the nutrient claims that may be used.

In the USA, nutrition labeling and nutrient claims requirements are codified in Title 9 of the Code of Federal Regulations (CFR) in Sections 317.300–317.400 (Code of Federal Regulations, 1995a) for red meat products, in Title 9 CFR Sections 381.400–381.500 (Code of Federal Regulations, 1995b) for poultry products and Title 21 CFR Sections 101.9–101.13 and 101.42–101.65 (Code of Federal Regulations, 1995c) for game meats (ostrich, venison, alligator, etc.), seafood and fishery products. In the USA, labeling of meat and poultry products is regulated by the Food

Safety and Inspection Service of the US Department of Agriculture (USDA); labeling of game meats, seafood and fishery products is regulated by the Food and Drug Administration (FDA) of the US Department of Health and Human Services. The regulations for the two agencies are essentially the same but where the regulations differ, a manufacturer must use the regulations for the Agency having jurisdiction over the food product. FDA-regulated products processed under voluntary USDA inspection must meet USDA requirements. The regulations are complex, very detailed and subject to change. Any attempt to construct a label for a lower fat and/or sodium food product should occur only after a thorough reading of the latest edition of the regulations.

3.2 Mandatory labeling requirements

In constructing the label for any food product, several features must always be present regardless of the nutritional profile of the product. These features are the product name (Code of Federal Regulations, 1995d), the list of ingredients (if more than one ingredient is used to make the product) (Code of Federal Regulations, 1995e), the manufacturer's or distributor's name and address (Code of Federal Regulations, 1995f), the net weight or quantity of contents (Code of Federal Regulations, 1995g), a handling statement (e.g. 'keep refrigerated') if the product is not shelf stable (Code of Federal Regulations, 1995h) and nutrition information (Code of Federal Regulations, 1995i). Nutrition labeling is not mandatory if the product qualifies for one of the exemptions from nutrition labeling (Code of Federal Regulations, 1995j). In addition, if the product falls under the regulatory jurisdiction of the FSIS (a meat or poultry product), the label must also bear the USDA mark of inspection and, unless the product is fully cooked or otherwise ready-to-eat, safe handling instructions are also required (Code of Federal Regulations, 1995h,k). All other information, including preparation instructions, is considered voluntary information. Labels should be designed so that all mandatory information is present before voluntary information is added.

3.2.1 Product name

The product name is a basic description of the product and usually includes a reference to all major and characterizing ingredients or components. Major ingredients and components are those that comprise the greatest proportion of the product; characterizing ingredients and components are those that, although they may be used in smaller amounts, have a significant flavor or texture impact on the product.

If the product meets a regulatory standard of identity or composition, the standardized product name must be used. The US product standards for red meat and poultry products are codified in Title 9 of the Code of Federal Regulations (1995l), and informal product standards for both red meat and poultry products are listed in the *Standards and Labeling Policy Book* (Food Labeling Division, 1995). Standards of identity for fishery and seafood products are codified in Title 21 of the Code of Federal Regulations (1995m). For some products, if the product fails to meet a standard of identity only because it has been specially formulated to be lower in fat or sodium, the standardized product name can still be used as long as it is properly modified to indicate how the product does not meet the standard.

(a) Fish and Shellfish regulated by FDA. Fishery and shellfish products, as well as other foods regulated by FDA, that do not conform to a standard of identity only because they have been specially formulated to be lower in fat and/or sodium may be labeled with the appropriate nutrient claim and the standardized product name (Code of Federal Regulations, 1995n). In the label's list of ingredients, those ingredients not permitted by the product standard, or used in amounts exceeding the limits in the product standard, must be marked by asterisks referencing a statement at the end of the ingredients list. This statement explains that the asterisked ingredient is not normally found in that product or that the ingredient has been used at levels above what is normally found in that product.

(b) FSIS labeling of low fat or low sodium products. The FSIS uses a slightly different labeling scheme for standardized products that have been specially formulated to be lower in fat. Uncooked meat and poultry products must be labeled with a product name that explains how they differ from the standardized product. For example, fresh pork sausage may be formulated with a binding agent and more water than the 3% permitted by the standard of identity (Code of Federal Regulations, 1995o) if the binder/water mixture is used to replace fat. The current labeling of this type of product would be 'Lower Fat Pork Sausage, Water and Soy Protein Isolate Product' (Wade, 1995a). Certain ready-to-eat meat or poultry products and breakfast sausage may be labeled according to the FDA labeling scheme (Wade, 1995b), and regulations are in progress to extend this scheme to raw products.

3.2.2 Other required labeling features

(a) List of ingredients. The list of ingredients must include each ingredient used to make the product in order of predominance by weight. Some ingredients, such as enriched flour, are themselves made from several ingredients and these sub-ingredients must also be declared, usually in a parenthetical listing.

(b) Manufacturer's or distributor's name and address. The manufacturer's or distributor's name and address must also appear on the label. If the address is for a distributor, the name and address must be preceded by 'packed for', 'distributed by' or some other phrase to indicate that some other company actually manufactured the product. If the company is listed in a telephone directory, only the city, state and zip code must be listed as the address.

(c) Net weight or quantity of contents. The net weight or quantity of contents must be declared in avoirdupois and metric units for products under FDA's jurisdiction and only in avoirdupois units for products under USDA's jurisdiction. This information must be printed in the lower 30% of the principal display panel, and in type sizes dictated by the area of the principal display panel. In addition, if the net weight is between 1 and 4 lb, the weight is declared first in ozs. then, parenthetically, in lb and oz [e.g. net wt 20 oz (1 lb 4 oz)]. The metric units are expressed after (to the right or below) the avoirdupois units.

(d) Handling statement. Unless the product is shelf stable (e.g. canned beef stew or oysters), some kind of handling statement, such as 'keep refrigerated' or 'keep frozen', must also appear on the label. Poultry products are permitted to bear handling statements that cover both options ('keep refrigerated or frozen'), but red meat products must be labeled with only one option. A third kind of handling statement, 'previously handled frozen for your protection – refrigerate or refreeze', may be used if the product is frozen at packaging and is permitted to thaw in transit or at the retail store.

Safe handling instructions must appear on all meat and poultry products that are not ready-to-eat. Their use is encouraged, but not required, on uncooked seafood and fishery products. Safe handling instructions on seafood and fishery products may not state that the product has been prepared from inspected and passed meat and/or poultry.

The instructions must be set off from all other information on the label by a thin line. The instruction must contain the title 'Safe Handling Instructions', a paragraph explaining why special handling is required and four sets of instructions for proper handling of the food. Each instruction must be placed adjacent to a pictorial representing the instruction. The arrangement of the required features may vary, but the correct pictorial must always appear next to the complementary instruction. Figure 3.1 shows one arrangement for the safe handling label. Other pictorials may be used, but the text may not vary from the example, except if the product is sold frozen and is to be cooked from the frozen state. For these products the first instruction may be 'Keep Frozen'.

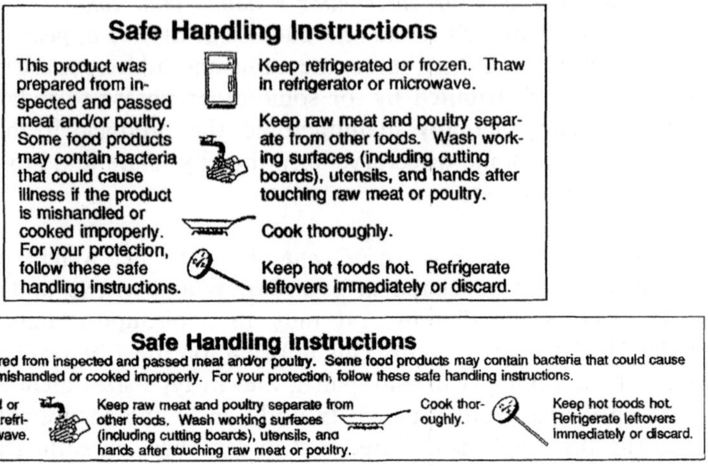

Figure 3.1 Safe food handling instructions.

(e) Inspection legend. The USDA inspection legend, or the inspection insignia of the country from which the product is exported, must also appear on the label. In addition to identifying that the product has been inspected, an official plant number appears within the legend so that any problems with the product can be traced back to the manufacturer. In some cases, the number may be embossed on the side of the package or provided on a bag clip.

3.2.3 Nutrition labels

Unless a product qualifies for a specific exemption from nutrition labeling, the nutrition label must appear on all labels of food products sold in the USA. Specific exemptions exist for small businesses, for products sold to food service establishments or another food manufacturer, for raw, single-ingredient products, for very small packages of foods and for products that contain no nutritional value. (Note that small businesses are defined differently by FSIS and FDA. For seafood and fishery companies, small businesses have less than 300 employees. For these companies, any product sold at a level below 100 000 units per year is exempt. If a company has less than 200 employees, a short-term exemption applies until May 1997 that any product sold at a level below 200 000 units per year is exempt. For meat and poultry product companies, small businesses have no more than 500 employees and any product produced at a level below 100 000 lb per year is exempt.) Even though an exemption from nutrition labeling may exist for a specific food product, if the label, labeling or advertising of the product bears a nutrient or health claim, the exemption is nullified.

Table 3.1 Nutrients that may be declared in nutrition labels[a]

Nutrient	Daily Value	Rounded to the nearest
Calories	n/a	5 if a serving contains less than 50; 10 if a serving contains more than 50
Kilojoules	n/a	
Calories from fat	n/a	
Calories from saturated fat	n/a	
Total fat (g)	65	0.5 g if a serving contains less than 5 g; whole gram if a serving contains more than 5 g; if less than 0.5 g, **must** be declared as 0.
Saturated fat (g)	20	
Polyunsaturated fat (g)	n/a	
Monounsaturated fat (g)	n/a	
Cholesterol (mg)	300	5 mg
Sodium (mg)	2400	5 mg if a serving contains less than 140 mg; 10 mg if a serving contains more than 140 mg
Potassium (mg)	3500	
Total carbohydrate (g)	30	Whole gram
Dietary fiber (g)	25	
Soluble fiber (g)	n/a	
Insoluble fiber (g)	n/a	
Sugars (g)	n/a	
Sugar alcohols (g)	n/a	
Other carbohydrates (g)	n/a	
Protein (g)	50	
Vitamin A (IU)	5000	2% DV if a serving contains less than 10% DV; 5% DV if a serving contains from 10–50% DV; 10% DV if a serving contains more than 10% DV
Vitamin C (mg)	60	
Calcium (mg)	1000	
Iron (mg)	18	
Vitamin D (IU)	400	
Vitamin E (IU)	30	
Thiamine (mg)	1.5	
Riboflavin (mg)	1.7	
Niacin (mg)	20	
Vitamin B_6 (mg)	2.0	
Folic acid (mg)	0.4	
Vitamin B_{12} (μg)	6	
Biotin (mg)	0.3	
Pantothenic acid (mg)	10	
Phosphorus (mg)	1000	
Iodine (μg)	150	
Magnesium (mg)	400	
Zinc (mg)	15	
Copper (mg)	2	

[a]DV - Daily Value. Nutrients in bold type are required declarations.

(a) Nutrients to be declared. Nutrition labels must contain declarations of the total calorie content, amount of calories from fat, total fat, saturated fat, cholesterol, sodium, total carbohydrate, dietary fiber, sugars, protein, vitamin A, vitamin C, calcium and iron in a serving of the food. Other nutrients may be declared voluntarily (Code of Federal Regulations, 1995c,i). Table 3.1 lists all the nutrients that may be declared within nutrition labels, and provides the rounding rules and daily values for each nutrient. With the exception of total calories, total fat, sodium, total car-

Table 3.2 Some selected reference amounts customarily consumed (RACC)

Product category	Ready-to-serve RACC	Ready-to-cook RACC
Bacon and breakfast strips	15 g	Poultry bacon = 26 g Meat bacon = 54 g Breakfast strips = 30 g
Dried, e.g. jerky, dried meat, dried sausages	30 g	n/a
Luncheon meats, Canadian style bacon, meat or poultry pattie crumbles, deviled ham, sandwich spreads, Lebanon bologna, potted meat food product, taco and pie fillings	55 g	n/a
Linked sausage products, smoked or pickled meat/poultry, pickled pigs feet	55 g	Poultry = 69 g Meat = 75 g
Entrées without sauce, meat cuts and poultry parts, pre-basted turkeys, meat patties, croquettes, fritters, ham, corned beef, country ham, meatballs, puréed adult foods	85 g	Fish/seafood = 110 g Meat/poultry = 114 g
Entrées with sauce, barbecue beef in sauce, minced chicken barbecue (see also the next two categories)	140 g	n/a
Mixed dishes not measurable with a cup, e.g. burritos, enchiladas, egg rolls, pizzas, pizza rolls, quiches, sandwiches, cracker and meat lunch-type packages, gyros, stromboli, burger or frank on a bun, calzones, tacos, stuffed pockets, foldovers, stuffed vegetables with meat/poultry, shish kebabs, empanadas	140 g; 195 g if topped with sauce	n/a
Mixed dishes measurable with a cup, e.g. casseroles, pot pies, pasta with meat/poultry/seafood and sauce, meat/poultry chili or chili with beans, hashes, stews, goulashes, Brunswick stew, ragouts, lasagnas	1 cup	n/a

bohydrate and protein, a required nutrient may be dropped from the nutrition label if the amount of the nutrient would be declared as zero. Dropped nutrients usually must be referenced elsewhere in the nutrition label by a statement 'Not a significant source of (nutrient).'

(b) Serving sizes. Serving sizes are expressed both in common household units and in metric units. One serving of the food is the amount, expressed in the appropriate household unit, whose weight or volume comes closest to a 'Reference Amount Customarily Consumed' (RACC) printed in the regulations (Code of Federal Regulations, 1995a–c). In selecting the appropriate household unit, household measuring cups and spoons are the preferred expression (usually used for meat, poultry, or seafood casseroles, stews and soups); number of pieces (for discrete

smaller pieces of foods such as fish sticks or chicken wings), fractions of the whole (e.g. ¼ pizza or ⅙ crab quiche) may be used if cup and spoon measurements are not applicable to the food; and ounces may be used if no other unit is applicable. Table 3.2 lists some of the more commonly used RACC amounts for meat, poultry, seafood and fishery products.

(c) Formats for nutrition labels. There are several available formats for nutrition labels. Figure 3.2 shows the regular format and a 'simplified' format. The simplified format may be used on fishery and seafood products when the food contains insignificant amounts of at least seven of the mandatory nutrients (other than calories from fat), and may be used on meat or poultry products when the food contains insignificant amounts of at least one mandatory nutrient (other than calories, total fat, sodium, total carbohydrate or protein). Figure 3.3 shows some of the formats that may be used on smaller packages having no more than 40 square inches of total surface area that could be used for labeling. Other nutrition label formats may be used to meet specific package constraints (Code of Federal Regulations, 1995a–c).

Regular Format
(product is a seafood sausage)

Nutrition Facts	
Serving Size 1 link (56g)	
Servings Per Container varied	
Amount Per Serving	
Calories 110 Calories from Fat 50	
	% Daily Values*
Total Fat 6g	**9%**
Saturated Fat 1.5g	**8%**
Cholesterol 50mg	**17%**
Sodium 240mg	**10%**
Total Carbohydrate 3g	**1%**
Dietary Fiber 0g	**0%**
Sugars 0g	
Protein 10g	
Vitamin A 4% • Vitamin C 0%	
Calcium 2% • Iron 2%	
*Percent Daily Values are based on a 2,000 calorie diet. Your Daily Values may be higher or lower depending on your calorie needs.	

	Calories:	2,000	2,500
Total Fat	Less than	65g	80g
Sat. Fat	Less than	20g	25g
Cholest.	Less than	300mg	300mg
Sodium	Less than	2,400mg	2,400mg
Total Carb		300g	375g
Fiber		25g	30g

Simplified Format
(product is an ostrich roast)

Nutrition Facts	
Serving Size 4 oz. (112g)	
Servings Per Container varied	
Amount Per Serving	
Calories 140 Calories from Fat 30	
	% Daily Values*
Total Fat 3g	**5%**
Saturated Fat 1g	**6%**
Cholesterol 85mg	**29%**
Sodium 70mg	**3%**
Total Carbohydrate 0g	**0%**
Protein 29g	
Iron 15%	
Not a significant source of dietary fiber, sugars, vitamin A, vitamin C, and calcium.	
*Percent Daily Values are based on a 2,000 calorie diet.	

Figure 3.2 Examples of regular and simplified nutrition label formats.

Columnar Format
(product is a
chicken pattie)

Nutrition Facts	
Serv Size 1 patty (76g)	
Servings 100	
Amount Per Serving	
Calories 140	Fat Cal 50
	% Daily Value*
Total Fat 6g	9%
Sat Fat 1.5g	8%
Cholest 95mg	32%
Sodium 390mg	16%
Total Carb 7g	2%
Fiber 1g	4%
Sugars 5g	
Protein 15g	
Vitamin A 2% • Vitamin C 0%	
Calcium 2% •	Iron 10%
*Percent Daily Values are based on a 2,000 calorie diet.	

Tabular Format
(product is Italian sausage)

Nutrition Facts	Amount/serving	%DV*	Amount/serving	%DV*
Serv Size 1 link (91g)	**Total Fat** 7g	11%	**Total Carb** 1g	07%
Servings varied	Sat Fat 2g	10%	Fiber 0g	0%
Calories 140	**Cholest** 60mg	19%	Sugars 0g	
Fat Cal. 60	**Sodium** 890mg	37%	**Protein** 17g	
*Percent Daily Values (DV) are based on a 2,000 calorie diet.	Vitamin A 0% • Vitamin C 0% • Calcium 0% • Iron 6%			

Linear Format
(product is a fish jerky)

Nutrition Facts Serv. Size 1 pkg. (45g) Amount/Serving: Calories 140, Fat Cal 0, Total Fat 0g (0% DV), Sat Fat 0g (0% DV), Cholest. 0mg (0% DV), Sodium 1100mg (46% DV), Total Carb 27g (9% DV), Fiber 3g (13% DV), Sugars 22g, Protein 7g, Vitamin A (6% DV), Vitamin C (10% DV), Calcium (0% DV), Iron (0% DV). Percent Daily Values (DV) are based on a 2,000 calorie diet.

Figure 3.3 Examples of nutrition label formats for use on packages having 40 square inches or less of usable label space.

3.3 Nutrient claims

3.3.1 Types of nutrient claims

Nutrient claims are not required on product labels, but they highlight the special nutritional attributes of a product and make it easy for consumers to identify products that can be most easily incorporated into a lower fat and/or sodium diet. There are three types of nutrient claims that may be made on labels: expressed nutrient claims, which are mainly adjectival terms used in association with a nutrient (e.g. 'Low Fat') and which are defined by regulation; quantitative statements of the amount or percent of a nutrient; and implied nutrient claims. Any reference to a nutrient on product labels must either be part of a nutrition label or be made in accordance with the regulations for one of these types of claims.

(a) Expressed nutrient claims. These nutrient claims are adjectival in nature (e.g. 'cholesterol free', 'low fat', 'very low sodium') and are defined by the amount of the nutrient in the RACC for the product (see Table 3.2). For products that are meals or main dishes, the claim is defined as the amount of the nutrient found in one serving (if the adjective in the claim is 'free') or in 100 g of the product (if other adjectives are used in the claim). [FDA defines meals as products weighing at least 10 oz, con-

Table 3.3 Definitions for nutrient claims[a]

Nutrient claim	Definition
'High'	⩾20% of the daily value for that nutrient
'Good source'	10–19% of the daily value for that nutrient
'More'	⩾10% more of the daily value than the reference food
'Light' or 'Lite'	⩾1/3 decrease in calories or ⩾50% reduction in fat (if percentage of calories from fat is <50%), or ⩾50% reduction in fat (if percentage of calories from fat is ⩾50%)
'Light/lite in sodium'	⩾50% decrease in sodium
'Calorie free'	<5 calories
'Low calorie'	⩽40 calories
'Sugar free'	<0.5 g sugars (the label must also bear a 'low' or 'reduced' claim or state that the product is not 'low calorie')
'Sodium free'	<5 mg sodium
'Very low sodium'	⩽35 mg sodium
'Low sodium'	⩽140 mg sodium
'Salt free'	<5 mg sodium, or label must state that the product is not sodium free
'Fat free'	<0.5 g total fat
'Low fat'	⩽3 g total fat
'Saturated fat free'	<0.5 g saturated fat; ⩽1% of fatty acids in *trans* configuration
'Low saturated fat'	⩽1 g saturated fat and ⩽15% of calories from saturated fat
'Cholesterol free'	<2 mg cholesterol; ⩽2 g saturated fat
'Low cholesterol'	<20 mg cholesterol; ⩽2 g saturated fat
'Lean'	<10 g total fat, <4 g saturated fat and <9 mg cholesterol (per RACC and per 100 g of food)
'Extra lean'	<5 g total fat, <2 g saturated fat and <9 mg cholesterol (per RACC and per 100 g of food)
'*X*% of fat free'	⩽3 g total fat

[a]Source: Code of Federal Regulations (1995a–c).

taining at least 40 g each of ingredients from three of the old, basic four food groups (meat, poultry, fish, egg and legume group, dairy group, fruit and vegetable group and grains/cereal group) and being labeled as intended to be consumed as a meal. USDA defines meals as products weighing at least 6 oz, containing ingredients from at least two of the old, basic four food groups, and being labeled as intended to be consumed as a meal. FDA defines main dishes as products weighing at least 6 oz, containing at least 40 g each of ingredients from at least two of the old, basic four food groups (meat, poultry, fish, egg and legume group, dairy group, fruit and vegetable group and grains/cereal group) and being labeled as intended to be consumed as the entree or main dish portion of a meal. USDA includes main dishes within the definition of meals.] A list of nutrient claim definitions is given in Table 3.3.

(b) Comparative nutrient claims. If the product does not meet one of the definitions for an adjectival nutrient claim, it may still be possible to make a comparative claim, such as 'Less Fat' or 'Lower Sodium' on the label.

These claims compare the amount of the nutrient in the food with the amount found in a comparable product. The comparable product is called the 'reference food', and there must be at least a 25% difference in the amount of the nutrient in the labeled food and the reference food to make a comparative claim. To assure that nutrient reductions are meaningful, the reference food may not meet the 'low' definition for the nutrient in the claim. Even a 50% reduction in fat would be meaningless if the reference food only contained 2 g of fat. In addition, to avoid comparisons to obscure products, the reference food must be one of the following:

- a food entry from a representative valid data base, or
- an average calculated from the nutrient contents of the top three national or regional brands for the product, or
- a market leader for the product, or
- your company's regular product, or
- a competitor's product.

The last option may only be used for labels of game meat, seafood and fishery products because USDA regulations do not allow this option on meat and poultry products (Code of Federal Regulations, 1995a–c).

Any comparative claim must identify what food the product is compared with (the reference food), the percentage or fractional difference between the amount of the nutrient in the labeled product and the amount in the reference food, and the amount of the nutrient in one serving of the labeled product and in one serving of the reference food. A typical comparative claim about fat would be '30% Less Fat than Our Regular Brand' accompanied by the explanatory phrase 'Fat reduced from 10 g to 7 g per serving.' The explanatory phrase may either be adjacent to the main part of the nutrient claim or may be printed adjacent to the nutrition label.

Lite/light claims. 'Light' or 'lite' is also a comparative claim, usually about fat content, but use of this claim requires at least a 50% reduction in fat instead of the 25% reduction required for other comparative claims (Code of Federal Regulations, 1995a–c). If the food product derives less than 50% of its calories from fat, however, 'lite/light' may be a one third reduction in calories. The explanatory language required for 'lite/light' claims must include the percentage or fractional reduction in both calories and fat and also the amount per serving in the labeled food and the reference food for these two nutrients. The reference food for 'lite/light' claims may not be a company's regular product or a competitor's product. 'Lite/light in sodium' claims require a 50% reduction in sodium.

(c) Quantitative statements about nutrients. Other nutrients claims may be factual quantitative statements of the amount or percent of a nutrient

in the food (Code of Federal Regulations, 1995a–c). The product does not have to comply with any of the nutrient claim definitions to be labeled with this type of statement. If the statement is phrased in such a way as to imply that the level of the nutrient is either a low (in the case of undesirable nutrients, such as fat or sodium) or high (in the case of desirable nutrients, such as protein or iron) level, and if the food does not meet the definition for a low or high level of that nutrient, the statement must be qualified to clarify this (e.g. 'Only 5 g of fat – Not a low fat food'). Quantitative statements may not be used to label a product as '*X*% Fat Free' or '*X*% Lean' unless the product meets the definition for 'low fat'.

(d) Implied nutrient claims. A third type of claim, implied nutrient claims, is usually a statement about the use or non-use of a food ingredient that is understood to be low or high in a particular nutrient. For example, a statement on a label for beef burritos of 'No Lard' implies that the burrito is low in saturated fat and cholesterol. These implied claims may be made on labels if the product meets the definition for the claim that is implied – low cholesterol and low in saturated fat in the case of the beef burrito label. In another example, a crab pattie label may bear the implied nutrient claim 'Browned in 100% Canola Oil', implying that the product is low in saturated fat. As in the beef burrito example, this implied claim may be made on the label if the crab pattie meets the definitions for a 'low in saturated fat' food, but FDA requires the expressed nutrient content claim to also appear on the label (Code of Federal Regulations, 1995c).

(e) 'Healthy' claims. Another type of nutrient content claim is a claim that the product is 'healthy', or uses any derivative of the word 'healthy' (Code of Federal Regulations, 1995a–c). This claim differs from the others in that it involves maximum levels of four nutrients for most foods and a minimum level of a fifth nutrient. The definition for 'healthy' raw, single-ingredient products, however, differs from the definition for other 'healthy' foods. 'Healthy' raw, single-ingredient products must meet the definition for 'extra lean' and must contain at least 10% of the Daily Value for protein, dietary fiber, vitamin A, vitamin C, calcium or iron. A maximum sodium content of these foods is not defined, but is indirectly controlled by being limited to the inherent sodium in the raw, single-ingredient food.

'Healthy' cooked and/or multiple-ingredient foods must meet the definitions for 'low fat' and 'low saturated fat', may contain no more than 60 mg of cholesterol and 480 mg of sodium per RACC and per serving and must contain at least 10% of the Daily Value for protein, dietary fiber, vitamin A, vitamin C, calcium or iron. The sodium content in this definition will change in the year 2000 for both USDA-regulated foods and FDA-regulated foods to be a maximum sodium content of 360 mg.

3.3.2 General rules for nutrient claims

(a) Nutrition labeling is mandatory if a nutrient claim is made. Although the product may be exempt from nutrition labeling requirements because the manufacturer is a small business, the product is destined for food service, or the product is not sold in the USA, if a claim about fat, sodium or any other nutrient is made on the label, labeling or advertising of the product, nutrition labeling is mandatory. 'Labels' are written and/or graphic information affixed to the product, 'labeling' is the same kind of information that accompanies the product and 'advertising' is information about the product that is distributed to the general public.

(b) Nutrient claims inherent to the food. There is a general labeling rule that prohibits special claims appearing on a label unless the converse of the claim would be likely. For this reason, any claim made about a food that is inherently true for that food must be phrased to assure that consumers understand this fact (Code of Federal Regulations, 1995a–c). For example, all shrimp is low in fat so a 'low fat' claim on a boiled shrimp package would be phrased 'Shrimp is a Low Fat Food'. If the food is specifically processed (e.g. by special trimming) or formulated (made with leaner meat or a high fat usual ingredient has been omitted from the formula) to make it eligible to bear the nutrient claim, the nutritional attributes in the claim are not considered inherent.

(c) Foods with small reference amounts customarily consumed. If the RACC for a food is relatively small (no more than two tablespoons or no more than 30 g), a 50 g portion of the food must meet the nutrient criteria in the defined claim. The rationale for this is that consumers may believe that they can safely consume larger than usual amounts of foods that are low in fat and/or sodium. Foods that meet a nutrient claim definition simply because the food is usually consumed in small amounts may not be labeled with the defined nutrient claim.

(d) Referral statements. If the nutrition label appears on a different side of the food package from a nutrient claim, FDA requires a referral statement to be placed adjacent to the claim telling consumers where to find the nutrition label (Code of Federal Regulations, 1995c). The USDA does not require this referral statement. The wording for the

referral statement is 'See ______ for nutrition information', and the blank is filled in by the side of the package where the nutrition label appears (e.g. 'back', 'right side', 'bottom'). The statement must be printed in boldface type that distinctly contrasts from other printed material on the label, and in a type size no smaller than the size required for the net weight statement.

To assure that consumers are alerted to any adverse nutritional attributes of a food product bearing a claim about a different nutrient, if the food contains moderately high levels of total fat, saturated fat, cholesterol or sodium, this is indicated in the referral statement. If the total fat content is more than 13 g, the saturated fat content is more than 4 g, the cholesterol content is more than 60 mg or the sodium content is more than 480 mg per serving or per RACC, the nutrients exceeding this level are included as part of the referral statement (e.g. 'See back for information about sodium and other nutrients').

3.4 Compliance

Because food comes from living organisms and because no two living organisms are exactly the same in composition, there are compliance tolerances for nutrition labeling and nutrient claims. Added nutrients, either by use of fortification or enrichment, may not be present at levels lower than those stated on the label (Code of Federal Regulations, 1995a–c). The USDA does not allow fortification or enrichment of meat or poultry products, and so the USDA does not have a special compliance tolerance for added nutrients.

3.4.1 Over- or under-declared nutrients

Nutrients that are inherent in the food or food ingredients may be over-declared or under-declared on the label as long as the declarations are within good manufacturing practices. For nutrients that consumers may wish to limit in their diets (e.g. calories, calories from fat, total fat, saturated fat, cholesterol, sodium and sugars), the analytical amount of the nutrient in the amount of food declared in the metric serving size may not exceed 120% of the label declaration. For nutrients that consumers may wish to increase in their diets (e.g. total carbohydrate, dietary fiber, protein, vitamins and minerals other than sodium), the analytical amount of the nutrient in the amount of food declared in the metric serving size may not be less than 80% of the amount declared on the label (Code of Federal Regulations, 1995a–c). The same tolerances are applied to the nutrient content claims.

3.5 International policies

3.5.1 Policies outside the USA

Although Canada requires a declaration of the meat product protein content on labels of products that contain phosphate salts or water as ingredients (Canadian Food and Drug Regulations, 1994), only the USA requires nutrition labeling to be provided on the labels of any products sold in commerce. When nutrition information is provided, it must be in the language of that country and comply with any requirements for declaration of nutrients, format, etc. Canada's nutrition labeling rules are much like those used in the USA prior to August 1994 (Agriculture and Agri-Food Canada, 1992). Unfortunately, it is not possible to satisfy both US and Canadian requirements with one nutrition label. This type of disparity between countries' rules for labeling causes products to be labeled differently when intended for export. Other required labeling features in other countries essentially mirror the US requirements, and a translation into the language of the country in which the product is sold is usually all that is needed. (Information about a specific country's labeling requirements may be down-loaded from the USDA International Program's Export Library System. Information on accessing the system may be obtained by contacting the Program Development Division, IP, FSIS, USDA, Room 3718 Franklin Court, Suite 3700W, Washington, DC 20250.) Many food packages are labeled with one country's requirements on one side of the package and a different country's requirements on another side. These types of food packages can be sold in either country.

3.5.2 Exemptions to nutritional labeling and claims

Exemptions from US nutrition and nutrient claim regulations apply to labels of products sold only outside the USA. Although it is possible to have different labels for the same products, the added expense of multiple labels for a single product may discourage exports. As we work toward lessening non-tariff trade barriers among different countries, there may be a greater willingness to accept other labeling schemes that contain the same information as is required in the country in which the product is sold.

3.6 Summary

The minimum requirements for labeling of lower fat and/or sodium meat, poultry, game meat, seafood and fishery products is the same as for other food products. These minimum requirements include a product name, list of ingredients (if the product contains two or more ingredients), the manufacturer's or distributor's name and address, a statement of the net weight or quantity of contents and nutrition labeling unless the product or manufacturer is specifically exempt from providing nutrition information on the label. In addition, if the product is not shelf stable, the label must bear a statement about proper storage, such as 'keep refrigerated' or 'keep frozen'. Products under the regulatory jurisdiction of the USDA must also be labeled with the official USDA inspection legend.

Voluntary information, such as nutrient claims, can be used on these products to enhance the marketing of these foods to consumers who will value their special nutritional attributes. There are a variety of nutrient claims that may be considered, and most products that have been specially formulated to reduce the fat, saturated fat, cholesterol or sodium content will be eligible to be labeled with some kind of claim. The rules for nutrition labeling and nutrient claims are extensive and complex; care should be taken in designing labels for these products.

References

Agriculture and Agri-Food Canada (1992) *Nutrition Labeling Handbook*, Food Division, Consumer Products Branch, Consumer and Corporate Affairs Canada, Hull, Quebec.

Canadian Food and Drug Regulations (1994) *Packaging and Labeling*, Agriculture Canada, Ottawa, Ontario, Chapter 7.

Code of Federal Regulations (1995a) *Title 9, Section 317, Subpart B*, US Government Printing Office, Washington, DC.

Code of Federal Regulations (1995b) *Title 9, Section 381, Subpart Y*, US Government Printing Office, Washington, DC.

Code of Federal Regulations (1995c) *Title 21,. Subparts A–C*, US Government Printing Office, Washington, DC.

Code of Federal Regulations (1995d) *Title 9, Sections 317.2, 381.117*, US Government Printing Office, Washington, DC.

Code of Federal Regulations (1995e) *Title 9, Sections 317.2, 381.118*, US Government Printing Office, Washington, DC.

Code of Federal Regulations (1995f) *Title 9, Sections 317.2, 381.122*, US Government Printing Office, Washington, DC.

Code of Federal Regulations (1995g) *Title 9, Sections 317.2, 381.121*, US Government Printing Office, Washington, DC.

Code of Federal Regulations (1995h) *Title 9, Sections 317.2, 381.125*, US Government Printing Office, Washington, DC.

Code of Federal Regulations (1995i) *Title 9, Sections 317.309, 381.409*, US Government Printing Office, Washington, DC.

Code of Federal Regulations (1995j) *Title 9, Sections 317.400, 381.500, Title 21, Section 101.9*, US Government Printing Office, Washington, DC.

Code of Federal Regulations (1995k) *Title 9, Sections 317.2, 381.123*, US Government Printing Office, Washington, DC.

Code of Federal Regulations (1995l) *Title 9, Sections 319 and 381, Subpart P*, US Government Printing Office, Washington, DC.

Code of Federal Regulations (1995m) *Title 21, Parts 102, 161*, US Government Printing Office, Washington, DC.

Code of Federal Regulations (1995n) *Title 21, Section 101.30*, US Government Printing Office, Washington, DC.

Code of Federal Regulations (1995o) *Title 9, Section 319.141*, US Government Printing Office, Washington, DC.
Food Labeling Division (1995) *Standards and Labeling Policy Book*, US Government Printing Office, Washington, DC.
Wade, C. (1995a) *Policy Memorandum 121B, Labeling of Modified, Substitute Versions of Fresh (Species) Sausage, Hamburger, or Ground Beef Products with Added Ingredients Used to Replace Fat that Qualify for Use of Certain Nutrient Contents Claims Associated with a Reduction in Fat Contents*, Food Labeling Division, FSIS, USDA, Washington, DC.
Wade, C. (1995b) *Policy Memorandum 123, Modified Breakfast Sausage, Cooked Sausage, and Fermented Sausage Products Identified by a Nutrient Content Claim and a Standardized or Traditional Name*, Food Labeling Division, FSIS, USDA, Washington, DC.

4 Principles and applications in production of reduced and low fat products

A.M. PEARSON

4.1 Introduction

As indicated in Chapter 1, there has been an increase in consumer demand for reduced and low fat meat, poultry and fish products. Consumer awareness of the influence of dietary fat levels on health and fitness and the relationship between the fat content of the diet and certain types of degenerative diseases have had a profound effect on the demand for reduced and low fat products, as explained in Chapter 1.

The demand for low fat or reduced fat products has resulted in a divergence of opinions as to the meaning of these terms. Chapter 3 discusses the definitions and regulations pertaining to both low fat and reduced fat products, but also to reduced and low salt products. The definitions are variable in different countries, as might be expected. For example, in many countries of the world where energy is in short supply, it may not be necessary to consider fat content as fat actually could be desirable. On the other hand, the industrialized countries may develop labeling requirements so that consumers can choose low fat or low salt products. Thus, the labeling laws should reflect the needs of consumers in the country of concern and may not apply to other countries. Chapter 3 does, however, explain the basis for labeling regulations and could serve as a guide in developing regulations.

Reduced and low fat products have constituted a growing segment of the food industry and demand for such products has contributed to the production of low fat meat, poultry and fish products and their increasing proportion of the market. The trend for a greater demand for low fat products by a health conscious public, although often not always well informed, seems likely to continue. Hence, this chapter will briefly review some of the health related issues and production and processing procedures that can be utilized in reducing the fat content of such products in order to meet consumer demands for low fat meat, poultry and fish products.

4.2 Rationale for reducing the fat-energy levels in meat, poultry and fish

The US Senate Select Committee on Nutrition and Human Needs (1977) emphasized the high incidence of deaths from degenerative disease, such as coronary heart disease (CHD), stroke and cancer, and specifically recommended a reduction in red meat consumption as a preventative procedure. Their recommendations also included a decrease in the proportion of saturated fatty acids in the diet. These recommendations, no doubt, have led to changes in eating habits of the American public, with a subsequent decline in meat consumption and an increase in per capita intake of poultry, as pointed out by Briggs and Schweigert (1990). The Surgeon General's Report on Nutrition and Health (1988), on the other hand, has recommended moderate amounts of meat in the diet as being healthful, as have a number of other nutritionists (Schweigert *et al.*, 1985; American Dietetic Association, 1986; Hansen *et al.*, 1986; Cronin *et al.*, 1987; Iacono, 1987; Cronin and Shaw, 1988; Watts *et al.*, 1988; Frantz *et al.*, 1989). The recent Surgeon General's Report (1995) on Dietary Guidelines for Americans has recommended 'choose a diet with most of the calories from grain products, vegetables, fruits, low fat milk products, lean meats, fish, poultry and dry beans. Choose fewer calories from fats and sweets'. Thus, the question is not whether meat is a desirable dietary constituent, but rather whether the amount of fat contributed by meat, poultry and fish products can be reduced by various procedures and still maintain the desirable palatability attributes.

4.2.1 Health issues associated with excess fat intake

The major health concerns from excess fat-energy intake include obesity, cardiovascular disease, stroke and cancer (US Senate Select Committee on Nutrition and Human Needs, 1977). Each of these issues will be summarized briefly.

(a) Obesity. One of the main health concerns in affluent populations is obesity, which is the result of energy intake in excess of needs (American Dietetic Association, 1986; Abrams, 1987; Dreon *et al.*, 1988; Surgeon General's Report on Nutrition and Health, 1988; American Heart Association, 1988). Obesity is associated with an increased risk for certain types of cancer, i.e. breast cancer in women and colon cancer in both men and women. The incidence of diet-associated cancers is not necessarily related to intake of fat, but may be due to excess energy in the diet regardless of the source (Dreon *et al.*, 1988).

The contribution of fat to obesity is due to the well known fact that fat is energy dense providing 2.25 times as much energy per unit as carbohydrates and protein (Maynard and Loosli, 1962). This has led to the Surgeon General's (1988) recommendation that less than 30% of the calories in the diet be from fat, which is based in part on the fact it is difficult to obtain the essential nutrients within the specified calorie restriction. The recommendation is impossible to achieve using high fat meat products in which as much as 70% of the calories are derived from fat. Excessive energy intake from fat or any other source can lead to obesity in both humans and other animals. Although fattening may be desirable in meat animals to improve both the yield and eating qualities of the meat, excessive fat adds calories and beyond the amount needed is undesirable. From the human nutrition standpoint, however, fat is undesirable only if intake is in excess of energy expenditures, in which case it may contribute to obesity (American Dietetic Association, 1986).

(b) Cardiovascular Disease (CVD). CVD is known to be associated with high blood cholesterol, which may be deposited in the blood vessels, where it may partially or completely block the blood vessels to the heart. Although blood serum cholesterol is closely related to CVD, dietary levels of cholesterol *per se* do not have a major effect on most humans, as pointed out by McNamara (1990).

Although saturated fatty acids have been implicated as being hypercholesterolemic, neither stearic (saturated) nor oleic (monounsaturated) acids are cholesterolemic (Hegsted *et al.*, 1965; Keys, Anderson and Grande, 1965). These are the main fatty acids in animal fats, and actually may decrease blood levels of cholesterol (Grundy, 1986a,b; Bonanome and Grundy, 1988). Reiser and Shorland (1990), in reviewing the relationship between dietary fatty acids and blood serum cholesterol levels, concluded 'The dietary cholesterol and saturated fat theory of hypercholesterolemia is simplistic, ignoring normal (nonpathological) variations in levels of plasma cholesterol'. They further pointed out that such information misinterprets 'to the public the statistics of epidemiological and research data as applying to all members of a population, but which, in fact, apply only to a small genetically aberrant minority'.

Fish oils contain considerable quantities of n-3 fatty acids, which help to reduce the cholesterolemic effects of other fatty acids, as shown by von Lossonczy *et al.* (1978). Thus, n-3 fatty acids may contribute to a reduction in CVD. The fatty acids from animal sources, particularly from fish, therefore, may have a beneficial effect on CVD (Reiser and Shorland, 1990).

(c) Stroke. Stroke or apoplexy is the sudden loss of consciousness caused by arterial occlusion of the blood supply to the brain and its consequences, as indicated by the US Senate Select Committee on Nutrition and Human Needs (1977) and Surgeon General's Report on Nutrition and Health (1988). As explained, stroke is usually accompanied by paralysis, loss of coordination and other symptoms, which are the direct result of the lack of blood to different areas of the brain. The decreased blood supply to the brain is probably due to an embolism or blood clot in the blood vessels supplying the brain. It is frequently associated with high

blood pressure and complete or partial blockage of the blood vessels to the brain by cholesterol deposits.

Although dietary factors may be involved in the etiology of stroke, there is no concrete evidence that meat, poultry or fish are contributors. There is, in fact, evidence that the monounsaturated fatty acids, which are major constituents of animal fats, either lower serum cholesterol levels or are neutral (Grundy, 1986a,b; Bonanome and Grundy, 1988). The relationship of meat fats to serum cholesterol levels has been discussed in greater detail by Reiser and Shorland (1990), while McNamara (1990) concluded that there is no good evidence that dietary intervention is desirable for the public as a whole. This does not mean, however, that diet is without effect on the incidence of stroke. It does support the concept that persons with atherosclerosis and high blood pressure should be under the care of a competent physician (McNamara, 1990; Reiser and Shorland, 1990).

Fish and fish oils contain appreciable levels of ω-3 fatty acids, which are known to have a cholesterol lowering effect, as explained by von Lossonczy *et al.* (1978). Meat also contains low levels of n-3 fatty acids (Reiser and Shorland, 1990). The beneficial effects of ω-3 fatty acids presumably may decrease the incidence of stroke.

(d) Cancer. Although meat diets have been suggested to contribute to the incidence of cancer in humans, the evidence is circumstantial and could well be due to other dietary factors, as explained by Kritchevsky (1990). As already mentioned, the increase in cancer may well be due to excess levels of energy or dietary fat from both plant and animal sources according to Briggs and Schweigert (1990), and not due to meat *per se*.

4.3 Problems encountered in producing low fat products

A number of problems frequently occur in the production of reduced and low fat products, which affect important sensory properties (Pearson *et al.*, 1987). The specific areas in which reduction of fat content may influence acceptance include flavor, tenderness, juiciness and color. Although related to the above items, even more important may be the dryness and rubberiness that is often associated with reduced and low fat meat products (Valvano, 1983; Pearson *et al.*, 1987).

4.3.1 Flavor

Flavor is undoubtedly the most important palatability attribute of all muscle foods (Reineccius, 1994), although tenderness is more variable (Dikeman, 1987). Humans eat meat, poultry and fish products because of their desirable flavor and their satiety factor, which in part may be associated with a minimum fat content (Pearson, 1976). Although a number of researchers (Koohmaraie, 1992; Savell, 1992) have stated that tenderness is the single most important component of meat quality, this applies only to steaks and oven roasts in which flavor is readily apparent and already has been deemed to be acceptable. Taste is one of the key sensory attributes that makes foods desirable and satisfying to the appetite of humans (Beauchamp and Brand, 1994). Consumers prefer muscle foods because of their desirable flavor.

Numerous studies (Bender, Wood and Palgrave, 1958; Hornstein and Crowe, 1960a,b, 1963; Kramlich and Pearson, 1960; Santoro and Landmann, 1962; Wasserman, 1979; MacLeod and Seyyedain-Ardebili, 1981) have determined some of the chemical components that contribute to meat flavor, but to date researchers have been unsuccessful in duplicating the flavor of meat, poultry and fish. This suggests that meat flavor is due to a large number of components in proper balance, both water- and fat-soluble substances, to produce the desirable flavors that are characteristic of meat, poultry and fish products (Reineccius, 1994), and their reactions on the chemical sense of humans (Beauchamp and Brand, 1994). The compounds responsible for meat flavor *per se* are water soluble (Kramlich and Pearson, 1960), while the species specific flavors are due to the fat-soluble components (Hornstein and Crowe, 1960a,b; Wasserman, 1972, 1979). Anything that causes a diminution or alteration in the desirable flavor characteristics of the muscle foods is obviously a problem. Flavor problems and procedures for overcoming those encountered in low fat products will be covered in Chapter 13.

Low fat or reduced fat products are frequently criticized for their lack of flavor. The problem could be a direct result of reduction in flavor compounds or be due to dilution and masking of flavor by additives that are added to replace fat and bind water. Gray, Pearson and Monahan (1994) have discussed some flavor and aroma problems and the measurement of the problems in meat, poultry and fish products. The effects of reducing the fat content on flavor will be discussed in greater detail in Chapter 13.

4.3.2 Tenderness

Dransfield (1994) has reviewed some of the factors that influence tenderness, including preslaughter influences, slaughtering, rigor development, muscle shortening, the effects of pH, post-rigor handling and artificial tenderization, and their impact on ultimate tenderness. Chrystall (1994) has discussed the various methods utilized to measure tenderness (texture), including subjective and objective assessments, structural changes, chemical measurements and other possible methods of assessment.

It is well established that there is a low but important relationship between marbling and tenderness for intact meat cuts (Wellington and

Stouffer, 1959; Blumer, 1963). On summarizing a number of reports in the literature, Blumer (1963) concluded that marbling or intramuscular fat accounted for less than 10% of the variation in meat tenderness. Dikeman (1987) demonstrated that Warner–Bratzler shear values decreased on average as external fat thickness increased. Although there was considerable overlap, on average each increase in fat thickness by groups resulted in an improvement in tenderness. Similarly, Dikeman (1987) found as marbling increased that the panel scores for tenderness improved within the A maturity group, and there was a similar trend in the A + B maturity group. Savell *et al.* (1987) found a positive relationship between marbling and consumer tenderness ratings for beef shortloin steaks, although there was some overlap between groups. These studies demonstrate that fat content, especially marbling, tends to improve tenderness.

A number of studies with ground beef have shown that a certain amount of fat is needed to give maximum consumer responses (Cole, Ramsey and Odum, 1960; Cross *et al.*, 1975; Drake *et al.*, 1975). Pearson (1976) reviewed consumer studies on effects of fat levels upon the acceptability of ground beef and concluded that about 20–25% fat is needed to give the maximum consumer response. This could, of course, be due to the dryness and toughness that develops as cooking takes place in low fat products. Conversely, it could be associated with the dryness and lack of juiciness, which will be discussed below.

4.3.3 Juiciness

Juiciness assessment can only be consistently achieved using sensory methods (Winger and Hagyard, 1994). Early research indicated that juiciness could be separated into two components. The first is the impression of wetness during the first few chews, which is produced by the rapid release of meat fluids, while the second is due to the sustained juiciness that is apparently associated with the stimulation of salivary flow and the coating of fat on the teeth, tongue and mouth (Gaddis, Hankins and Hiner, 1950; Weir, 1960). Harries, Rhodes and Chrystall (1972) and Dransfield, Jones and Robinson (1984) later demonstrated that initial wetness and overall juiciness could be combined into a single attribute, which simplifies the overall analysis, yet juiciness is still the least understood sensory trait of meat.

Recently, Winger and Hagyard (1994) have reviewed the literature on the importance of juiciness and some contributing factors. Honikel and Hamm (1994) have discussed measurement of water-holding capacity and its relationship to juiciness. As indicated in the above reviews, juiciness and its relationship to the water-holding capacity of muscle foods are important characteristics of acceptability, although they are poorly understood.

Both Wellington and Stouffer (1959) and Blumer (1963) found a low but statistically significant relationship between juiciness and marbling. Although the relationships between marbling and juiciness and tenderness are both low, the fact that both of these important sensory traits are positively related to intramuscular fat content emphasizes the importance of fat for maximum consumer acceptance. This does not necessarily mean that high fat levels are needed for acceptance since some meat, poultry and fish products of relatively low fat content are highly acceptable. Examples of this can be found in fillet mignon steaks from beef, ham from pork, chicken and turkey breasts and a number of species of fish that all are low in fat. All of these products are in high demand.

The trend in producing low fat products is not new, although it has become of greater interest with consumer demands for low fat products taking over a greater portion of the market in recent years. John Morrell and Company produced a reduced fat frankfurter in the early 1970s that was not successful because it was ahead of its time and cost more to produce (W.E. Kramlich, personal communication). It is common knowledge that McDonald's McLeanburger, which has a reduced fat content, has not taken over the market as expected, perhaps because of cost or even more importantly as a result of requiring special cooking and handling to maintain acceptance.

4.3.4 Color

Color in meat, poultry and fish products is an important sensory trait. Cornforth (1994) has reviewed the importance of color and some factors that influence its acceptability, while MacDougall (1994) has discussed its measurement. These two reviews are timely and explain both the role of color and methods of measuring it in meat, poultry and fish products.

Raw meat acceptance is based upon the desirable bright red color that is characteristic of the presence of oxymyoglobin. Greene, Hsin and Zipser (1971) concluded that when the proportion of brown metmyoglobin reaches or exceeds 30–40% of the total pigments consumers find it objectionable and will not purchase it. Estimates indicate that 4–10% of all meat in retail markets must be discounted, processed into hamburger or else discarded as a result of its brown color (Cornforth, 1994). Other color problems encountered include pale, soft and exudative (PSE) and dark, firm and dry (DFD) pork, dark cutting beef and cooked meat defects, such as brown color in hamburger, and pink color in cooked turkey rolls and pre-cooked bratwurst, which are discussed by Cornforth (1994). These problems can all occur in low or reduced fat products and need to be understood by processors of these products.

Color and appearance of low fat products also constitute an important problem. As other additives are used to replace fat, they tend to dilute the color of meat, poultry and fish items. This is especially true on using water binding agents, which reduce the myoglobin concentration in low or reduced fat products. Frequently particles of water binding agents can be seen in the low fat products, making them unattractive to consumers (Dickson *et al.*, 1995). These effects, which have been shown to be objectionable to consumers, are common knowledge to food scientists, although they are not well documented in the scientific literature. Nevertheless, color and appearance are important problems in some low fat or reduced fat products and need careful consideration by processors, especially in their choice of water binding agents. Overcoming color problems in low fat products will be discussed in more detail in Chapter 13.

4.3.5 Dryness–rubberiness

As pointed out in a review by Pearson *et al.* (1987), the most serious problem encountered in the production of reduced or low fat meat products is the dryness and/or rubberiness. The dryness and rubberiness of low fat products are apparently associated with the lack of juiciness and the toughness in low fat products and can be overcome by adding water binders. As indicated in the above review, addition of substances that bind water can improve the juiciness and prevent the rubberiness that is common in low fat products. Various fat substitutes include a number of synthetic fat replacers, such as Olestra, which was developed by Proctor and Gamble and recently approved for use in snack foods. In addition, there are a considerable number of other fat mimetics. The structure and mechanism of action of these compounds are reviewed in Chapter 9. In addition, various synthetic and naturally occurring fat replacers will be reviewed in Chapters 9 and 10. These compounds include the gums, starches and hemicelluloses, the hydrocolloids and alginates and the carrageenans. These substances have all been investigated as fat replacers for meat, poultry and fish products. Apparently these compounds improve the gel matrix so the products become more juicy and have a softer mouth-feel.

Still, two other groups of products that can be utilized to bind water or mimic fat include additives from plant and animal sources, which have been widely used in meat, poultry and fish processing. The plant products include the soy proteins, glandless cottonseed meal, sunflower seed proteins, canola seed proteins, starches and flours from the cereal grains and potatoes and konjac flour. These plant products will be discussed in Chapter 10, as will various animal products. The animal proteins include deboned poultry and fish and have been widely utilized in processed products from muscle foods. Other animal proteins such as collagen, blood proteins, milk proteins and surimi also have been used or show promise in extending or maintaining the quality of low or reduced fat muscle foods.

4.4 Reducing fat content by production practices

A number of production methods can be utilized for reducing the fat content of meat, poultry and fish. These include breeding and selection, feeding and management and trimming of excess fat from carcasses and/or cuts. Each of the above procedures will be discussed briefly here, with some of them being discussed in greater detail in other chapters.

4.4.1 Breeding and selection

Kirkpatrick and Rutledge (1991) have reviewed the effects of genetics upon growth and the composition of gains in domestic farm animals. They pointed out that breeding and selection of leaner animals can greatly alter the composition of the carcasses and cuts produced. It is recognized that breeding leaner animals may provide a solution to excessive fatness, but genetic improvement requires time and is not a short-term solution to the problem. It is recognized that there are marked differences in the value of different carcasses, both within and between breeds. This is true for cattle, pigs and sheep, and perhaps to a lesser extent for poultry and fish. Although breeding and selection offer a solution to the problems associated with excessive fat production, they are not without their own problems, e.g. fatness is known to be directly related to dressing percentage or live to carcass yield. Some breeds that are lean are slow gainers or less efficient in converting feed to carcass gains. Nevertheless, progress has been made in overcoming some of these problems and has resulted in improvement in carcass leanness.

Chapter 7 will cover reducing the fat content by breeding and selection with cattle, sheep and pigs. Double muscling and its potential for improvement of lean yields will be discussed, which has been successfully used in commercial production of broad-breasted turkeys. The potential of utilizing the callipyge gene in production of leaner lambs will be covered, as will the possibility of improvement through utilizing transgenic animals (Pursel *et al.*, 1988, 1989). These tools will be reviewed in Chapter 7.

4.4.2 Feeding and management

Production practices, such as limited feeding, have been widely used in Europe and the UK as a means of controlling excessive fat deposition in pigs (McMeekan, 1940; Widdowson and Lister, 1991) and cattle (Berg and Butterfield, 1976). Limited feeding, however, has not been in favor in the USA. The composition of the ration can also be utilized to limit intake, with high fiber, bulky feeds reducing the rate of gains and carcass fatness of pigs (Hochstetler *et al.*, 1959; Hoefer *et al.*, 1963).

One method of altering composition in beef cattle has been the selection of larger breeds and slaughtering them at an earlier stage of their growth curve. This procedure has been discussed by Tulloh (1964) and Dickerson (1978). Slaughter at an earlier stage of the growth curve produces leaner carcasses and has resulted in the use of large continental breeds of cattle in US feedlots to help in production of leaner carcasses and cuts. Crossbreeding utilizing large terminal sire breeds is also a common practice, in which both the male and female offspring are all marketed. This is common in cattle, sheep, broilers and turkeys. Another means of increasing leanness is by raising intact (uncastrated) males, which has resulted in a large number of young bulls going to market, since uncastrated males are generally leaner than their castrated mates. The effects of sex condition on leanness has been reviewed by Purchas (1991). Chapter 8 will review the influence of production practices in reducing the fat content of meat, poultry and fish products.

There are a number of other production practices that may influence fatness, which include the use of hormones, both estrogens and androgens (Hancock, Wagner and Anderson, 1991), antibiotics (Hays, 1991), ionophores (Owens, Zorilla-Rios and Dubeski, 1991), somatotrophin (Beermann and DeVol, 1991) and β-adrenergic agonists (Moloney *et al.*, 1991). Some of these compounds are approved, whereas others are not, and their effects are often variable. Chapter 8 will discuss the effects of these compounds on composition of the products produced.

4.4.3 Trimming of excess fat

Still another procedure that can greatly reduce the fat content of overfat carcasses and cuts is the removal of excess fat by trimming. This is commonly practiced by meat packing companies or supermarkets in the USA (Savell *et al.*, 1987). Meat packers today commonly remove the kidney fat at their plants, which earlier was left in the carcasses. Retail stores and supermarkets then usually remove some of the excess fat from the retail cuts, which renders them more desirable to purchasers. It is a common practice in US supermarkets to reduce the fat covering on retail cuts to ¼ inch (6.5 mm) or less (Savell *et al.*, 1987).

Although trimming or removal of excess fat from carcasses and cuts provides a temporary or short-term solution, the production of excess fat is a wasteful process as it requires 2.25 times as much energy to produce a unit of fat as it does to produce an equivalent amount of protein (Maynard and Loosli, 1962). Hence the long-term solution to production of excess fat should be by breeding or else by other production practices.

4.5 Reducing fat content by processing procedures

Reduction of the fat content can also be achieved during processing. Commonly, the reduction in fat content is accomplished by adding substances that bind water, which overcomes the dryness and rubberiness and improves juiciness (Pearson *et al.*, 1987). The additives that bind water can be classified as protein additives, nonprotein additives and other procedures, all of which will be briefly discussed.

A promising but not widely used processing method for reducing the fat and cholesterol contents of meat, poultry and fish products, namely supercritical fluid extraction (Clarke, 1991), will be briefly reviewed. More details on the procedure will be presented in Chapter 6.

4.5.1 *Addition of protein additives*

As already briefly mentioned, the use of protein additives has been utilized not only to improve the water binding capacity of muscle foods, but also actually to increase the protein content. Protein additives can be from either animal or plant sources and will be discussed in greater detail in Chapter 10.

(a) Animal proteins. Among the animal protein sources are the milk proteins, including nonfat dry milk, calcium-reduced nonfat dry milk, dried whey, whey protein concentrate, buttermilk proteins, the caseinates and skim milk coprecipitates. These milk by-products contain appreciable amounts of protein and have been discussed by Delaney (1976), Morr (1976), Hugunin and Ewing (1977) and Thomas *et al.* (1978). These dairy products are widely utilized in processed meat and poultry products because of their desirable effects in binding water and for their influence on texture and flavor (Pearson and Gillett, 1995).

Other animal proteins utilized in meat and poultry products include mechanically deboned meat and poultry, which are both widely used in processed products (Pearson and Gillett, 1995). Collagen (gelatin) is also commonly utilized as a binding agent since it absorbs and binds appreciable amounts of water (Stainsby, 1987). Collagen is also used in producing edible sausage casings (Hood, 1987). Other animal proteins that can be incorporated into processed meat, poultry and fish products include blood and blood protein derivatives that can be upgraded into food grade products. The production of blood proteins and their derivatives is described by Dill and Landmann (1988) and Knipe (1988). Blood and blood products are widely used as human food in Russia, as explained by Gorbatov (1988). Surimi is also commonly utilized in producing fish analogs and imitation seafoods.

(b) Plant proteins. The most widely used plant protein additives are soy protein products, which are available as flour, concentrates and isolates (Pearson, 1984). These products differ in the content of protein and in the amount of purification that takes place during processing, as explained by

Pearson (1984). Soy proteins in various forms both bind water and increase the protein content of the products in which they are incorporated (Huffman and Powell, 1970). The soy proteins are not only utilized in ground meat systems but are also common ingredients in processed meats (Pearson and Gillett, 1995).

Glandless cottonseed proteins have been shown by Rhee, Smith and Rhee (1983) to be suitable additives for meat and poultry products, and apparently have similar advantages to soy proteins. Terrell *et al.* (1981) have utilized cottonseed proteins in the production of frankfurters. However, to date they have not been used as widely as soy proteins. Although little work has been conducted on sunflower seed proteins, they should have properties that are similar to those of soy proteins. Canola seed proteins have not been exploited in muscle food formulations, but seem likely to have the same advantages as soy proteins.

4.5.2 *Addition of nonprotein additives*

Starches have been utilized in sausage products and bind water about as well as soy proteins (Pearson and Gillett, 1995). They may contain more impurities and are generally not highly refined. They may be produced from any of the cereal grains and have slightly different properties, depending on the source. The starches are especially common in fresh sausages produced in the UK and Europe and are often added to make the products lower in price. Thus, such products are often considered to be of lower quality.

Konjac flour (a hydrocolloid derived from the *Amorphophallis konjac* plant) is produced in Japan and other Asian countries, where it has been shown to have special water binding properties. It has been claimed to have special advantages in muscle foods, although it seems likely that its ability to bind water is its main advantage. It is sometimes used in combination with carrageenan or other gums. Dickson (1996) has reviewed the properties of konjac flour and its advantages and disadvantages in meat, poultry and fish products.

4.5.3 *Other techniques*

Any technique that can enhance water binding or improve the perception of juiciness would have advantages in meat, poultry and fish products. These would include synthetic fat replacers and fat mimetic systems, which are discussed in greater detail in Chapter 9. Several commercial products have been developed, but they must be approved by regulatory agencies before they can be incorporated into foods. Recently the FDA approved the use of Olestra for use as a fat replacer in snack foods. Some of the fat mimetic systems have already been discussed in this chapter, but their structures and mechanisms of action are covered in more detail in Chapter 9.

(a) Supercritical fluid extraction (SFE). Supercritical fluid extraction is based upon altering the properties of a fluid by utilization of elevated pressure and temperature conditions that produces a supercritical phase, which gives the fluid unique extraction properties (Rizvi *et al.*, 1986a). For use in foods, CO_2 has commonly been utilized since it has a critical temperature and pressure of 31 °C and 7.38 MPa (Clarke, 1991), and is not toxic in foods (Rizvi *et al.*, 1986b). It has been used commercially for production of decaffeinated coffee (Rizvi *et al.*, 1986b) and for extraction of the flavor compounds from hop oil (Gardner, 1982). SFE has also been shown to be an effective procedure for extraction of fat from meat products (King, Johnson and Friedrich, 1989) and of lipid and cholesterol from ground beef (Chao *et al.*, 1991) and cholesterol and other lipids from dried egg yolk (Froning *et al.*, 1990).

The process of removing lipid and cholesterol from meat, poultry and fish products by SFE has been shown to be possible, as discussed in Chapter 6. However, the feasibility of applying SFE to these animal products will depend upon the costs and the demand for fat-free and cholesterol-free products. With the development of better equipment and reduction in costs, SFE may find a niche in the production of healthful products.

4.6 Research needs

4.6.1 Basic studies on mechanism and sensation of juiciness

Of particular importance are basic studies on the mechanism and sensation of juiciness, which were recently reviewed by Winger and Hagyard (1994). To exploit fully the utilization of low fat foods, the mechanism and sensation of juiciness need to be more fully understood. Only after the factors that contribute to juiciness are fully understood can the development of low fat products be fully exploited. Certainly the full understanding of juiciness and its related sensory properties is a primary need in order to make the best use of fat substitutes, whether they are synthetic compounds or fat mimetics. Understanding the basic sensations of taste and flavor is also central to progress in this important area.

4.6.2 Fat mimetics and substitutes

There are, no doubt, other food ingredients that have water binding and juiciness properties that show promise in low fat products. These products should be identified and characterized. They may be either fat mimetics or substitutes. There are also numerous possibilities to develop new substances that can be used. A search of GRAS (Generally Recognized As Safe) substances for new products that are approved for foods could be a profitable area for exploration and development.

The development of synthetics has already been shown to be an important area. This is true since chemical synthesis has unlimited possibilities. The major drawback to this approach, however, is the slow process involved in obtaining approval for their use in foods. Nevertheless, synthesis of new compounds and testing them for safety should have a high research priority.

4.6.3 *Improving leanness by breeding and genetics*

Research on improving leanness of beef cattle, hogs and sheep by breeding and genetics, although already under way (Chapter 7), should receive further emphasis to hasten the production of low fat meats. The possibility of utilizing specific genes that control leanness shows promise and should be exploited. Gene mapping of domestic meat-producing species should be encouraged and could offer a long-term solution to the problem of excessive fat production. Transgenic animals have developed some serious defects but the technique could still be useful in reducing fatness (Pursel *et al.*, 1988, 1989).

4.6.4 *Altering leanness by feeding and management*

Feeding and management can alter the lean content of carcasses (Chapter 8) and can alter composition more rapidly than genetics and breeding. Although some of the feeding and management techniques that can be utilized to alter composition are already understood, other possible methods should be examined. Hormonal means of altering growth and lean meat production should be investigated for their usefulness. Although repartitioning agents show promise for altering fatness/leanness, more research is needed in this area (Hancock, Wagner and Anderson, 1991). β-Agonists are known to influence leanness greatly, but research is still needed on their mechanism of action and the length of time that they need to be administered to gain maximum benefits. Of course, the safety of all of these new compounds and additives must be established.

4.6.5 *Protein and nonprotein additives*

Research is also needed on protein and nonprotein additives and how they alter the physical properties of meat, poultry and fish products. Research on their mechanisms of action and how they improve the sensory properties of the products is essential before these additives can be utilized fully in muscle foods. Finally, further research on supercritical fluid extraction is needed to make it a more attractive procedure for reducing the fat and cholesterol contents of meat, poultry and fish products.

4.7 Summary

Meat, poultry and fish products generally are desirable in human diets, but the relatively high fat content of some of these products is a barrier to their widespread acceptability. This chapter discusses some of the problems encountered in producing low fat meat, poultry and fish products that are acceptable to consumers. An overview of the decrease in juiciness as the fat content is lowered and the related problems of rubberiness/dryness is presented. Techniques for preventing these problems are reviewed, which include addition of water binding agents that may be either protein (animal or plant) or nonprotein additives.

Other procedures for producing low fat products are also discussed, which include trimming off excess fat, breeding and genetic improvement of animals, poultry and fishes and feeding and management. These topics are introduced in this chapter but will be covered in depth in other chapters.

Finally, some research needs are mentioned and the importance of further research in various areas is emphasized. In-depth discussion of such research needs will be covered in other chapters.

References

Abrams, H.L., Jr (1987) The preference for animal protein and fat: a cross-cultural survey, in *Food and Evolution, Toward a Theory of Human Food Habits* (eds M. Harris and E.B. Ross), Temple University Press, Philadelphia, PA, p. 207.

American Dietetic Association (1986) The American Dietetic Association's nutrition recommendations for women. *J. Am. Diet. Assoc.*, **86**, 1663.

American Heart Association (1988) Dietary guidelines for healthy American adults. Arteriosclerosis, **8**, 218A.

Batzer, O.F., Santoro, A.T. and Landmann, W.A. (1962) Identification of some beef flavor precursors. *J. Agric. Food Chem.*, **10**, 94.

Beauchamp, G.K. and Brand, J.G. (1994) The chemical senses. *Adv. Meat Res.*, **9**, 162.

Beermann, D.H. and DeVol, D.L. (1991) Effects of somatoprophin, somatotrophin releasing factor and somatostatin on growth. *Adv. Meat Res.*, **7**, 373.

Bender, A.E., Wood, T. and Palgrave, J.A. (1958) Analysis of tissue constituents. Extract of fresh ox muscle. *J. Sci. Food Agric.*, **9**, 812.

Berg, R.T. and Butterfield, R.M. (1976) *New Concepts of Animal Growth*, Sydney University Press, Sydney.

Blumer, T.N. (1963) Relationship of marbling to the palatability of beef. *J. Anim. Sci.*, **22**, 771.

Bonanome, A. and Grundy, S.M. (1988) Effect of dietary stearic acid on plasma cholesterol and lipoprotein levels. *N. Engl. J. Med.*, **318**, 1244.

Briggs, G.M. and Schweigert, B.S. (1990) An overview of meat in the diet. *Adv. Meat Res.*, **6**, 1.

Chao, R.R., Mulvaney, S.J., Bailey, M.E. and Fernando, L.N. (1991) Supercritical CO_2 conditions affecting extraction of lipid and cholesterol from ground beef. *J. Food Sci.*, **56**, 183.

Chrystall, B. (1994) Meat texture measurement. *Adv. Meat Res.*, **9**, 316.

Clarke, A.D. (1991) Supercritical fluid extraction. Technology for fat reduction. *Proc. Recip. Meat Conf.*, **44**, 101.

Cole, J.W., Ramsey, C.B. and Odum, L.O. (1960) What effect does fat content have on palatability of broiled ground beef? *Tenn. Farm Home Sci.*, **36**, 5.

Cornforth, D. (1994) Color – Its basis and importance. *Adv. Meat Res.*, **9**, 34.

Cronin, F.J. and Shaw, A.M. (1988) Summary of dietary recommendations for healthy Americans. *Nutr. Today*, **23**(6), 26.

Cronin, F.J., Shaw, A.M., Krebs-Smith, F.M. *et al.* (1987) Developing a food guidance system to implement the dietary guidelines. *J. Nutr. Educ.*, **19**, 281.

Cross, H.R., Stanfield, M.S., Griene, E.C. *et al.* (1975) Effect of fat and textured soy protein content on consumer acceptance of ground beef. *J. Food Sci.*, **40**, 1331.

Delaney, R.A.M. (1976) Composition, properties and uses of whey protein concentrates. *J. Soc. Dairy Technol.*, **29**, 91.

Dickerson, G.E. (1978) Animal size and efficiency. Basic concepts. *Anim. Prod.*, **27**, 367.

Dickson, R.L. (1996) Konjac flour gel as a suitable replacement for fat in a ground meat system. *MS Thesis*, Oregon State University, Corvallis, OR.

Dickson, R.L., Pearson, A.M., Anglemeir, A.F. and Holmes, Z. (1995) Konjac flour/carrageenan gel as a suitable fat replacer in a ground meat system. *Proc. Int. Cong. Meat Sci. Technol.*, **41**(II), 433.

Dikeman, M.E. (1987) Fat reduction in animals and the effects on palatability and consumer acceptance of meat products. *Proc. Recip. Meat Conf.*, **40**, 93.

Dill, C.W. and Landmann, W.A. (1988) Food grade proteins from edible blood. *Adv. Meat Res.*, **5**, 127.

Drake, S.R., Hinnergardt, L.C., Kluter, R.A. and Prell, P.A. (1975) Beef patties: the effect of textured soy protein and fat levels on quality and acceptability. *J. Food Sci.*, **40**, 1065.

Dransfield, E. (1994) Tenderness of meat, poultry and fish. *Adv. Meat Res.*, **9**, 289.

Dransfield, E., Jones, R.C.E. and Robinson, J.M. (1984) Development and application of a texture profile for UK beefburgers. *J. Texture Stud.*, **15**, 337.

Dreon, D.M., Frey-Hewitt, B., Ellsworth, N. *et al.* (1988) Dietary fat:carbohydrate ratio and obesity in middle-aged men. *Am. J. Clin. Nutr.*, **47**, 995.

Frantz, L.D., Jr, Dawson, E.A., Ashman, P.L. *et al.* (1989) Test of effect of lipid lowering by diet on cardiovascular risk. *Arteriosclerosis*, **9**, 129.

Froning, G.W., Wehling, R.L., Cuppett, S.L. *et al.* (1990) Extraction of cholesterol and other lipids from dried egg yolk using supercritical carbon dioxide. *J. Food Sci.*, **55**, 95.

Gaddis, A.M., Hankins, O.G. and Hiner, R.L. (1950) Relationships between the amount and composition of press fluid, palatability and other factors of meat. *Food Technol.*, **4**, 498.

Gardner, D.S. (1982) Industrial scale hop extraction with liquid CO_2. *Chem. Ind. (London)*, **19**, 402.

Gorbatov, V.M. (1988) Collection and utilization of blood and blood proteins for edible purposes in the USSR. *Adv. Meat Res.*, **5**, 167.

Gray, J.I., Pearson, A.M. and Monahan, F.J. (1994) Flavor and aroma problems and their measurement in meat, poultry and fish products. *Adv. Meat Res.*, **9**, 250.

Greene, B.E., Hsin, I. and Zipser, M.W. (1971) Retardation of oxidative color changes in raw ground beef. *J. Food Sci.*, **36**, 940.

Grundy, S.M. (1986a) Comparison of monounsaturated fatty acids and carbohydrates for lowering plasma cholesterol. *N. Engl. J. Med.*, **314**, 745.

Grundy, S.M. (1986b) Cholesterol and coronary heart disease. *J. Am. Med. Assoc.*, **256**, 2849.

Hancock, D.L., Wagner, J.F. and Anderson, D.B. (1991) Effects of estrogens and androgens on animal growth. *Adv. Meat Res.*, **7**, 255.

Hansen, R.G., Windham, C.T., Wittwer, C.T. and Wyse, B.W. (1986) Placing the role of meat in human nutrition in perspective. *J. Anim. Sci.*, **62** (Suppl. 1), 7.

Harries, J.M., Rhodes, D.N. and Chrystall, B.B. (1972) Meat texture I. Subjective assessment of the texture of cooked beef. *J. Texture Stud.*, **3**, 101.

Hays, V.W. (1991) Effects of antibiotics. *Adv. Meat Res.*, **7**, 299.

Hegsted, D.M., McGandy, R.B., Meyers, M.L. and Stare, F.J. (1965) Quantitative effects of dietary fat on serum cholesterol in man. *Am. J. Clin. Nutr.*, **17**, 281.

Hochstetler, L.N., Hoefer, J.A., Pearson, A.M. and Luecke, R.W. (1959) Effect of varying levels of fiber of different feed sources upon growth and carcass characteristics of swine. *J. Anim. Sci.*, **18**, 1397.

Hoefer, J.A., Pearson, A.M., Stevenson, J.W. and Luecke, R.W. (1963) Effect of fibrous feedstuffs fed during the finishing period on gain, feed efficiency and carcass characteristics of swine. *Mich. Agric. Expt. Stn. Q. Bull.*, **45**, 480.

Honikel, K.O. and Hamm, R. (1994) Measurement of water-holding capacity and juiciness. *Adv. Meat. Res.*, **9**, 125.

Hood, L.L. (1987) Collagen in sausage casings. *Adv. Meat Res.*, **4**, 109.

Hornstein, I. and Crowe, P.F. (1960a) Constituents of meat flavor: beef. *J. Agric. Food Chem.*, **8**, 65.

Hornstein, I. and Crowe, P.F. (1960b) Flavor studies on beef and pork. *J. Agric. Food Chem.*, **8**, 494.

Hornstein, I. and Crowe, P.F. (1963) Meat flavor: lamb. *J. Agric. Food Chem.*, **11**, 147.

Huffman, D.L. and Powell, W.E. (1970) Fat content and soya level effect on tenderness of ground beef patties. *Food Technol.*, **24**, 1418.

Hugunin, A.G. and Ewing, N.L. (1977) *Dairy Based Ingredients for Food Products.* Dairy Research, Rosemont, IL.

Iacono, J.M. (1987) Dietary intervention studies to reduce risk factors related to cardiovascular diseases and cancer. *Prev. Med.*, **16**, 516.

Keys, A., Anderson, J.T. and Grande, F. (1965) Serum cholesterol response to changes in the diet. IV. Particular saturated fatty acids in the diet. *Metabolism*, **14**, 776.

King, J.W., Johnson, J.H. and Friedrich, J.P. (1989) Extraction of fat tissue from meat products with supercritical carbon dioxide. *J. Agric. Food Chem.*, **37**, 961.

Kirkpatrick, B.W. and Rutledge, J.J. (1991) Genetics and transgenetics of growth. *Adv. Meat Res.*, **7**, 47.

Knipe, C.L. (1988) Production and use of animal blood and blood proteins for human food. *Adv. Meat Res.*, **5**, 147.

Koohmaraie, M. (1992) Role of the neutral proteinases in postmortem muscle protein degradation and meat tenderness. *Proc. Recip. Meat Conf.*, **45**, 63.

Kramlich, W.E. and Pearson, A.M. (1960) Separation and identification of cooked beef flavor components. *Food Res.*, **25**, 712.

Kritchevsky, D. (1990) Meat and cancer. *Adv. Meat Res.*, **6**, 89.

MacDougall, D.B. (1994) Color of meat. *Adv. Meat Res.*, **9**, 79.

MacLeod, G. and Seyyedain-Ardebili, M. (1981) Natural and simulated meat flavors (with particular reference to beef). *CRC Crit. Rev. Food Sci. Nutr.*, **14**, 309.

Maynard, L.A. and Loosli, J.K. (1962) *Animal Nutrition*, 5th edn, McGraw-Hill, New York.

McMeekan, C.P. (1940) Growth and development in the pig with special reference to carcass quality characters. III. Effect of plane of nutrition on form and composition of the bacon pig. *J. Agric. Sci. Camb.*, **30**, 276.

McNamara, D.J. (1990) Relationship between blood and dietary cholesterol. *Adv. Meat Res.*, **6**, 63.

Moloney, A., Allen, P., Joseph, R. and Tarrant, V. (1991) Influence of β-adrenergic agonists and similar compounds on growth. *Adv. Meat Res.*, **7**, 455.

Morr, C.V. (1976) Whey protein concentrates: an update. *Food Technol.*, **30**, 18.

Owens, F.N., Zorilla-Rios, J. and Dubeski, P. (1991) Effects of ionospheres on metabolism, growth, body composition and meat quality. *Adv. Meat Res.*, **7**, 321.

Pearson, A.M. (1976) The consumer desire for animal products, in *Symposium on the Fat Content and Composition of Animal Products*, held December 12–13, 1974, National Academy of Sciences, Washington, DC, pp. 45–79.

Pearson, A.M. (1984) Soy proteins, in *Developments in Food Proteins*, Vol. 2 (ed. B.J.F. Hudson), Applied Science, London, pp. 67–108.

Pearson, A.M. and Gillett, T.A. (1995) *Processed Meats*, 3rd edn, Chapman & Hall, New York.

Pearson, A.M., Asghar, A., Gray, J.I. and Booren, A.M. (1987) Impact of fat reduction on palatability and consumer acceptance of processed meat. *Proc. Recip. Meat Conf.*, **40** 105.

Purchas, R.W. (1991) Effect of sex and castration on growth and composition. *Adv. Meat Res.*, **7**, 203.

Pursel, V.G., Rexroad, C.E., Jr, Palmiter, R.D. *et al.* (1988) Gene transfer for increased animal growth. Biomechanisms regulating growth and development. *Symp. Agric. Res.*, **12**, 77.

Pursel, V.G., Pinkert, C.A., Miller, K.F. *et al.* (1989) Genetic engineering of livestock. *Science*, **244**, 1281.

Reineccius, G. (1994) Flavor and aroma chemistry. *Adv. Meat Res.*, **9**, 184.

Reiser, R. and Shorland, F.B. (1990) Meat fats and fatty acids. *Adv. Meat Res.*, **6**, 21.

Rhee, K.S., Smith, G.C. and Rhee, K.C. (1983) Retardation by glandless cottonseed flour of lipid oxidation and discoloration in raw ground beef containing salt. *J. Food Sci.*, **48**, 351.

Rizvi, S.S.H., Benado, A.L., Zollweg, J.A. and Daniels, J.A. (1986a) Supercritical fluid extraction: fundamental principles and modeling methods. *Food Technol.*, **40**(6), 55.

Rizvi, S.S.H., Daniels, J.A., Benado, A.L. and Zollweg, J.A. (1986b) Supercritical fluid extraction: Operating principles and food applications. *Food Technol.*, **40**(7), 57.

Savell, J.W. (1992) Significance of tenderness to the meat industry. *Proc. Recip. Meat Conf.*, **45**, 43.

Savell, J.W., Branson, R.E., Cross, H.R. *et al.* (1987) National consumer retail beef study: palatability evaluation of beef loin steaks that differed in marbling. *J. Food Sci.*, **52**, 517.

Schweigert, B., Kamin, H., Kritchevsky, D. *et al.* (1985) *Dietary Guidelines for Americans. Home and Garden Bulletin No. 232*, 2nd edn, US Department of Agriculture, Washington, DC.

Smith, G.C., Carpenter, Z.L., Cross, H.R. *et al.* (1984) Relationship of USDA marbling groups to palatability of cooked meat. *J. Food Qual.*, **7**, 289.

Stainsby, G. (1987) Gelatin gels. *Adv. Meat Res.*, **4**, 209.

Surgeon General's Report on Nutrition and Health (1988) Summary of recommendations. *Nutr. Today*, **23**(5), 22. Full report available from Superintendant of Documents, Stock No. 017-001-00465-1, US Government Printing Office, Washington, DC.

Surgeon General's Report (1995) *Nutrition and Your Health: Dietary Guidelines for Americans*, 4th edn, US Department of Agriculture and US Department of Health and Human Services, Washington, DC.

Terrell, R.N., Swasdee, R.L., Wan, P.J. and Lusas, E.W. (1981) Cottonseed proteins in frankfurters: effects on pH, cured color, and sensory and physical properties. *J. Food Sci.*, **46**, 845.

Thomas, M.A., McBride, R., Turner, A.D. and Ahab, G.H. (1978) Effect of extending beef patties with milk co-precipitates. *J. Food Sci.*, **43**, 162.

Tulloh, N.M. (1964) The carcase compositions of sheep, cattle and pigs as functions of body weight, in *Carcase Composition and Appraisal of Meat Animals* (ed. D.E. Tribe), CSIRO, Melbourne, pp. 5.1–5.16.

US Senate Select Committee on Nutrition and Human Needs (1977) *Dietary Goals for US*, Index No. 052-070-04376-8, US Government Printing Office, Washington, DC.

Valvano, R.A. (1983) The effects of preblending and lowered fat content on processed meats, *MS Thesis*, University of Nebraska, Lincoln, NE.

von Lossonczy, T.O., Ruiter, A., Bronsgeest-Schoute, H.C. *et al.* (1978) The effect of a fish diet on serum lipids in healthy human subjects. *Am. J. Clin. Nutr.*, **31**, 1340.

Wasserman, A.E. (1972) Thermally produced flavor components in the aroma of meat and poultry. *J. Agric. Food Chem.*, **20**, 737.

Wasserman, A.E. (1979) Chemical basis of meat flavor: a review. *J. Food Sci.*, **44**, 6.

Watts, G.F., Ahmed, W., Quincy, J. *et al.* (1988) Effective lipid lowering diets including lean meat. *Br. Med. J.*, **296**, 235.

Weir, C.E. (1960) Palatability characteristics of meat, in *The Science of Meat and Meat Products*, Freeman, San Francisco, pp. 212–233.

Wellington, G.H. and Stouffer, J.R. (1959) Beef marbling. Its estimation and influence on tenderness and juiciness. *Cornell Univ. Agric. Exp. Stn. Bull.*, 941.

Widdowson, E.M. and Lister, D. (1991) Nutritional control of growth. *Adv. Meat Res.*, **7**, 67.

Winger, R.J. and Hagyard, C.J. (1994) Juiciness. Its importance and some contributing factors. *Adv. Meat Res.*, **9**, 94.

5 Scientific basis for reducing the salt (sodium) content in food products

T.F.T. ANTONIOS and G.A. MacGREGOR

5.1 Introduction

5.1.1 Importance of blood pressure

Cardiovascular disease, through the strokes and heart attacks that it causes, is now the most common cause of death in developed countries. Large-scale population follow-up studies have identified three major risk factors for cardiovascular disease: high blood pressure, high blood cholesterol and smoking. Of these three, high blood pressure is now recognized as the most potent predictor of life expectancy, a finding which life insurance companies have recognized for many years (Metropolitan Life Insurance Company, 1961). The risk of cardiovascular disease from high blood pressure is directly related to the height of blood pressure with no evidence of a cut-off point. Conventionally, high blood pressure is designated as a systolic pressure greater than 140 mmHg or a diastolic pressure greater than 90 mmHg.

5.1.2 Proportion of population affected

Based on the above definition, 5–10% of most western populations have high blood pressure. It is known that high blood pressure tends to run in families and tends to be inherited, but there is as yet no clear cut indicator as to who will develop high blood pressure at a later stage of life (Pickering, 1968). As it is not possible to identify a particular section of the population who may be at risk, the only strategy in trying to prevent the development of high blood pressure is to apply this strategy to the whole population. Because the risk of high blood pressure is distributed throughout the range, it has been calculated that only very small falls in the average population blood pressure, e.g. 2 mmHg, would carry immense benefits. Indeed, it is almost equivalent to that of the current blood pressure lowering therapy that is being given to patients with high blood pressure (Rose, 1981).

In considering the relationship between salt and blood pressure there are two questions. One is whether our current high salt intake in any way causes, precipitates or aggravates the development of high blood pressure.

The second and very separate question is whether a short-term reduction in salt intake lowers raised blood pressure when it has already developed. The purpose of this chapter is to review briefly the normal physiology of sodium chloride and the overwhelming evidence that links our high sodium chloride intake to the development of high blood pressure. At the same time, we also review some of the other adverse effects of a high salt intake, other than that on blood pressure.

With the increasing realization that salt may not be such an innocuous substance, many people have stopped adding salt to their food and their cooking. However, because of the increase in the consumption of processed food products, most of which have a very high salt content, salt intake has not fallen over the last 10–20 years and indeed there is some evidence suggesting that it is increasing. Over 80% of our current salt intake now comes from salt added by food manufacturers (James, Ralph and Sanchez-Castillo, 1987).

5.2. Importance of sodium and chloride

5.2.1 Physiological needs for sodium and chloride

Sodium and chloride are the two predominant ions in the fluid surrounding the tissues, the extracellular space. The amount of sodium and chloride in the body, therefore, plays a very important role in regulating the volume of extracellular fluid. There is a very carefully regulated balance between the amount of salt eaten and the amount excreted in the urine. The physiological need for sodium is extremely small. Studies of herbivorous animals living away from the sea and studies of humans have shown that the body has very powerful mechanisms for conserving both sodium and chloride in urine and in sweat. In humans during evolution and even in some primitive societies today, salt intake only varies between 1 and 20 mmol per day (Oliver, Cohen and Neel, 1975; Denton, 1982). These amounts of sodium and chloride are found in natural foods and there is no physiological need for additional salt in our diet.

5.2.2 Historical background and consumption patterns

During evolution, particularly in northern climates, there was a great need to preserve food during the winter, particularly as communities started to develop. Probably the Chinese first discovered that salt had the magical property of preserving food. Salt then became of great economic importance, as those communities that had access to salt were able to survive during the winter. Because of its white colour and that when food was not fresh salt made it edible, salt was thought to be a purifier of food and became of considerable significance in most religions. With its growing economic importance, governments throughout the world used salt as a main source of tax revenue (Multhauf, 1978). In most societies salt was extremely expensive, but when it became cheap, salt was then added to natural food to bring it towards the same concentration as in preserved food. Salt consumption, therefore, increased from 1–20 to 100–400 mmol per day, that is, less than 1 g of salt per day, to our current level of consumption, which varies between 6 and 20 g per day.

5.3 Salt and high blood pressure

5.3.1 Early observations

The first suggestion that a high salt intake gives rise to high blood pressure goes back many years. The Chinese Yellow Emperor Huang Ti wrote several thousand years ago, 'If too much salt is used in the food, the pulse hardens, tears make appearance and the complexion changes'. Ambard and Beaujard (1904) were the first to show that salt restriction lowers high blood pressure in patients with severe hypertension and renal disease in the early 1900s. Allen and Sherril (1922) confirmed these findings in both patients with renal disease and those with high blood pressure without renal involvement, but these findings were largely ignored. It was only in the mid-1940s that Kempner (1948) gave patients a very low protein rice and fruit diet, which then led to lower blood pressure because it was very low in salt.

5.3.2 Epidemiological studies

There are now a large number of studies which have related salt intake (measured by the excretion of sodium in the urine over 24 h, all of the salt that we eat is absorbed and comes out in the urine) to the level of blood pressure in different communities. However, it must be realized that dietary assessments of sodium are not accurate. In particular, the collection of 24 h urine samples can be difficult, especially in more primitive communities and in western countries, since sodium intake varies from day to day. This means that it is difficult to characterize for one individual their usual sodium intake from one measurement. One difficulty with these studies is that there may well be other factors in the undeveloped countries, which have the low sodium intakes, that also may account for the blood pressure differences so that it may not necessarily be due to the difference in salt intake. However, Law, Frost and Wald (1991) recently re-analysed all the epidemiological studies, and when they plotted sodium intake against blood pressure in developed countries, there was a very close relationship. There was a similar close relationship for undeveloped countries, but it was set at a slightly lower blood pressure. In other words, salt intake was closely related to blood pressure levels, but there were other factors in developed countries which were also causing a rise in blood pressure. These studies agree with a large standardized and systematic international study, the INTERSALT (1988) study, where urinary sodium excretion and blood pressure were measured in over 10 000 people from 52 centres throughout the world. The results showed a direct relationship between urinary sodium and blood pressure, and clearly demonstrated that the rise in blood pressure with age was very closely related to sodium intake. A recent re-analysis of this study has confirmed that salt intake is one of the most important determinants of blood pressure differences, both between and within populations, as well as the rise in blood pressure that occurs with age in those countries that are exposed to a high salt intake (Elliott *et al.*, 1996).

Recently, Uzodike (1993), in a study comparing the urinary excretion of sodium and blood pressures in two rural communities in Nigeria without ethnic differences but with different levels of salt intake, found that the community with access to a salt lake and a thriving local salt-making industry had the highest mean urinary sodium excretion, the highest mean blood pressure and the highest prevalence rate of hypertension. Similar findings were observed in children and adolescents living in that community, suggesting that the relationship between sodium and blood pressure starts early in childhood and continues in adult life.

5.3.3 Intervention studies

Stronger evidence comes from intervention studies, such as a carefully controlled 6 month, double-blind study in newborn babies by Hoffman, Hazebroek and Valenburg (1983). Just under 500 newborn babies were randomized into two groups, one of which was placed on a reduced salt intake, while the other group had a normal salt intake. The study was done double-blind, in other words neither the mother nor the investigators who were measuring the blood pressure knew which salt intake the baby was consuming. The babies were followed up for a period of 6 months with the difference in diet being maintained for this time. Spot urine checks showed that the urinary sodium concentration was reduced in the restricted salt intake group compared with the normal salt group. Over the 6 months of follow-up there was a gradual and progressive increasing difference in blood pressure, so that by the end of 6 months the babies with the restricted salt intake had a systolic pressure 2.1 mmHg lower than those on the usual salt intake ($p < 0.001$). As the sodium restriction was discontinued at 6 months it is not known whether this difference would have increased, but a reduction of even 2 mmHg in the blood presblood pressure. This has been shown both in epidemiological studies and also in treatment trials (MacGregor and Cappuccio, 1993; Elliott, 1994).

5.5.2 Obesity and other factors

There is good evidence to suggest that obesity is also an important factor causing higher blood pressure (Elliott *et al.*, 1990), although there are some methodological problems in measuring blood pressure in fatter arms and, unless the right techniques are used, these can over-estimate blood pressure. Excessive alcohol consumption can also increase blood pressure, but this appears mainly to be due to withdrawal from alcohol and stimulation of the sympathetic nervous system, and there is little evidence to suggest that it causes long-term changes in blood pressure (Lip and Beevers, 1995). Studies suggesting that calcium might be related to blood pressure are probably incorrect, or have greatly exaggerated this effect, as pointed out by MacGregor (1985).

5.6 Other adverse effects of excessive salt intake

High salt intake may, however, have more subtle effects independent of its influence on blood pressure, as concluded by Antonios and MacGregor (1996). For instance, it is now realized that salt intake may have a direct effect on strokes, independent of blood pressure, and may also be important in the development of left ventricular hypertrophy and progression of kidney disease. At the same time, salt intake plays an important role in conditions where there is sodium and water retention, such as heart failure, liver cirrhosis and nephrotic syndrome, where dietary salt restriction is essential in their management. Also, salt restriction has been shown to prevent the so-called idiopathic oedema and cyclical oedema. The important role of restricting salt intake in these conditions has largely been forgotten. Perhaps the area of greatest concern is the increasing evidence that a high salt intake may cause an increase in urinary calcium excretion, and in the long term may be associated with bone demineralization (MacGregor and Cappuccio, 1993; Devine *et al.*, 1995).

5.6.1 Dietary salt and stroke

Cerebrovascular disease is the third commonest cause of death in the developed world and the commonest cause of chronic neurological disability. Epidemiological studies have shown that mortality rates from stroke have declined steadily in the USA and most of western Europe since the 1950s, with an accelerated rate of decline since the early 1970s for reasons which are still not very clear. Undoubtedly, treatment of hypertension and the control of other cerebrovascular risk factors, such as tobacco smoking, have contributed to the decline, but there is evidence that dietary factors may have a role in stroke mortality (Perry, 1990). Joossens (1980) argued that salt intake may be a major determinant of stroke rates in the population, and he suggested that the decline of salt intake in the western world since 1945 is a major contributing factor in the decline in stroke incidence. He also provided impressive data on the decline of salt intake in Belgium, which has been clearly accompanied by a decline in mortality from stroke (Joossens and Geboers, 1985).

Perry and Beevers (1992) performed an ecological analysis of the relationship between regional mortality from cerebrovascular disease in western Europe and regional data on urinary sodium excretion, systolic blood pressure and relevant confounding variables. They used published WHO cerebrovascular disease mortality rates and data provided by the INTERSALT (1988) study. They found that stroke mortality was positively and significantly related to median sodium excretion, but not to median systolic blood pressure. They concluded that a high intake of salt may exert an adverse effect on the risk of stroke that is independent of the effects on blood pressure.

5.6.2 Dietary salt and cardiac hypertrophy

Left ventricular hypertrophy is related to cardiovascular morbidity and mortality independently of other risk factors, including hypertension. Data from the Framingham study showed that left ventricular hypertrophy as established by electrocardiographic evidence cannot be considered a compensatory process only, but should be viewed as a powerful independent risk factor for congestive heart failure, coronary artery disease and sudden death (Kannel, Gordon and Offut, 1969). The pathogenesis of left ventricular hypertrophy in essential hypertension remains a subject of major controversy, since the increase in afterload caused by arterial hypertension does not consistently produce the same degree of left ventricular hypertrophy. In addition to pressure overload, several clinical, hemodynamic and endocrine factors have been identified as pathogenetic factors in the process of structural cardiac adaptation in arterial hypertension (Frolich and Tarazi, 1979).

Ferrara *et al.* (1984) evaluated the effects of short-term (12 week) salt restriction on left ventricular mass in hypertensive patients. They showed that either low sodium intake or diuretic treatment produced a significant reduction in non-invasively determined left ventricular mass in a period as brief as 6 weeks. They concluded that factors other than hypertension, namely salt intake, have been involved in the regression of left ventricular mass. Schmieder *et al.* (1988) reported that salt intake estimated by sodium excretion is one of the strongest predictors of posterior wall thickness and relative wall thickness independent of other variables. In a further study by Schmieder *et al.* (1990) performed on 66 middle-aged never-treated mild hypertensive white male patients, they found significant correlations between sodium excretion and diastolic diameter of the left ventricle and with left ventricular mass. Again, dietary sodium intake was identified as the strongest determinant of cardiac structural adaptation to a persistent increase in arterial pressure. They concluded that dietary sodium intake must be regarded as a blood pressure-independent determinant of left ventricular hypertrophy in patients with essential hypertension.

In another study done on adolescents, Daniels, Meyer and Loggie (1990) reported that salt intake was a powerful, blood pressure-independent determinant of left ventricular hypertrophy. In this study dietary salt intake was not only assessed by measurement of 24 h sodium excretion but also estimated via observation by a trained dietician of foods consumed and calculation of sodium intake from the known sodium composition of each food. Du Cailar *et al.* (1992) studied the effect of salt intake on left ventricular mass. In both normotensives and never-treated patients with hypertension, left ventricular mass was positively correlated with urinary sodium excretion as a measure of salt intake. They also noticed that for a given level of salt intake, hypertensive patients have a higher left ventricular mass index than normotensive subjects. Data from the Treatment of Mild Hypertension Study (TOMHS) group showed a significant correlation between reduced salt intake and left ventricular mass in the entire study group. Furthermore, a moderate reduction in salt intake has been shown to cause regression of left ventricular hypertrophy (Liebson *et al.*, 1992).

All the available evidence points to a close relationship between the development and persistence of left ventricular hypertrophy and sodium homeostasis. The higher the sodium intake, the greater was left ventricular mass in normotensive and, most relevant, in hypertensive subjects. At the present state of knowledge, high salt intake should be considered as a possible reason for persisting left ventricular hypertrophy in the presence of good blood pressure control.

5.6.3 Dietary salt and renal injury

An increase in salt intake causes an increase in renal blood flow and glomerular filtration rate, and in animals an increase in glomerular pressure. In animal models of hypertension and renal impairment, a high salt intake has been shown to increase the rate of deterioration of renal function while salt restriction slowed the deterioration to a greater extent than thiazide diuretics (Daniels and Hostetter, 1986; Benstein *et al.*, 1990).

5.6.4 Dietary salt and bronchial asthma

There is evidence of a positive association between increasing sodium intake and worsening of asthma. Epidemiological studies show that there is an increased prevalence of asthma in communities adopting a western life style, while a low prevalence is found in some of the world's poorest areas (van Niekerk *et al.*, 1979). Furthermore, regional mortality data for England and Wales show a strong correlation between asthma mortality and regional per person purchases of table salt for men (Burney, 1987). Burney *et al.* (1986), in a study on 138 men, found a close relationship between the results of a histamine challenge test and 24 h urinary sodium excretion. In a larger randomized controlled study by the same group, slow sodium showed an increase in bronchial reactivity in men but not in women (Burney *et al.*, 1989).

Javaid, Cushley and Bone (1988), in a study on 10 asthmatic and five healthy volunteers, also found that increased salt intake produced a significant increase in bronchial reactivity to histamine in both male and female asthmatics but not in healthy control subjects. Recently, Carey, Locke and Cooksen (1993) studied 27 asthmatic men established on a low sodium diet before receiving 200 mmol of slow sodium or placebo in random order. A change from a high to a low sodium intake was associated with lower methacholine reactivity, consumption of bronchodilators and symptom scores and higher peak flow and forced expiratory volume in 1 s.

5.6.5 Dietary salt and cancer

Most of the available knowledge about the relationship between cancer and salt intake comes from epidemiological observations. Epidemiological studies using data on mortality provided by the World Health Organization showed a strong relationship between cerebrovascular mortality and stomach cancer mortality from 1950 to 1977, and this was true for both sexes (Joossens, 1980). From 1973 data, it seems that stroke is decreasing much faster than stomach cancer, at least in countries such as England and Wales, the USA and Germany. This new deviation from the linearity observed from 1955 to 1968 is probably due to treatment of hypertension (Garraway *et al.*, 1979). Although antihypertensive drugs have been available since the end of the 1940s, they have not been used on a population-based level, but only in selected cases. Treatment of hypertension is unlikely to improve stomach cancer, and therefore, if effective, would only influence stroke. From the known risk factors of stroke and stomach cancer, salt appears to be the linking factor. Various studies in Japan and elsewhere, have linked gastric cancer and dietary factors, especially salted food (Sato, Fukuyama and Suzuki, 1959; Hirayama, 1967).

Nasopharyngeal carcinoma is the second cancer which appears to be strongly related to salt intake. It is a disease with a remarkable racial and geographical distribution. It is very rare in most parts of the world, and only a handful of populations are known to deviate from this low-risk profile, including people of southern China, Eskimos, natives of Southeast Asia and Arab populations of north Africa and Kuwait. Case-control studies conducted in various Chinese populations with distinct risks of nasopharyngeal carcinoma have suggested that ingestion of salted fish and other kinds of preserved foods by the Chinese constitutes the most important cause of nasopharyngeal carcinoma development among these people (Geser, Charney and Day, 1987).

5.6.6 Dietary salt and osteoporosis

Dietary salt is one of the major determinants of urinary calcium excretion: the higher the salt intake, the greater is the urinary calcium excretion (McCarron *et al.*, 1981; Sabto *et al.*, 1984). Salt is a more important determinant of urinary calcium excretion than calcium intake within the range of salt and calcium intake that is currently consumed. The rise in urinary calcium excretion with salt had been assumed to be compensated for by an identical increase in calcium absorption with no reabsorption of calcium from bone, but this assumption is incorrect. A 2 year longitudinal study of bone density in 124 postmenopausal women designed to look for predictive factors that would relate to bone loss showed a significant inverse relationship between urinary sodium excretion and a reduction in hip bone density, as reported by Devine *et al.* (1995). No bone loss occurred at the hip at a urinary sodium excretion of less than 90 mmol per day. The authors concluded that a reduction of salt intake by approximately half would have the same effect on bone mineral density as an increase of calcium intake of approximately 900 mg.

When salt intake was increased in postmenopausal women there was not only an increase in urinary calcium excretion, but also an increase in parathyroid hormone levels and hydroxyproline excretion, indicating mobilization of calcium from bone (McParland, Goulding and Campbell, 1989). In a separate study by Matkovic *et al.* (1995), when sodium intake was reduced in postmenopausal women from 170 to 70 mmol of sodium per day, there was a large reduction in calcium excretion with a significant reduction in both 1,25-dihydroxyvitamin D and serum osteocalcin. It has also been suggested that in young females, a low calcium intake and a relatively high obligatory calcium loss in the urine caused by a high salt intake during rapid growth may reduce calcium retention in the skeleton, with a concomitant reduction in peak bone mass.

Patients with idiopathic hypercalciuria are known to have a decrease in bone mineral mass compared with controls and this loss of bone mass has been shown to be worse the higher is the salt intake (Silver *et al.*, 1983). It has been known for many years that a moderate reduction in salt intake in hypercalciuria is as effective in reducing calcium excretion as a thiazide diuretic (Pak *et al.*, 1984). Patients with high blood pressure have an increase in urinary calcium excretion for a given urinary sodium excretion, as well as a tendency for reduced ionized calcium, an increase in parathyroid hormone and increase in 1,25-dihydroxyvitamin D, indicating that the increased urinary calcium excretion is not compensated for by an increase in calcium absorption and that there is mobilization of calcium from bone. Patients with essential hypertension are potentially at greater risk of bone demineralization for a given sodium intake than normotensive subjects.

Thiazide diuretics are known to decrease urinary calcium excretion and also to cause retention of calcium, increase in bone density and reduction in the number of hip fractures, in both normotensive and hypertensive subjects (LaCroix *et al.*, 1990). A moderate reduction in salt intake is likely to have the same effect as a thiazide diuretic in increasing bone density and preventing long-term bone demineralization.

5.7 Research needs

The preponderance of evidence suggests that reduction of salt intake would have a beneficial effect upon blood pressure, yet some of those relationships need further clarification and extension of the data to mortality as related to stroke, cardiovascular disease and other conditions that appear to be associated with high salt intake. Where possible, epidemiological studies that demonstrate a relationship between salt intake and hypertension should be extended to include morbidity rates from stroke and cardiovascular disease.

Further research is also needed to elucidate the direct effect of salt intake on the incidence of stroke, an effect which Antonios and MacGregor (1996) have observed to be independent of hypertension. Such information may help to explain the cause of stroke in normotensive individuals.

Although Midgley *et al.* (1996) have shown that normotensives responded to short-term salt restriction less than hypertensives, as pointed out earlier in this chapter salt restriction was only for 10 days, and thus the results could be flawed. Studies on salt restriction in the diet of infants by Hoffman, Hazebroek and Valenburg (1983) and a later follow-up study by Geleijnse *et al.* (1996) showed that those on the higher salt intake still had a higher blood pressure even after 15 years. These studies demonstrate the importance of long-term studies to determine the effects of salt intake. Studies on the same children throughout their lifetimes would be of great value in determining the influence of salt intake on the causes of death.

One of the problems in trying to determine the relationship of salt intake to high blood pressure is the fact that blood pressure generally increases with age. Many normotensive individuals become hypertensive later in life. Although there is currently no test to determine one's likelihood of changing from normotensive to hypertensive, it would be useful to try to develop such a test in order to find true hypertensives.

The blood pressure lowering effects of potassium have been reported by MacGregor and Cappuccio (1993) and Elliott (1994). Data are needed on the usefulness of potassium intake on lowering blood pressure, the levels needed and the mechanism of action.

One area where research is greatly needed is the effect of salt intake on urinary calcium and bone demineralization (McParland, Goulding and Campbell, 1989; Matkovic *et al.*, 1995). Since calcium losses are a contributing cause to demineralization and bone fractures in women, establishment of dietary recommendations for increasing bone density and avoiding demineralization are badly needed, including the influence of lowering dietary salt levels in improving bone density in men and more especially in pre- and postmenopausal women.

Other areas in which research is needed include the relationship of salt intake to renal injury, cancer and bronchial asthma. High salt intake also appears to be a contributing factor to the development of left ventricular hypertrophy (Kannel, Gordon and Offut, 1969), in which further research could lead to a decrease in thickening of the ventricular wall by decreasing salt intake.

Encouragement should be given to meat, poultry and fish processors to decrease levels of salt without compromising safety. Chapters 12 and 14–17 discuss the production and processing of low salt products and should contribute to our understanding of the effectiveness of different processing procedures in reducing salt levels.

5.8 Summary

There is much circumstantial and now direct evidence to suggest that our current high consumption of salt predisposes individuals and communities to the development of high blood pressure and also causes a rise in pressure throughout the range of blood pressure. There is also increasing evidence that salt has other adverse effects apart from its effect on blood pressure. Of particular importance is its potential to aggravate bone demineralization, an increasing problem as the population tends to live longer. Nearly all expert committees in most western countries have now recommended a reduction in salt consumption for the whole population.

In spite of this advice, however, there has been little attempt to reduce sodium intake in most developed countries. Indeed, the salt and processed food industry has conducted a largely successful campaign over the past decade, particularly in the USA and UK and elsewhere, to convince the rest of the food industry and suppliers, politicians and health care workers that the evidence for reducing salt intake is insufficient for any action to be taken. This raises the question of why the food industry is so defensive of the high salt content of processed foods. According to the Salt Data Centre, a centre set up by a public relations company for the salt manufacturers in the UK, a major justification for the use of salt in processed food is that salt makes unpalatable food edible. What is unclear from this remarkable statement is why we need to eat unpalatable food. The sensitivity of the receptors in the mouth for salt are dependent on the usual salt intake of that individual.

Studies have clearly demonstrated that when salt intake is reduced, the salt taste receptors become much more sensitive, and highly salted food then becomes very unpleasant. This is one of the major reinforcing factors that allows people to stick to a low salt diet, because they are unable to eat the highly salted foods made by many food manufacturers. Whilst it is probably correct that many processed foods would initially not be acceptable if the salt concentration was suddenly reduced, it would be acceptable if this was done gradually. This has already occurred with bread, where there have been substantial reductions in the salt content over the last 20 years, and indeed in certain areas of Italy, e.g. Florence, bread has not had salt added since the eighteenth century because of the very high salt taxes.

Perhaps of more relevance to the food industry is the fact that adding sodium and chloride to processed food does allow more water to be added to the product, allowing the price of the processed food to be reduced, or more profit to be made. Another very relevant consideration is the well known finding that the higher the salt intake the greater is the consumption of fluid, and it is interesting in this regard that many soft drink manufacturers also own companies that produce and market salted snacks.

Clearly, the above is very relevant to meat products, many of which contain a very high salt content, often equivalent to that of sea water, i.e. 0.9 g of sodium per 100 g, or 2.5 g of sodium chloride per 100 g. At present most of the population is unaware of the very high salt content of processed food, which is hidden in the foods. However, as the population becomes more used to not adding salt at the table and not adding salt in their own cooking, it is likely that many of these foods will be rejected as they taste unpleasant. A gradual reduction in salt content of manufactured foods would not be noticed by the general public if it was done slowly and universally, and is likely to have immense public health benefits in reducing blood pressure and preventing strokes and heart attacks and also would have a beneficial effect on the prevention of bone demineralization.

References

Allen, F.M. and Sherril, J.W. (1922) The treatment of arterial hypertension. *J. Metab. Res.*, **2**, 429.

Ambard, L. and Beaujard, E. (1904) Causes de l'hypertension arterielle. *Arch. Gen. Med.*, **1**, 520.

Antonios, T.F. and MacGregor, G.A. (1996) Salt – more adverse effects. *Lancet*, **348**, 250.

Benstein, J.A., Feiner, H.D., Parker, M. and Dworkin, L.D. (1990) Superiority of salt restriction over diuretics in reducing renal hypertrophy and injury in uninephrectomized SHR. *Am. J. Physiol.*, **258**, F1675.

Burney, P.G.J. (1987) A diet rich in sodium may potentiate asthma: epidemiological evidence for a new hypothesis. *Chest*, **91**, 143S.

Burney, P.G.J., Britton, J.R., Chinn, S. *et al.* (1986) Response to inhaled histamine and 24 hour sodium excretion. *Br. Med. J.* **292**, 1483.

Burney, P.G.J., Neild, J.E., Twort, C.H. *et al.* (1989) Effect of changing dietary sodium on the airway response to histamine. *Thorax*, **44**, 36.

Carey, O.J., Locke, C. and Cooksen, J.B. (1993) Effect of alterations of dietary sodium on the severity of asthma in men. *Thorax*, **48**, 714.

Daniels, B.S. and Hostetter, T.H. (1986) Influence of dietary salt on chronic renal injury. *Kidney Int.*, **29**, 244.

Daniels, S.D., Meyer, R.A. and Loggie, M.H. (1990) Determinants of cardiac involvement in children and adolescents with essential hypertension. *Circulation*, **82**, 1243.

Denton, D. (1982) *The Hunger for Salt: an Anthropological, Physiological and Medical Analysis*. Springer, Berlin.

Denton, D., Weisinger, R., Mundy, N.I. *et al.* (1995) The effect of increased salt intake on blood pressure of chimpanzees. *Nature Med.*, **1**, 1009.

Devine, A., Cridle, A.R., Dick, I.M. *et al.* (1995) A longitudinal study of the effect of sodium and calcium intakes on regional bone density in postmenopausal women. *Am. J. Clin. Nutr.*, **62**, 740.

Du Cailar, G., Ribstein, J., Daures, J.P. and Mimran, A. (1992) Sodium and left ventricular mass in untreated hypertensive and normotensive subjects. *Am. J. Physiol.*, **263**, H177.

Elliott, P. (1994) Nutritional factors in blood pressure. *J. Hypertension*, **8**, 595.

Elliott, P., Rogers, S., Scally, G. *et al.* (1990) Sodium, potassium, body mass, alcohol and blood pressure in three United Kingdom centres (the INTERSALT study). *Eur. J. Clin. Nutr.*, **44**, 637.

Elliott, P., Stamler, J., Nichols, R. *et al.* (Intersalt Cooperative Research Group) (1996) Intersalt revisited: further analyses of 24 hour sodium excretion and blood pressure within and across populations. *Br. Med. J.*, **312**, 1249.

Ferrara, L.A., De Simone, G., Pasanisi, F. and Mancini, M. (1984) Left ventricular mass reduction during salt depletion in arterial hypertension. *Hypertension*, **6**, 755.

Forte, J.G., Pereira Miguel, J.M., Pereira Miguel, M.J. *et al.* (1989) Salt and blood pressure: a community trial. *J. Hum. Hypertension*, **3**, 179.

Frolich, E.D. and Tarazi, R.C. (1979) Is arterial hypertension the sole factor responsible for hypertensive cardiac hypertrophy? *Am. J. Cardiol.*, **44**, 959.

Garraway, W.M., Whisnant, J.P., Furlan, A.J. *et al.* (1979) The declining incidence of stroke. *N. Engl. J. Med.*, **300**, 449.

Geleijnse, J.M., Hofman, A., Witteman, J.C.M. *et al.* (1996) Long-term effects of nonatal sodium restriction on blood pressure. *J. Hypertension*, **14**, S210.

Geser, A., Charney, N. and Day, N.E. (1987) Environmental factors in the etiology of nasopharyngeal carcinoma. Report on a case-control study in Hong Kong. *IARC Sci. Publ.*, **20**, 213.

Haddy, F.J. (1980) Mechanisms, prevention and therapy of sodium dependent hypertension. *Am. J. Med.*, **69**, 746.

Hirayama, T. (1967) The epidemiology of cancer of the stomach in Japan with special reference to the role of diet, in *Proceedings of the 9th International Cancer Congress*, (ed. R.J.C. Harris), Springer, Tokyo, pp. 37–49.

Hoffman, A., Hazebroek, A. and Valenburg, H.A. (1983) A randomised trial of sodium intake and blood pressure in newborn infants. *J. Am. Med. Assoc.*, **250**, 370.

INTERSALT (1988) Intersalt Cooperative Research Group. An international study of electrolyte excretion and blood pressure. Results of 24 hour urinary sodium and potassium excretion. *Br. Med. J.*, **297**, 319.

James, W.P., Ralph, A. and Sanchez-Castillo, C.P. (1987) The dominance of salt in manufactured food in the sodium intake of affluent societies. *Lancet*, **i**, 426.

Javaid, A., Cushley, M.J. and Bone, M.F. (1988) Effect of dietary salt on bronchial reactivity to histamine in asthma. *Br. Med. J.*, **297**, 454.

Joossens, J.V. (1980) Stroke, stomach cancer and salt: a possible clue to the prevention of hypertension, in *Epidemiology of Arterial Blood Pressure* (eds H. Kesteloot and J.V. Joossens), Martinus Nijhoff, The Hague, pp. 489–508.

Joossens, J.V. and Geboers, J. (1985) Control of hypertension in Belgium, in *Handbook of Hypertension* (eds W.H. Birkenhager and C.J. Bulpitt), Elsevier, Amsterdam, pp. 424–440.

Kannel, W.B., Gordon, T. and Offut, D. (1969) Left ventricular hypertrophy by electrocardiogram: prevalence, incidence and mortality in the Framingham study. *Ann. Intern. Med.*, **71**, 89.

Kempner, W. (1948) Treatment of hypertensive vascular disease with rice diet. *Am. J. Med.*, **4**, 545.

LaCroix, A.Z., Wienpahl, J., White, L.R. *et al.* (1990) Thiazide diuretic agents and the incidence of hip fracture. *N. Engl. J. Med.*, **322**, 286.

Law, M.R., Frost, C.D. and Wald, N.J. (1991) By how much does dietary salt reduction lower blood pressure? III. Analysis of data from trials of salt reduction. *Br. Med. J.*, **302**, 819.

Liebson, P., Prineas, R., Grandits, G. *et al.* (1992) Variables associated with regression of left ventricular mass in the Treatment of Mild Hypertension Study (TOMHS). *Circulation*, **86**, 599.

Lip, G.Y. and Beevers, D.G. (1995) Alcohol, hypertension, coronary disease and stroke. *Clin. Exp. Pharmacol. Physiol.*, **22**, 189.

MacGregor, G.A. (1985) Sodium is more important than calcium in essential hypertension. *Hypertension*, **7**, 628.

MacGregor, G.A. and Cappuccio, F.P. (1993) The kidney and essential hypertension: A link to osteoporosis? *J. Hypertension*, **11**, 781.

MacGregor, G.A., Markandu, N.D., Best, F.E. *et al.* (1982) Double-blind randomised crossover trial of moderate odium restriction in essential hypertension. *Lancet*, **i**, 531.

MacGregor, G.A., Markandu, N.D., Singer, D.R.J. *et al.* (1987) Moderate sodium restriction with angiotensin converting enzyme inhibitor in essential hypertension: A double blind study. *Br. Med. J.*, **294**, 531.

Matkovic, V., Ilich, J.Z., Andon, M.B. *et al.* (1995) Urinary calcium, sodium, and bone mass of young females. *Am. J. Clin. Nutr.*, **62**, 417.

McCarron, D.A., Rankin, L.I., Bennett, W.M. *et al.* (1981) Urinary calcium excretion at extremes of sodium intake in normal man. *Am. J. Nephrol.*, **1**, 84.

McParland, B.E., Goulding, A. and Campbell, A.J. (1989) Dietary salt affects biochemical markers of resorption and formation of bone in elderly women. *Br. Med. J.*, **299**, 834.

Metropolitan Life Insurance Company (1961) *Blood Pressure: Insurance Experience and Its Implications*, Metropolitan Life Insurance Company, New York.

Midgley, J.P., Mattew, A.G., Greenwood, C.M.T. and Logan, A.G. (1996) Effect of reduced dietary sodium on blood pressure. *J. Am. Med. Assoc.*, **275**, 1590.

Ministry of Health and Welfare (1980) *National Survey on Circulatory Disorders*, Koseisho, Japan.

Multhauf, R.B. (1978) *Neptune's Gift: a History of Common Salt*, Johns Hopkins University Press, Baltimore.

Oliver, W.J., Cohen, E.L. and Neel, J.V. (1975) Blood pressure, sodium intake and sodium related hormones in the Yanomamo Indians, a "no-salt" culture. *Circulation*, **52**, 146.

Pak, C.Y.C., Smith, L.H., Resnick, M.I. and Weinerth, J.L. (1984) Dietary management of idiopathic calcium urolithiasis. *J. Urol.*, **131**, 850.

Perry, I. (1990) What has caused the widespread decline in stroke mortality? *J. Ir. Coll. Physicians Surg.*, **19**, 257.

Perry, I.J. and Beevers, D.G. (1992) Salt intake and stroke: a possible direct effect. *J. Hum. Hypertension*, **6**, 23.

Pickering, G. (1968) *High Blood Pressure*, Churchill Livingstone, London.

Poulter, N., Khaw, K.T., Hopwood, B.E.C. *et al.* (1984) Salt and blood pressure in various populations. *J. Cardiovasc. Pharmacol.*, **6**, S197.

Rose, G. (1981) Strategy of prevention: lessons from cardiovascular disease. *Br. Med. J.*, **282**, 1847.

Sabto, J., Powell, M.J., Breidahl, M.J. and Gurr, F.W. (1984) Influence of urinary sodium on calcium excretion in normal individuals. A redefinition of hypercalciuria. *Med. J. Austr.*, **140**, 354.

Sato, T., Fukuyama, T. and Suzuki, T. (1959) Studies of the causation of gastric cancer 2. The relation between gastric cancer mortality rate and salted food intake in several places in Japan. *Bull. Inst. Publ. Health*, **8**, 187.

Schmieder, R.E., Messerli, F.H., Garavaglia, G.E. and Nunez, B.D. (1988) Dietary salt intake: a determinant of cardiac involvement in essential hypertension. *Circulation*, **78**, 951.

Schmieder, R.E., Grube, E., Impelmann, V. *et al.* (1990). Determinants of myocardial hypertrophy in mild essential hypertension. Impact of dietary salt intake on left ventricular hypertrophy. *Z. Kardiol.*, **79**, 557.

Silver, J., Rubinger, D., Friedlaeder, M.M. and Popovtzer, M.M. (1983) Sodium-dependent idiopathic hypercalciuria in renal-stone formers. *Lancet*, **ii**, 484.

Uzodike, V.O. (1993) Epidemiological studies of arterial blood pressure and hypertension in relation to electrolyte excretion in three Igbo communities in Nigeria. MD Thesis, London University, London.

van Niekerk, C.H., Weinberg, E.C., Shore, S.C. *et al.* (1979) Prevalence of asthma: a comparative study of urban and rural Xhosa children. *Clin. Allergy*, **9**, 319.

Watkin, D.M., Froeb, H.F., Hatch, F.T. and Gutman, A.B. (1950) Effects of diet in essential hypertension. Results with unmodified Kempner rice diet in fifty hospitalised patients. *Am. J. Med.*, **9**, 441.

6 Reduction of cholesterol levels in meat, poultry and fish products

A.D. CLARKE

6.1 Introduction

As mentioned in Chapter 4, a relationship between elevated plasma cholesterol and cardiovascular disease has been known for some time. While this became known from the Framingham Study (McNamara, 1985), it is questionable whether the relationship between dietary cholesterol and heart disease is significant. McNamara (1991) has also said that the effect of dietary cholesterol on plasma cholesterol is very small and a problem with some dietary changes is that they may lower high-density lipoprotein (HDL) cholesterol more than low-density lipoprotein (LDL) cholesterol. Besides, typical dietary cholesterol intake (0.5 g per day) is often less than the amount synthesized (0.7 g per day) by the body (McNamara, 1995). Thus, efforts to reduce the cholesterol content in meat, poultry or fish may have little impact on the health of the general public.

Whether physicians and consumers believe that cholesterol reduction will have an impact on human health or not, interest in developing low cholesterol or reduced cholesterol products has arisen among livestock producers and meat processors. Altering a natural component of animal tissue presents a significant challenge. However, there are several approaches for cholesterol reduction. Methods include modifying the animal before harvest, altering the composition of meat products or direct removal of cholesterol by supercritical fluid extraction. This chapter presents some of the recent efforts made to establish cholesterol reduction by each of these methods.

6.2 Reduction of cholesterol by animal modification

Reducing the cholesterol content of meat, poultry or fish at the source is an appealing concept, but one that is the most complex and appears the least successful. Selecting animals with the right genetics, feeding diets rich in unsaturated lipids and treating with growth or repartitioning drugs are the most common choices for modification of the cholesterol content before animal slaughter. Most have a negligible impact, but all can involve animal producers in the effort to modify cholesterol content.

6.2.1 Genetic approach

It is sometimes believed that particular breeds are lower in cholesterol content, but results are mixed. For example, Wheeler *et al.* (1987) analyzed tissue and serum cholesterol in fullblood Chianina and crossbred cattle but reported that days on feed, breed and sex had no effect on tissue cholesterol values. Serum cholesterol, however, was affected by days on feed. On the other hand, breeding beef animals (*Bos taurus*) with bison (*Bison bison*) resulted in a cholesterol content of the crossbred animals that was greater in the longissimus muscle (59.4 mg per 100 g), but lower in the subcutaneous fat (88.3 mg per 100 g) than the *Bos taurus* breeds (54.8 and 100.8 mg per 100 g, respectively) (Koch *et al.*, 1995).

Selection of animals for reproduction and meat production depends on knowledge of measurable traits, and it is possible to collect plasma cholesterol information to assist in the decision process. Unfortunately, plasma cholesterol variation in turkeys was reported by Smith and Savage (1992) to be due to environmental variation, and the genetic component was largely nonadditive. In Wheeler *et al.*'s (1987) study, the fluctuation of serum cholesterol was related to feeding and lipid metabolism of the animal and thus considered to be a poor indicator of tissue cholesterol content.

6.2.2 Dietary alteration

The concept that feeding animals a diet rich in unsaturated fatty acids would lower cholesterol levels mimics the attempts of humans to alter their cholesterol levels by diet. However, only a few studies provide support for the theory. Ajuyah *et al.* (1991) studied broiler chicks on diets with 10–20% full-fat flax seed or full-fat canola seed versus a corn and soybean meal diet and reported that the flax seed reduced total carcass fat and increased levels of the nutritionally desirable ω-3 fatty acids. Perhaps not surprisingly, the tissue cholesterol and lipid content of the white meat was lower than for dark meat in all experimental diets. In further work, Ajuyah *et al.* (1992) determined that spent hens fed flax seed at either 8 or 16% of the diet resulted in dark meat with a significantly lower cholesterol content (72 mg per 100 g) than the control diet (100 mg per 100 g). It is unclear whether the dietary alteration caused a meaningful reduction in tissue cholesterol, but at least there is evidence that tissue and plasma cholesterol may be reduced by some dietary interventions.

Not all methods of dietary modification are successful, however. Lough *et al.* (1992) reported that the cholesterol content of longissimus muscle and subcutaneous fat from ram lambs fed canola and soy lecithin were not different. When the cholesterol content of the longissimus muscle of pigs fed extruded full-fat soybeans or 4% tallow were compared with normal corn–soybean diets (Leszczynski *et al.*, 1992), the cholesterol was actually increased by the inclusion of 20% extruded full-fat soybeans. Furthermore, Brandt *et al.* (1992) determined that crossbred steers had a significantly greater cholesterol content in the longissimus muscle when fed a diet supplemented with 4% yellow grease, but they acknowledged that the difference was minor (52 vs 49 mg per 100 g).

While feeding unsaturated fatty acid diets may not always reduce cholesterol, some diets with elevated saturated fatty acid content would be undesirable. In an attempt to measure this effect, Lough *et al.* (1994) modified the diets of ram lambs with 10.7% palm oil and reported that the plasma concentration of cholesterol was elevated and the carcasses were all significantly fatter than lambs on the control diet.

The most promising method for dietary modification resulting in cholesterol reduction is by altering the mineral supplement in the diet. Bakalli *et al.* (1995) reported that supplementation of broiler chicken diets with 250 mg copper per kg diet for 35–42 days decreased breast muscle cholesterol content by 20.4%. As hypercholesterolemia is a symptom of copper deficiency, their experiments helped to demonstrate that copper in excess of dietary requirement can reduce the cholesterol content.

6.2.3 Pharmaceutical agents

Several pharmaceutical agents are commercially available or have been evaluated experimentally to alter fat deposition in livestock and poultry. According to Mills and Mersmann (1995), the β-adrenergic agonists ractopamine, clenbuterol, cimaterol and others have been reviewed and summarized as to their efficacy in promoting growth and improving protein to fat ratios in livestock and poultry species. Trenbolone acetate is another drug that has an influence on carcass composition, and dichloroacetate is an inhibitor of cholesterol synthesis.

While all of these pharmaceuticals have an impact on carcass composition, the results in reduction of cholesterol are mixed. Because drugs such as ractopamine are expected to reduce fat deposition and encourage protein accretion, this process could be expected to reduce cholesterol levels by dilution. The fatty acid and cholesterol contents of ractopamine treated pigs were evaluated by Perkins *et al.* (1992), and they reported that cholesterol content of longissimus dorsi muscle was reduced by 9% compared with controls. Ractopamine only had a limited effect in altering the fatty acid profile of the muscle and subcutaneous fat samples analyzed. Wander *et al.* (1993) determined that roasts from porcine somatotropin treated pigs had lower lipid content than roasts from placebo-treated pigs and that the cholesterol content decreased slightly in roasts stored frozen for 4 months.

Less successful examples of pharmaceutical alteration of animal composition have been reported. Lough *et al.* (1993) found ram and ewe lamb plasma concentrations of total cholesterol were not affected by trenbolone acetate treatment when compared with untreated animals. Beyer and Jensen (1993) reported that dichloroacetate was not effective in altering the plasma cholesterol levels in laying hens. Implementation of pharmaceutical intervention in light of mixed results remains unresolved and further research needs to be carried out.

6.2.4 *Immunization*

Immunizing animals against their adipocyte cell membrane proteins has been reported by Brodie and Hu (1995) and has met with some success in reducing fat deposition in lambs, pigs and rabbits, but not in chickens. The concept of the procedure is to create passive immunity to the adipocytes, and thus limit adipocyte number so leaner products would result. The impact on cholesterol level was not reported, but if lean tissue has less cholesterol than adipose tissue, immunization should be effective in reducing cholesterol.

6.2.5 *Endpoint for marketing*

Some research indicates that a reduction in feeding of animals would be effective in reducing cholesterol content. The basis of this concept is that adipose tissue in many studies has less cholesterol than muscle tissue. Wheeler *et al.* (1987) stated that the average value for beef longissimus muscle cholesterol (63 mg per 100 g) was lower than that for subcutaneous fat (99 mg per 100 g). A study by Hoelscher *et al.* (1988) concluded that raw beef adipose tissue had a greater concentration of cholesterol (>110 mg per 100 g) than muscle tissue (<65 mg per 100 g). Lough *et al.* (1992) reported the muscle tissue of lambs had approximately 62 mg cholesterol per 100 g, whereas adipose tissue had 76 mg per 100 g. VanKoevering *et al.* (1995) reported that cholesterol and total lipid concentrations in longissimus muscle increased linearly with time on feed for British and continental crossbred steers.

In contrast, some reports exist that offer conflicting values for cholesterol content and, therefore, less justification for reducing the degree of finish for livestock.

In one such case, Berg *et al.* (1985) found that longissimus muscle with the lowest degree of marbling (practically devoid) had the highest cholesterol content (211 mg per 100 g), whereas the highest marbling degrees (modest and moderate) were the lowest in cholesterol content (180 mg per 100 g). Brandt *et al.* (1992) also stated that there was no correlation between degree of marbling and cholesterol content of beef muscle tissues. Comparisons of early data were summarized by Pearson *et al.* (1987); data for rats and chickens showed that adipose tissue could have a much lower cholesterol content than lean tissue whereas veal examples were shown to have more cholesterol per 100 g than some beef samples.

Despite the possibility that the amount of cholesterol may or may not increase with increase in degree of marbling, there is enough proof that reducing the degree of finish has an impact on meat composition. With the opportunities available for selecting more efficient animals, control of their diet and length of feeding, it is clear that producers can use the option of marketing leaner animals as a means of altering cholesterol content.

6.3 Reduction of cholesterol by product modification

After harvest, the composition of raw materials can be analyzed, modified and manufactured into meat products for consumer enjoyment. With the interest in reducing dietary cholesterol high among the goals of many consumers, methods to decrease cholesterol content are of interest to researchers and processors. A considerable amount of information is available about two common approaches thought to affect the cholesterol content of meat products, although some methods are not as effective as expected.

6.3.1 Cooking or rendering

It is reasonable to assume that cooking would render fat and presumably reduces the cholesterol from meat products, but surprisingly little evidence exists to support this method for cholesterol reduction. One case that provided support for this concept showed that beef longissimus muscle samples cooked after the subcutaneous fat was trimmed off the steak resulted in lower cholesterol levels (Berg *et al.*, 1985). Most other studies revealed little or no change in cholesterol content. As an example, no difference in cholesterol content was reported for pork retail cuts containing different levels of subcutaneous fat trim prior to cooking (Morgan, Calkins and Mandigo, 1988). The four methods of cooking evaluated did not result in any statistically significant differences, but it was noted that microwaved chops retained more cholesterol. In another instance, the effect of endpoint cooking temperature of pork roasts was determined by Heymann *et al.* (1990). They found cholesterol levels were unaffected by the endpoint temperature, although both lipid and protein were concentrated by higher temperatures. In related research, Swize *et al.* (1992) reported that cooking steaks with or without external fat cover had no effect on the cholesterol content of the separable lean for longissimus and semimembranosus muscles of beef or pork. In contrast, longissimus muscle of lamb had a higher cholesterol content when the external fat remained attached during cooking. Akinwumni, Thompson and Ramsey (1993) cooked beef steaks of slight and modest marbling and reported higher cholesterol levels in the modestly marbled products, but neither trimming nor degree of doneness of the steaks had any effect on cholesterol levels. It seems that processors and consumers would have little chance of altering cholesterol content by merely cooking the product.

Unlike whole-muscle products, the fat content of ground beef can have considerably greater variation, and it is possible to alter the cholesterol content slightly by cooking. Rhee and Smith (1983) demonstrated that retention of cholesterol in cooked patties was favored in low fat (8%) products in comparison with high fat (27%) products. Raw samples containing the highest amounts of fat resulted in the lowest retention of cholesterol for ground beef cooked by three methods (Ono, Berry and Paroczay, 1985). The cholesterol content of ground beef patties increased with higher endpoint cooking temperature but differences in cholesterol levels could not be attributed to the fat content of the raw product (Kregel, Prusa and Hughes, 1986). The consistency among reports suggests that a higher cholesterol content is associated with higher fat as compared with low fat patties (Ono, Berry and Paroczay, 1985; Kregel, Prusa and Hughes, 1986). However, even this information is contested because Johnson *et al.* (1994) reported that the lipid and cholesterol contents of ground beef of regular, lean and extra-lean types were not different in a survey of samples collected in 12 cities.

The most successful method for reducing cholesterol by cooking is by draining and rinsing the cooked meat. Rhee, Griffith-Bradle and Ziprin (1993) reported that fat and cholesterol concentrations were reduced by the brown and drain process, particularly from ground meats with higher initial fat levels when based on a percentage of the amount in the raw product. In a variation of this procedure, Bonsell, Andersen and Rule (1993) cooked ground beef in a variety of vegetable oils and then rinsed the meat with hot water. Their results included a significant reduction of cholesterol and an alteration of fatty acid composition consistent with the characteristics of the oils used for cooking. On the other hand, some authors reported no change in the cholesterol content in their evaluation of the process (Love and Prusa, 1992; Snyder *et al.*, 1994), even though lipid reduction was successful.

Cooking, without intervention by significant trimming or extraction by hot water, seems to be an ineffective method for altering the cholesterol content of most meats. While the outcome is often a decrease in lipid content, without a net change of the cholesterol it is considered a poor choice. However, considerable public interest was generated by the cook, drain and rinse approach for ground beef and it remains one of the viable options in the effort to reduce the lipid and cholesterol content of meat products.

6.3.2 *Dilution of cholesterol*

Dilution of cholesterol is one of the most practical methods of product modification and is a technique that can easily be utilized by meat processors. Substituting materials that are void of cholesterol, particularly plant products or vegetable oils, in further processed products has been widely practiced. Not surprisingly, there is diminished overall acceptability with the use of many cholesterol-free extenders.

Researchers have usually tried diluting the cholesterol content of meat products with various grains. Rhee and Smith (1983) reported that dilution of cholesterol in ground beef patties was achieved by including increasing amounts of textured soy protein. Isolated soy protein at 8–12% in ground beef patties can reduce the cholesterol content by 16–19% (McMindes, 1991). Cooked wild rice at levels up to 30% have been tested by Minerich *et al.* (1992) in ground beef mixtures and the cholesterol content was reduced in proportion to the added grain. Soy hulls in restructured pork (Ibrahim *et al.*, 1990) and pork patties (Goldman and Brown, 1992) also had an impact on cholesterol content. In addition, wheat gluten (Miller *et al.*, 1986), oat bran (Anon., 1991), wheat flour (Ba-Jaber *et al.*, 1992), rice or corn flour (Flores, Clarke and Hsieh, 1992) and wheat germ (Gnanasambandam and Zayas, 1994) are all examples of grain-based extenders that have been used to alter the composition of meat products.

Two other groups of diluting ingredients are vegetable oils and gums, both of which have appeal to processors trying to reduce the cholesterol content of meat products. Paneras and Bloukas (1994) produced frankfurters with olive, corn, sunflower or soybean oils and reported a significant reduction in cholesterol content. Several gums were evaluated in structured beef rolls by Shand *et al.* (1993) and they were able to alter the net cholesterol content of products through dilution. Much of the dilution effect is achieved through the yield enhancement or improved water binding imparted by the nonmeat ingredients.

Many grain products have little or no impact themselves on cholesterol reduction when included in the diet. Oat gum (Beer, Arrigoni and Amando, 1995) and wheat or oat bran (Uusitupa *et al.*, 1992) were shown to have no influence on blood cholesterol levels, even after several weeks of significant intake. Psyllium, however, was reported to have some benefit in reducing blood cholesterol levels (Anderson *et al.*, 1992) and barley bran had a similar effect (Lupton, Robinson and Morin, 1994). The chance that a cholesterol-lowering effect could exist for some of these items will interest processors and enhance the appeal for using these and other extenders in meat products.

6.4 Supercritical fluid extraction (SFE)

One of the more innovative techniques for fat and cholesterol reduction may be the use of supercritical fluid extraction (SFE). Researchers began the use of SFE as a method to alter the composition of meat (King, Johnson and Friedrich, 1989; Chao *et al.*, 1991) and meat processors became interested. Animal products such as dried egg yolk (Froning *et al.*, 1990) and butter (Shishikura *et al.*, 1986; Anon., 1989) have also been modified with SFE. Despite interest from the animal product industry, decaffeination of coffee and extraction of flavor compounds from hops (Gardner, 1982; Morris, 1982; Dziezak, 1986) remain the principal uses for SFE on a commercial scale.

6.4.1 Fundamentals of SFE

Supercritical fluids are a phase that results when pressure and temperature conditions exceed a critical point unique to the fluid. Above the critical temperature, no amount of pressure will force the gas into a liquid, and therefore the fluid remains in a supercritical phase with properties somewhat like those of both a gas and a liquid. Phase diagrams, such as that shown in Figure 6.1, reflect the relationship between supercritical fluids and gas, liquid or solid phases. The interest in supercritical fluids arises

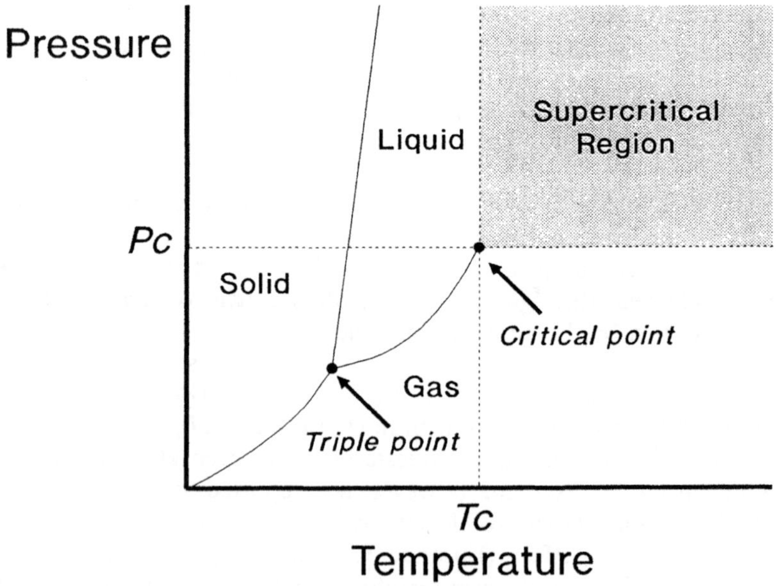

Figure 6.1 Typical pressure/temperature phase diagram with supercritical region. P_c is critical pressure and T_c is critical temperature. (Source: Clarke, 1991.)

because they have unique density and diffusivity values that are usually between those of the gas and liquid phase (Rizvi *et al.*, 1986a). The supercritical region encompasses a wide range of temperatures and pressures and this feature may be harnessed to enhance the selectivity of the supercritical solvent.

The properties of supercritical carbon dioxide and other solvents are well documented (Brogle, 1982; Hyatt, 1984; Rizvi *et al.*, 1986a). The critical temperature and pressure for the fluid are important because these conditions influence the requirements for equipment to process materials. For example, the critical temperature and pressure of carbon dioxide are 31 °C and 7.38 MPa (Brogle, 1982; Rizvi *et al.*, 1986b). The ability to use supercritical carbon dioxide at temperatures below 40 °C is most fortunate for maintaining the quality of raw materials, such as meats. Although it is evident that the critical pressure is attainable for many solvents, the critical temperature has a significant impact on the quality of foods and must be considered.

Selection of a solvent for SFE also depends on the sample and the intended solute. Since any product that is processed by SFE is totally immersed in the fluid, factors such as cost, toxicity, flammability and availability must be considered (Friedrich and Pryde, 1984; Hyatt, 1984; Rizvi *et al.*, 1986b). For these reasons, carbon dioxide is the preferred solvent for any meat product application.

A schematic representation of an SFE system with the essential components is presented in Figure 6.2. The four principal parts of the system are the solvent compressor, the extraction vessel, the separation vessel and the temperature/pressure control system. Extraction systems may have additional pumps for the introduction of entrainers (cosolvents) or multiple separatory vessels (with independent temperature/pressure control) to enhance the fractionation of mixtures. Other refinements may include using adsorbents (e.g. activated alumina) before the separation vessel or the recycling of solvent. Descriptions for these configurations may be found in various sources (Friedrich and Pryde, 1984; Rizvi *et al.*, 1986b; Huang, 1990).

6.4.2 Applications of SFE

Most research involving the reduction of cholesterol content in meat has been performed with beef and pork, but some reports also mention poultry or fish. Evidence that SFE can remove lipids from muscle foods is primarily based on extractions of materials with moisture levels below 10%.

(a) Fresh meats. The application of SFE to fresh meat products has not been overly successful with respect to fat and cholesterol reduction. One

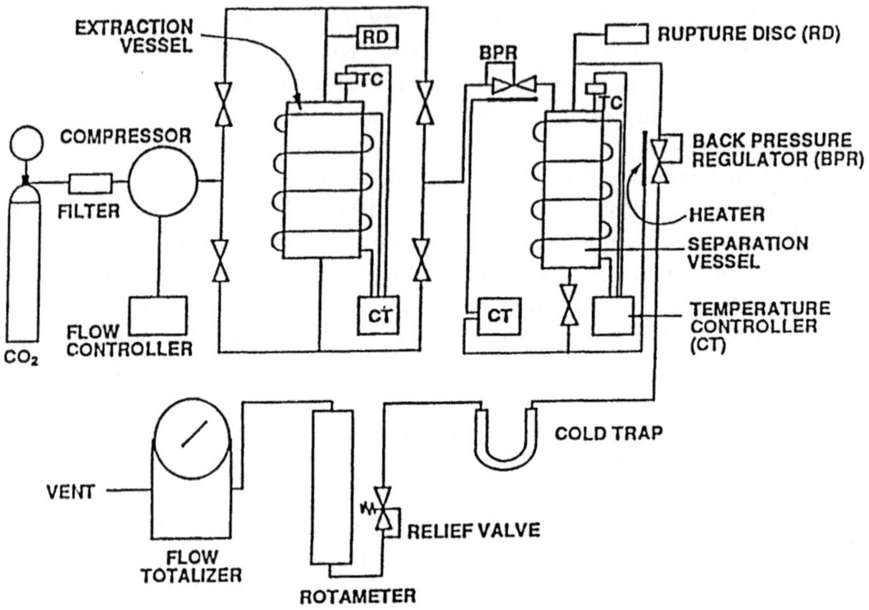

Figure 6.2 Supercritical fluid extraction system with a single separation vessel. (Source: Clarke, 1991.)

of the few successful efforts dealt with the extraction of fresh ground beef and was reported by Chao *et al.* (1991), who found that approximately 20–40% of the cholesterol could be removed at the rate of 2% per kg of carbon dioxide. King *et al.* (1993) found that without freeze-drying, raw or cooked ground beef patties had either no or limited reduction of cholesterol content. With freeze-drying they found that up to 103 mg per 100 g could be removed from the cooked patties compared with only 63 mg per 100 g for the raw ones.

Efforts to establish the efficiency of SFE conditions for fresh porcine muscle tissue have had poor results. A comparison of two extraction temperatures (35 and 50 °C) and three extraction pressures (17.3, 24.2 and 31.1 MPa) did not result in a significant reduction in total lipid or cholesterol (Clarke, 1991). Extraction of fresh ground pork with wide differences in composition [40–70% crude fat, dry weight basis (dwb)] were performed at 34.5 MPa and 40 °C, but essentially no cholesterol was removed (Clarke, 1991).

(b) Dried meats. The SFE process has been tested with several forms of dried meat, poultry and fish to determine the effectiveness of the process. Some of these results are summarized in Table 6.1. In most of the studies, the crude fat and cholesterol contents were reduced by the

Table 6.1 Comparison of fat and cholesterol removal from fresh and dried muscle foods by supercritical fluid extraction

Species (form)	Moisture (%)	Fat extracted or (gained) (%)	Cholesterol extracted or (gained) (%)	Pressure (MPa)	Temperature (°C)
Beef (ground)[a]	64.8	71.2	36.9	31.1	50
Beef (ground)[b]	63.1	2.5	(19.9)	54.9	40
Pork (ground)[c]	66.3	(1.7)	(1.3)	31.1	50
Beef (powder)[d]	1.4	87.7	87.8	38.5	55
Beef (chunk)[d]	1.7	97.6	85.4	38.5	55
Beef (ground)[b]	4.6	26.7	38.5	54.9	40
Pork (ground)[c]	3.9	71.9	78.1	31.1	50
Chicken (powder)[e]	1.3	89.1	90.3	38.5	55
Chicken (chunk)[e]	1.9	95.1	89.1	38.5	55
Trout (ground)[f]	1.7	78.2	96.8	27.6	40

Sources: [a]Chao *et al.* (1991); [b]King *et al.* (1993); [c]Clarke (1991); [d]Wehling *et al.* (1992); [e]Froning *et al.* (1994); [f]Hardardottir and Kinsella (1988).

SFE process, but it is a drawback to require dehydration before extraction.

Successful results for krill (Yamaguchi *et al.*, 1986) and different fish (Hardardottir and Kinsella, 1988; Fujimoto *et al.*, 1989) probably stimulated further work on land-based species. For example, Clarke (1991) reported that fatty acids were reduced by over 70% and cholesterol was reduced by more than 78% to an average of 37.2 mg per 100 g dwb in freeze-dried ground pork. Wehling *et al.* (1992) extracted up to 87% of the cholesterol from dehydrated beef powder and chunks. This was followed by a similar study in which dehydrated chicken meat powder or chunks had 82–90% of the cholesterol removed (Froning *et al.*, 1994). These results are valuable for demonstrating the efficiency of SFE as a technique with potential to alter the composition of dried meats without the use of toxic organic solvents.

Only limited reports exist to indicate the functionality of extracted tissues as materials for further processed products. One case involved pork that was freeze-dried, extracted by SFE, rehydrated with three parts of distilled water and used at 0 or 100% of the product formulation (Clarke, 1991). The frankfurter-like product reflected a 76% reduction of both crude fat and cholesterol compared with a 100% fresh pork product. It was shown that shear force and product hardness diminished when the extracted material was used, which suggested that the performance of the rehydrated meat powder in emulsion-type products was affected. In another study, Fujimoto *et al.* (1989) found that sardine meat powder subjected to SFE had less fishy odor, which may increase the opportunities for using such material in sensitive products, such as surimi.

(c) Fats, lard and tallow. A few interesting applications for SFE include fractionation of fats and extraction of flavor compounds. Some studies on fractionation of tallow (Rizvi *et al.*, 1986b; Huang, 1990; Chao, Mulvaney and Huang, 1993) and lard (King, Johnson and Friedrich, 1989) indicate that SFE is useful for reducing the cholesterol concentration of these sources of animal fat.

In using SFE for preparing beef fat, Um *et al.* (1992) identified numerous volatile compounds in the extracted material. Others have confirmed this in the extraction of triglycerides from beef fat (Merkle and Larick, 1993) and chicken fat (Taylor and Larick, 1995). Although no mention of cholesterol was made in these reports, Kwon (1993) recognized that the solubility of cholesterol varies when mixed with triglycerides. He determined that the solubility of cholesterol generally increased in mixtures with trilauin or trimyristin, but decreased in the presence of tripalmitin or tristearin. Cholesterol solubility also increased with relative degree of unsaturation in combinations with mono-, di- or triolein mixtures.

The ability to alter the composition of various animal fats will permit interesting applications in meat products. The selection of specific flavors and reduction of cholesterol is an obvious outcome of SFE. There is also potential to extract desirable flavors from animal fats and combine them with vegetable oils to effect a reduction of cholesterol when the enhanced oils are used in place of animal fats.

6.4.3 Cosolvent extractions

Another application of the SFE process requires the use of a cosolvent, such as ethanol, to improve further the fat or cholesterol reduction in meat products. Freeze-dried pork tissue was the subject in a study reported by Clarke (1991) using a 3% (w/w) level of ethanol incorporated into the carbon dioxide stream prior to supercritical pressurization. The cosolvent extraction resulted in 33% greater removal of the cholesterol and an overall reduction of 89% (22.6 mg per 100 g dwb) compared with the control. Hardardottir and Kinsella (1988) had earlier reported the effect of ethanol as a cosolvent with a 97.1% reduction of lipids. The usefulness of cosolvent extraction is probably limited to analytical processing, since removal of cosolvent residues from biological materials is not easy, and therefore may defeat the idea of using carbon dioxide as a 'safe' solvent.

6.5 Research needs

As the most promising means for selectively reducing the cholesterol content of muscle tissue in a living animal appears to be through provision of elevated dietary copper, further work in this area may be useful.

The mechanisms by which copper deficiency can induce hypercholesterolemia and abundant copper results in hypocholesterolemia need clear elucidation. Investigation of dietary supplementation with copper for species other than chickens would be useful, especially if animal growth is not significantly altered, other than the cholesterol content.

Implementation of pharmaceutical intervention in the light of mixed results is indicative of another area suitable for further research. While several compounds have a clear impact on the reduction of lipid content and enhancement of muscle size, it is also evident that human understanding of tissue metabolism and growth regulation is not complete. New compounds that are selective in reducing the cholesterol content of lean tissues may be discovered and their relationship with other environmental influences could be studied.

Further research with food additives and processes which alter product composition may be easily justified. However, the incorporation of additives is best directed toward defining the functional roles achieved by the additive, beyond the simple dilution of the cholesterol content. It would be more useful to carry out investigations with new processes that physically remove the cholesterol from the tissues. Development of large-scale SFE operations may be achieved once thorough evaluation of the quality attributes of extracted meat products is completed.

6.6 Summary

Modification of animal diet and marketing of leaner livestock, usually coupled with judicious application of repartitioning agents, have the most reasonable chance of reducing cholesterol in meat, poultry and fish species. Changes in genetic selection will probably not permit significant progress because environmental factors may have a greater influence on cholesterol content.

Dilution is effective in reducing cholesterol and is easily practiced for meat products of any species where the opportunity exists to incorporate nonmeat ingredients. Cooking of meat is obviously necessary, but it is disappointing as a method for cholesterol reduction. However, combining dilution with cooking provides a meaningful way to reduce total fat content and should be perceived as a healthy and valuable option for processors and consumers.

Awareness that SFE has potential as a new procedure in food and flavor processing has been heightened by recent articles in trade and research journals. The principle behind SFE as a process is that supercritical fluids have solvent properties that are between those of gases and liquids; they have solubility characteristics similar to those of liquids, but can penetrate material matrices with the ease of gases. By changing the extraction conditions (temperature and/or pressure), the selectivity of the solvent can be modified, and thus extraction or fractionation of selected solutes may be controlled. Studies with muscle foods indicate that extracted ingredients for further processed products with significantly lower fat and cholesterol levels may be produced by SFE.

References

Ajuyah, A.O., Lee, K.H., Hardin, R.T. and Sim, J.S. (1991) Changes in the yield and in the fatty acid composition of whole carcass and selected meat portions of broiler chickens fed full-fat oil seeds. *Poult. Sci.*, **70**, 2304.

Ajuyah, A.O., Hardin, R.T., Cheung, K. and Sim, J.S. (1992) Yield, lipid, cholesterol and fatty acid composition of spent hens fed full-fat oil seeds and fish meal diets. *J. Food Sci.*, **57**, 338.

Akinwumni, I., Thompson, L.D. and Ramsey, C.B. (1993) Marbling, fat trim and doneness effects on sensory attributes, cooking loss and composition of cooked beef steaks. *J. Food Sci.*, **58**, 242.

Anderson, J.W., Riddell-Mason, S., Gustafson, N.J. *et al.* (1992) Cholesterol-lowering effects of psyllium-enriched cereal as an adjunct to a prudent diet in the treatment of mild to moderate hypercholesterolemia. *Am. J. Clin. Nutr.*, **56**, 93.

Anon. (1989) Supercritical extraction holds promise for cholesterol-free butter. *Food Eng.*, **61**, 83.

Anon. (1991) Oat-bran-based ingredient blend replaces fat in ground beef and pork. *Food Technol.*, **45** (11), 60.

Ba-Jaber, A.S., Sofos, J.N., Schmidt, G.R. and Maga, J.A. (1992) Extrusion cooking of chicken meat with various nonmeat ingredients. *Dev. Food Sci.*, **29**, 761.

Bakalli, R.I., Pesti, G.M., Ragland, W.L. and Konjurfca, V. (1995) Dietary copper in excess of nutritional requirement reduced plasma and breast muscle cholesterol of chickens. *Poul. Sci.*, **74**, 360.

Beer, M.U., Arrigoni, E. and Amando, R. (1995) The effects of oat gum on blood and cholesterol levels in healthy young men. *Europ. J. Clin. Nutr.*, **49**, 517.

Berg, P.T., Marchello, M.J., Erickson, D.O. and Slanger, W.D. (1985) Selected nutrient content of beef longissimus muscle relative to marbling class, fat status, and cooking method. *J. Food Sci.*, **50**, 1029.

Beyer, R.S. and Jensen, L.S. (1993) The hypocholesterolemic agent dichloroacetate increases egg cholesterol content of laying hens. *Poul. Sci.*, **72**, 1063.

Bonsell, T.D., Andersen, M.K. and Rule, D.C. (1993) Effect of cooking oil type on final cholesterol content and fatty acid composition of ground beef. *J. Food Qual.*, **16**, 383.

Brandt, R.T., Jr, Kuhl, G.L., Campbell, R.E. *et al.* (1992) Effects of steam-flaked sorghum grain or corn and supplemental fat on feedlot performance, carcass traits, longissimus composition, and sensory properties of steers. *J. Anim. Sci.*, **70**, 343.

Brodie, A. and Hu, C.Y. (1995) Fat reduction through the use of passive immunity, in *The Biology of Fat in Meat Animals* (eds S.B. Smith and D.R. Smith), American Society of Animal Science, Champaign, IL, pp. 70–77.

Brogle, H. (1982) CO_2 as a solvent: its properties and applications. *Chem. Ind. (London)*, **19**, 385.

Chao, R.R., Mulvaney, S.J., Bailey, M.E. and Fernando, L.N. (1991) Supercritical CO_2 conditions affecting extraction of lipid and cholesterol from ground beef. *J. Food Sci.*, **56**, 183.

Chao, R.R., Mulvaney, S.J. and Huang, H. (1993) Effects of extraction and fractionation pressures on supercritical extraction of cholesterol from beef tallow. *J. Am. Oil Chem. Soc.*, **70**, 139.

Clarke, A.D. (1991) Supercritical fluid extraction technology for fat reduction. *Proc. Recip. Meat Conf.*, **44**, 101.

Dziezak, J.D. (1986) Innovative separation process finding its way into the food industry. *Food Technol.*, **40**(6), 66.

Flores, N.C., Clarke, A.D. and Hsieh, F.H. (1992) Extrusion processing of rice or corn flour with mechanically deboned turkey. *J. Food Qual.*, **15**, 399.
Friedrich, J.P. and Pryde, E.H. (1984) Supercritical CO_2 extraction of lipid-bearing materials and characterization of the products. *J. Am. Oil Chem. Soc.*, **61**, 223.
Froning, G.W., Wehling, R.L., Cuppett, S.L. *et al.* (1990) Extraction of cholesterol and other lipids from dried egg yolk using supercritical carbon dioxide. *J. Food Sci.*, **55**, 95.
Froning, G.W., Fieman, F., Wehling, R.L. *et al.* (1994) Supercritical carbon dioxide extraction of lipids and cholesterol from dehydrated chicken meat. *Poult. Sci.*, **73**, 571.
Fujimoto, K., Endo, Y., Cho, S.-Y. *et al.* (1989) Chemical characterization of sardine meat powder produced by dehydration with high osmotic pressure resin and defatting with high pressure carbon dioxide. *J. Food Sci.*, **54**, 265.
Gardner, D.S. (1982) Industrial scale hop extraction with liquid CO_2. *Chem. Ind. (London)*, **19**, 402.
Gnanasambandam, R. and Zayas, J.F. (1994) Microstructure of frankfurters extended with wheat germ proteins. *J. Food Sci.*, **59**, 474.
Goldman, D.C. and Brown, N.E. (1992) Effects of fat level and addition of soy fiber on sensory and other properties of ground pork patties. *J. Foodserv. Syst.*, **7**(1), 1.
Hardardottir, I. and Kinsella, J.E. (1988) Extraction of lipid and cholesterol from fish muscle with supercritical fluids. *J. Food Sci.*, **53**, 1656.
Heymann, H., Hedrick, H.B., Karrasch, M.A. *et al.* (1990) Sensory and chemical characteristics of fresh pork roasts cooked to different endpoint temperatures. *J. Food Sci.*, **55**, 613.
Hoelscher, L.M., Savell, J.W., Smith, S.B. and Cross, H.R. (1988) Subcellular distribution of cholesterol within muscle and adipose tissues of beef loin steaks. *J. Food Sci*, **53**, 718.
Huang, H. (1990) Fractionation and cholesterol reduction of crude beef tallow using supercritical fluid CO_2 extraction. *MS Thesis*, University of Missouri, Columbia, MO.
Hyatt, J.A. (1984) Liquid and supercritical carbon dioxide as organic solvents. *J. Org. Chem.*, **49**, 5097.
Ibrahim, N., Unklesbay, N., Kapila, S. and Puri, R.K. (1990) Cholesterol content of restructured pork/soy hull mixture. *J. Food Sci.*, **55**, 1488.
Johnson, L.P., Williams, S.E., Neel, S.W. and Reagan, J.O. (1994) Foodservice industry market profile study: nutritional and objective textural profile of foodservice ground beef. *J. Anim. Sci.*, **72**, 1487.
King, J.W., Johnson, J.H. and Friedrich, J.P. (1989) Extraction of fat tissue from meat products with supercritical carbon dioxide. *J. Agric. Food Chem.*, **37**, 951.
King, J.W., Johnson, J.H., Orton, W.L. *et al.* (1993) Fat and cholesterol content of beef patties as affected by supercritical CO_2 extraction. *J. Food Sci.*, **58**, 950.
Koch, R.M., Jung, H.G., Crouse, J.D. *et al.* (1995) Growth, digestive capability, carcass, and meat characteristics of *Bison bison*, *Bos taurus*, and *Bos* × *Bison*. *J. Anim. Sci.*, **73**, 1271.
Kregel, K.K., Prusa, K.J. and Hughes, K.V. (1986) Cholesterol content and sensory analysis of ground beef as influenced by fat level, heating, and storage. *J. Food Sci.*, **51**, 1162.
Kwon, Y.A. (1993) Determination and prediction of the solubilities of cholesterol and triglycerides in supercritical carbon dioxide. *PhD Dissertation*, University of Missouri, Columbia, MO.
Leszczynski, D.E., Pikul, J., Easter, R.A. *et al.* (1992) Characterization of lipid in loin and bacon from finishing pigs fed full-fat soybeans or tallow. *J. Anim. Sci.*, **70**, 2175.
Lough, D.S., Solomon, M.B., Rumsey, T.S. *et al.* (1992) Effects of dietary canola seed and soy lecithin in high-forage diets on cholesterol content and fatty acid composition of carcass tissues of growing ram lambs. *J. Anim. Sci.*, **70**, 1153.
Lough, D.S., Kahl, S., Solomon, M.B. and Rumsey, T.S. (1993) The effect of trenbolone acetate on performance, plasma lipids, and carcass characteristics of growing ram and ewe lambs. *J. Anim. Sci.*, **71**, 2659.
Lough, D.S., Solomon, M.B., Rumsey, T.S. *et al.* (1994) The effects of high-forage diets with added palm oil on performance, plasma lipids, and carcass characteristics of ram lambs with initially high or low plasma cholesterol. *J. Anim. Sci.*, **72**, 330.
Love, J.A. and Prusa, K.J. (1992) Nutrient composition and sensory attributes of cooked ground beef: effects of fat content, cooking method, and water rinsing. *J. Am. Diet. Assoc.*, **92**, 1367.

Lupton, J.R., Robinson, M.C. and Morin, J.L. (1994) Cholesterol-lowering effect of barley bran flour and oil. *J. Am. Diet. Assoc.*, **94**, 65.

McMindes, M.K. (1991) Applications of isolated soy protein in low-fat meat products. *Food Technol.*, **45**(12), 61.

McNamara, D.J. (1985) Diet–heart disease risk. *Proc. Recip. Meat Conf.*, **38**, 149.

McNamara, D.J. (1991) Relationship between fat, cholesterol and coronary heart disease. *Proc. Recip. Meat Conf.*, **44**, 147.

Merkle, J.A. and Larick, D.K. (1993) Triglyceride content of supercritical carbon dioxide extracted fractions of beef fat. *J. Food Sci.*, **58**, 1237.

Miller, M.F., Davis, G.W., Seideman, S.C. *et al.* (1986) Extending beef bullock restructured steaks with soy protein, wheat gluten or mechanically separated beef. *J. Food Sci.*, **51**, 1169.

Mills, S. and Mersmann, H.J. (1995) Beta-adrenergic agonists, their receptors, and growth: special reference to the peculiarities in pigs, in *The Biology of Fat in Meat Animals* (eds S.B. Smith and D.R. Smith), American Society of Animal Science, Champaign, IL, pp. 1–34.

Minerich, P.L., Addis, P.B., Epley, R.J. and Bingham, C. (1992) Properties of wild rice/ground beef mixtures. *J. Food Sci.*, **56**, 1154.

Morgan, J.B., Calkins, C.R. and Mandigo, R.W. (1988) Effect of trim level, cooking method, and chop type on lipid retention, caloric content, and cholesterol level in cooked pork. *J. Food Sci.*, **53**, 1602.

Morris, C.E. (1982) New process tool: supercritical CO_2. *Food Eng.*, **54**, 89.

Ono, K., Berry, B.W. and Paroczay, E. (1985) Contents and retention of nutrients in extra lean, lean and regular ground beef. *J. Food Sci.*, **50**, 701.

Paneras, E.D. and Bloukas, J.G. (1994) Vegetable oils replace pork backfat for low-fat frankfurters. *J. Food Sci.*, **59**, 725.

Pearson, A.M., Ashgar, A., Gray, J.I. and Booren, A.M. (1987) Impact of fat reduction on palatability and consumer acceptance of processed meat. *Proc. Recip. Meat Conf.*, **40**, 105.

Perkins, E.G., McKeith, F.K., Jones, D.J. *et al.* (1992) Fatty acid and cholesterol changes in pork longissimus muscle and fat due to ractopamine. *J. Food Sci.*, **57**, 1266.

Rhee, K.S. and Smith, G.C. (1983) Effect of cooking on cholesterol content of patties containing different amounts of beef, textured soy protein and fat. *J. Food Sci.*, **48**, 268.

Rhee, K.S., Griffith-Bradle, H.A. and Ziprin, Y.A. (1993) Nutrient composition and retention in browned ground beef, lamb, and pork. *J. Food Compos. Anal.*, **6**, 268.

Rizvi, S.S.H., Benado, A.L., Zollweg, J.A. and Daniels, J.A. (1986a) Supercritical fluid extraction: fundamental principles and modeling methods. *Food Technol.*, **40**(6), 55.

Rizvi, S.S.H., Daniels, J.A., Benado, A.L. and Zollweg, J.A. (1986b) Supercritical fluid extraction: operating principles and food applications. *Food Technol.*, **40**(7), 57.

Shand, P.J., Sofos, J.N. and Schmidt, G.R. (1993) Properties of algin/calcium and salt/phosphate structured beef rolls with added gums. *J. Food Sci.*, **58**, 1224.

Shishikura, A., Fujimoto, K., Kaneda, T. *et al.* (1986) Modification of butter oil by extraction with supercritical carbon dioxide. *Agric. Biol. Chem.*, **50**, 1209.

Smith, E.J. and Savage, T.F. (1992) Genetic variation in total plasma and high density lipoprotein cholesterol and body weight in medium white turkeys. *Poult. Sci.*, **71**, 807.

Snyder, M.P., Obarzanek, E., Montgomery, D.H. *et al.* (1994) Reducing the fat content of ground beef in a school foodservice setting. *J. Am. Diet. Assoc.*, **94**, 1135.

Swize, S., Harris, K.B., Savell, J.W. and Cross, H.R. (1992) Cholesterol content of lean and fat from beef, pork and lamb cuts. *J. Food Compos. Anal.*, **5**, 160.

Taylor, D.L. and Larick, D.K. (1995) Volatile content and sensory attributes of supercritical carbon dioxide extracts of cooked chicken fat. *J. Food Sci.*, **60**, 1197.

Um, K.W., Bailey, M.E., Clarke, A.D. and Chao, R.R. (1992) Concentration and identification of volatile compounds from heated beef fat using supercritical CO_2 extraction–gas liquid chromatography/mass spectrometry. *J. Agric. Food Chem.*, **40**, 1641.

Uusitupa, M.I.J., Ruuskanen, E., Makinen, E. *et al.* (1992) A controlled study on the effect of beta-glucan-rich oat bran on serum lipids in hypercholesterolemic subjects: relation to apolipoprotein E phenotype. *J. Am. Coll. Nutr.*, **11**, 651.

VanKoevering, M.T., Gill, D.R., Owens, F.N. *et al.* (1995) Effect of time on feed on performance of feedlot steers, carcass characteristics, and tenderness and composition of longissimus muscles. *J. Anim. Sci.*, **73**, 21.

Wander, R.C., Clark, S.L., Hu, C.Y. *et al.* (1993) Interaction of porcine somatotropin administration to growing pigs and frozen storage of carcass on lipids and quality characteristics of roasts. *J. Food Compos. Anal.*, **6**, 62.

Wehling, R.L., Froning, G.W., Cuppett, S.L. and Niemann, L. (1992) Extraction of cholesterol and other lipids from dehydrated beef using supercritical carbon dioxide. *J. Agric. Food Chem.*, **40**, 1204.

Wheeler, T.L., Davis, G.W., Stoecker, B.J. and Harmon, C.J. (1987) Cholesterol concentration of longissimus muscle, subcutaneous fat and serum of two beef cattle breed types. *J. Anim. Sci.*, **65**, 1531.

Yamaguchi, K., Murakami, M., Nakano, H. *et al.* (1986) Supercritical carbon dioxide extraction of oils from Antarctic krill. *J. Agric. Food Chem.*, **34**, 904.

and beef cuts with 2 mm or devoid of fat cover rather than steaks with 4 mm or more (Hearnshaw *et al.*, 1995). In New Zealand, pork sales have been increased through the introduction of a 'trim pork' programme involving the display for sale of lean pork cuts and pieces. In Denmark and elsewhere, retail demand for leaner pork has resulted in national programmes for leaner pork production (Fredeen, 1958) in advance of similar fat reduction pressures for other species. Goutefongea and Dumont (1990) reported that the fat content of pork in Finland had dropped 30% over the past 20 years and in the USA the equivalent figure for pork was 15%. Development of a farmed deer industry (Fennessy and Drew, 1985) resulting in increasing sales of venison and also of goat meat (both low fat products) also indicates consumer demand for leaner meats as well as a search for new products. Undoubtedly, continued media publicity on the negative aspects of a high fat diet and advice from nutritionists and expert panels (Chapters 1 and 2) have led to consumer demand for leaner meat and have had effects in most countries on retail practices, resulting in less visible fat on meat cuts on display for sale.

7.2.3 Impact upon trade in the European Community, United Kingdom, United States and worldwide

The changing consumer and, as a result, retailer requirements for leaner cuts of meat have obviously had a major effect on the meat trade in the developed world (sections 7.2.1 and 7.2.2). This is a change from the fatter meat cuts that were historically available when more of the population was engaged in manual work, in which fat was a source of calories to support such activities, and before the potentially harmful effects of high fat diets had been discovered. Consumer requirements for lower fat meat can only be recognized through trade systems in which carcasses are classified for fatness levels to enable retailers to specify carcasses and cuts which contain less fat. Marketing procedures designed to change carcass composition were discussed by Kempster (1990). Worldwide, increased emphasis is being placed on the search for more accurate carcass classification systems so that those in the meat trade can specify the carcasses they require. This will encourage the production and trade in such carcasses and cuts.

7.3 Trimming excess fat from cuts and carcasses

7.3.1 Leaner cuts

Where carcasses have more fat than consumers find acceptable on displays of cuts (Savell *et al.*, 1991; Thatcher and Couchman, 1983; Hearnshaw *et al.*, 1995), the carcasses and/or cuts must be trimmed before the meat can be sold. There is a saving in both feed energy and labour if production systems supply animals with acceptable levels of fat that does not require trimming. Chapter 8 and Bass *et al.* (1990) discuss systems for reducing fat in animals through nutrition and other management factors, while section 7.5 in this chapter discusses genetic strategies.

7.3.2 Upgrading of fats

Depending on subsequent processing, surplus trimmed fat from edible carcasses can be converted through rendering into edible fat. It can also be processed with carcass material containing fat classified as inedible sources of material, such as slaughter wastes, feet and gut material, from live or even dead animals (Swan, 1992) to produce different grades of inedible tallows and greases. Edible tallow (sheep and cattle) and lard (pigs) can be used in margarine, shortenings and cooking fats. Inedible tallow is used for purposes such as soap manufacture, oleochemical production and an energy source for animal feedstuffs. Tallow and methyl esters can be used as fuel oil, while glycerine and fatty acids can be made into purified products. In some countries, only inedible products can be produced from raw materials classified as inedible whereas in others sterile, purified fat obtained from such raw materials is classified as edible.

As a general rule, edible fat is a lower priced product than saleable meat and, in turn, inedible tallow is lower priced than edible tallow. The prices of individual fats and oils are affected by the prices of all fats and oils traded internationally. Tallow is normally the lowest priced fat in world markets (J.E. Swan, personal communication). Because of similarity of composition, the price differential between palm oil and tallow has been low in recent years. The exception to this rule is where edible fat can be incorporated with lean meat into processed products, such as mince and sausages, where it can be sold at meat prices.

7.4 Selection

7.4.1 Leaner cuts and carcasses

Section 7.5 will discuss the genetic strategies that can be used for the production of leaner animals which yield leaner carcasses and cuts. Kempster, Cuthbertson and Harrington (1982) described the carcass classification systems (phenotypic) in operation in the main countries using them at that time and, in addition, Kirton (1989) and Price (1995) have discussed the factors considered in carcass grading and classification. Kempster (1990) also discussed the role of classification in marketing, which is necessary for the selection of leaner carcasses and cuts. When used properly, they can indicate to producers the qualities, including fatness, which best meet the requirements of meat processors and consumers. Fatness has always been one of the most important factors considered in all carcass classification systems which are often tied to producer payments. A feature of some carcass classification systems has been the belief that some appropriate level of fatness (external fatness or alternatively marbling/intramuscular fat) is needed to promote palatability. A recent review by Hearnshaw *et al.* (1995) examined the relationship between factors commonly considered in carcass grading and expected to affect the resulting eating quality of beef. They concluded that fat cover and marbling of the longissimus muscle were almost unrelated to the tenderness of the beef, whereas electrical stimulation of the carcasses, age and carcass weight were generally related to tenderness. The beef from lighter, younger animals whose carcasses had been electrically stimulated was the most tender. Tatum (1981) suggested that the main role of subcutaneous fat in tenderness may be to act as insulation in preventing cold shortening.

Most carcass classification systems for pigs to indicate fatness are based on objective measurements, and electronic probes, which measure fatness and longissimus muscle depth, are commonly used (Kempster *et al.*, 1982; Forrest, 1995). Pigs are anatomically designed for probing, with around 75% of their carcass fat located subcutaneously compared with cattle, which have 40–50% of their carcass fat in the subcutaneous depots (Goutefongea and Dumont, 1990). Sheep are similar to cattle. In addition, the skin is retained in pig carcasses at slaughter, preventing fat loss or displacement during the process of hair removal and providing a firm base against which to probe compared with sheep and cattle, where subcutaneous fat can be lost or deformed as the skin is removed. In some systems, each pork carcass is multiply probed to increase the accuracy of yield prediction. Most of the intensive studies on the predictive relationships of different indicators of carcass composition show that multiple indicators can be much more revealing of variations among animals than any single indicator. Few single trait indicators can be expected to do a very good job, even when used in association with carcass weight, and even for rather simplistic objectives such as predicting the total fat content of carcasses. This principle is implicit in the carcass classification practices undertaken by the Danish pig industry, where as many as 15 carcass measurements and some of their live animal counterparts are routinely taken using automated electronic methods to predict the yield and value of carcasses for various purposes (Jensen, 1991; Mannion, 1993; Fensvig, 1995). They provide valuable feedback signals to producers and sire breeders in economic terms.

Most of the technology commonly reported for use on sheep and cattle (and pigs) has tended to result from research which pointed to possible new directions for carcass fatness classification without producing equipment that could be operated at chain speed. Such equipment could not be purchased at an affordable price relative to the value of the product. Most classification systems for fatness are still largely subjective with, in some cases, printed standards for comparison. In the UK, sheep carcass classification is still subjectively based (MLC, 1994a). In New Zealand, an objective measurement (GR, defined as tissue depth 11 cm from the midline at the twelfth rib; Kirton, 1989) is specified for sheep classification but most carcasses are still classified subjectively. In Australia, sheep carcasses are largely classified subjectively, although there is now some use of the AUS-Meat grading probe to measure GR (Cabassi, 1990). It is hoped to introduce video image analysis (VIA) for sheep/lamb carcass classification (Ferguson, Thompson and Cabassi, 1995) if it is proved to be sufficiently accurate. VIA is now registered under the trade name Viascan and there are three Viascan systems available in Australia, one for measurement of carcass dimensions, fat cover and bruise score, a second for measurement of eye muscle area, fat cover, meat colour and marbling and the third for use by the food services sector for portion assessment. Subjective systems are still largely in use in Canada and the USA for sheep/lamb carcass classification, although in the latter country a yield grading option has been added (Price, 1995).

Beef classification in the UK is still largely subjective (MLC, 1994b), as is also the case in New Zealand. The UK system predicts yield (MLC, 1994b). European classification systems have been described by Kempster, Cuthbertson and Harrington (1982). In Australia, beef carcass classification is in a state of change with a P8 fat depth (described below) which may be used for carcass specification, a chiller assessment system to record meat colour, fat colour and marbling being introduced for beef shipped to Japan (Herrod and Treadwell, 1995) and a VIA system under consideration for predicting yield (which involves fatness) on carcasses and cuts (Ferguson *et al.*, 1995). The P8 rump fat depth is measured in the hanging carcass from 'the point of intersection of a vertical line from the dorsal tuberosity of the tripartite tuber ischii parallel with the chine, and a horizontal line from the crest on the spinal process of the third sacral vertebra' (Hopkins, 1989). The USA subjective system as carried out on cold, ribbed or quartered sides to reveal longissimus muscle cross-section and fat was explained by Kempster *et al.* (1982), along with an alternative yield grading system based mainly on fat measurements. The Canadian system of beef grading included a fat thickness measurement taken between the eleventh and twelfth ribs over the longissimus muscle (Kempster *et al.*, 1982).

Some of the options available for measuring body and carcass composition were covered in an NRC (1988) report. The automated measurement of beef was discussed by Brownlie, Hall and Fabiansson (1989). Since then, Allen (1990) reviewed methods of estimating the composition of live animals, while Fisher (1990) reviewed methods of predicting fat in carcasses. Subsequently, there have been developments, such as greater to a fixed age of 20 months, showed a range among 11 sire breeds of 3.1 mm or 1.6 phenotypic standard deviations (σ_P) for twelfth/thirteenth rib fat depth, when the mean was 3.6 mm. At 31 months when the mean was 6.5 mm, the range was higher at 5.3 mm, but it was similar ($1.8\sigma_P$) in standard units, i.e. in relation to variability within a breed. When the 31 month data were adjusted to a common carcass weight of 277 kg, the range of fat depths changed to 5.8 mm ($2.0\sigma_P$), suggesting that fat variation from breed to breed is not weight related.

In Clay Center, Nebraska, which is undertaking the largest of the beef cattle breed comparisons in the USA, the most recent trial is known as the Germ Plasm Utilization and includes purebreds and three very different advanced-generation crosses (Gregory *et al.*, 1994). All groups were compared at constant age, and there was a range in twelfth rib fat depth of 8.2 mm ($2.4\sigma_P$), with an overall mean of 6.5 mm at 333 kg carcass weight.

In sheep, Kirton *et al.* (1995b) published a comparison of 15 sire breeds in New Zealand, including data on various fat depths and chemical fat percentages in the left side. The halfbred lambs had a mean fat content of 27.8% and a range of 5.2 percentage points at a fixed age ($= 1.9\sigma_P$). Their results have direct relevance to the choice of breeds for use as terminal sires. This avenue of genetic change provides major flexibility in meeting particular production and market circumstances (McEwan *et al.*, 1995). Some of its features pertaining to the history of the New Zealand export lamb industry were discussed by Clarke and Meyer (1982) and Carter and Cox (1982). Generally, this industry appears to have given greatest attention to breed differences in live or carcass weight for age (i.e. growth rate), favouring larger over smaller terminal sire breeds, but whose progeny tended to be leaner at any given age as well. However, many other issues may affect the final choice of a terminal sire breed (Bennett and Kirton, 1983), not the least of which is production costs relative to market requirements. In particular, Clarke and Meyer (1982) stress crossbred differences in progeny survival and post-weaning lamb fleece weights, which influence lamb returns per breeding ewe. These can give rise to different breed rankings to those based on meat returns per lamb, and provide one likely explanation for the rather low current use of terminal sire breeds by the New Zealand lamb industry (McEwan *et al.*, 1995).

Thus, industry needs reliable animal production and carcass quality information for many performance features of its major breed resources if it is to retain a high level of flexibility to respond to emerging market requirements for meat. A summary of two important carcass features for the breeds considered by Kirton *et al.* (1995b) and of major relevance to this review is presented in Figure 7.1. It gives emphasis to crossbred age-adjusted variations in carcass fat percentage relative to carcass size for crossbred lambs aged between 3 and 6 months.

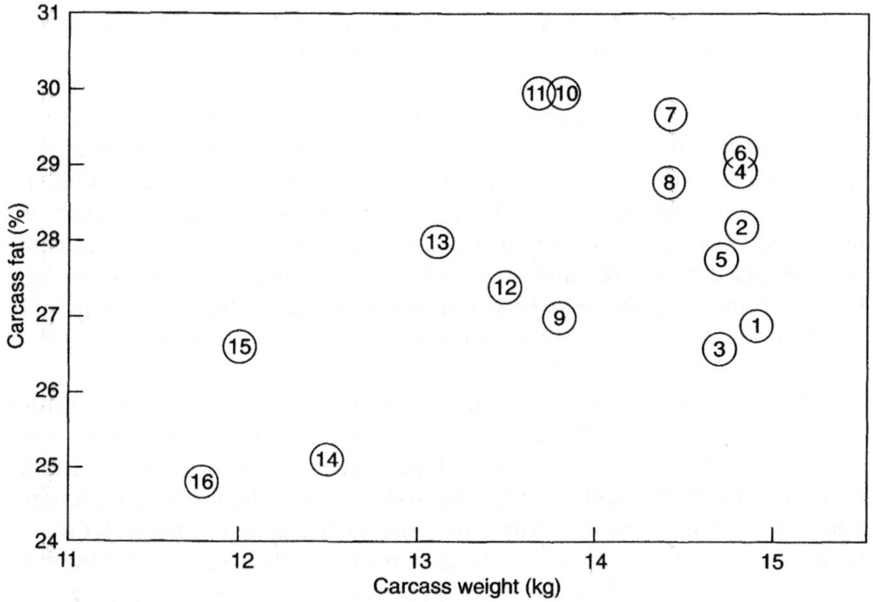

Figure 7.1 Average sire breed means for percentage chemical carcass fat and carcass weight for lambs by different sire breeds and from Romney dams. Breeds are coded in order of lean growth rate: 1, Dorset Horn; 2, Border Leicester; 3, Suffolk; 4, Poll Dorset; 5, Dorset Down; 6, Hampshire; 7, South Suffolk; 8, South Dorset Down; 9, Cheviot; 10, Southdown (unregistered); 11, Southdown (registered); 12, English Leicester, 13, Ryeland; 14, Lincoln; 15, Merino; 16, Romney. Source: Kirton *et al.* (1995b).

The data in Figure 7.1 illustrate that both fat content and carcass growth rates make important contributions to breed variations in lean growth rate to a given age. These two traits tend to be positively related overall, but negatively related within the most common of the Down breeds (i.e. excluding the Ryeland) of UK origin. Fat content was especially important in differentiating among Suffolks (3), Dorset Downs (5) and Hampshires (6), which had similar lean growth rates, and between Dorset Horns (1) and Poll Dorsets (4), which were also similar to one another. Both traits are important to the contrast of Southdowns (10 and 11) with Suffolks (3) and Dorset Downs (5), and provide some explanation of industry trends away from the former breed since the late 1950s. The high lean growth rate of Border Leicester × Romney crosses (2), coupled with their high wool production (Clarke and Meyer, 1982), provide good justification for the high industry use of these breeds in the synthesis of a recently derived synthetic breed, the Coopworth.

Other comparisons of different sheep breeds for fatness have involved breeds of mainly British origin as reviewed by Bass, Butler-Hogg and Kirton (1990). The review also includes the Texel – a breed of international interest because of its high lean and low fat content relative to carcass weight in comparison to many other breeds – and some of the more improved versions of them (Clarke and Kirton, 1990; Leymaster and Jenkins, 1993). Bass, Butler-Hogg and Kirton (1990) also included a number of native breeds farmed under less intensive conditions and likely to be particularly well adapted as well as more readily available for immediate commercial exploitation on their home ground. The potential use of new genotypes to the Australian and New Zealand lamb industries was discussed by Clarke and Banks (1995). They also considered new synthetic strains, specific crossbred combinations and strains developed by selection for specific productive attributes, marker traits, identifiable genes or gene complexes.

Cattle and sheep data have both demonstrated considerable variation among breeds in backfat depth or body fat percentage. Therefore, breed substitution or breed crossing could be used to reduce fat levels (if this was the only trait affecting farm profits). In pigs, there are few common domestic breeds to choose from, so most of the genetic manipulations to reduce fatness have been within breeds, which is described later. However, some of the various breed/cross comparisons have been summarized by Johnson (1981). He found a 7.6 mm range in backfat depth among seven breeds, or $2.0\sigma_p$. Thus, breed differences in σ_p units were similar in cattle, sheep and pigs.

7.5.2 Genetic selection

In cattle, sheep and pigs, genetic selection has achieved divergence between High and Low, or Control and Low groups, for fat depth or body fat percentage.

In cattle, few selection studies for fatness/leanness have been carried out. One example was reported by Bergfeld *et al.* (1995), where correlated responses to selection for or against marbling were studied. Since this was for a single generation, inevitably differences were small.

In sheep, a number of experiments on selection for leanness have been published, which have been reviewed by Simm (1992). He concluded that ultrasonic fat depth measured on live animals (usually fat depth adjusted for live weight) was heritable and responded to selection. Cameron and Bracken (1992) reported on early results of successful divergent selection for carcass leanness in Texel × Oxford sheep using ultrasonic measurements of backfat and eye muscle depth at 20 weeks of age. After 3 years of selection using an index designed to change body composition without any change in liveweight, the line divergence in estimated carcass lean and fat proportions was equal and opposite ($14\,g\,kg^{-1}$) in the direction sought. Slaughter data on crossbred progeny indicated that selection has established lines of animals with different rates of lean growth but similar rates of fat deposition (Cameron, 1992).

Since the time-span reviewed by Simm (1992), two other New Zealand selection experiments have been completed, one with Southdowns at Massey University (Solis-Ramirez, Blair and Purchas, 1993) and one with Coopworths at Invermay (Morris *et al.*, 1997). At Massey, a 1.7 mm divergence was achieved between the High and Low selection flocks after 8 years of selection (46% of the High mean), and at Invermay a 2.9 mm divergence in ultrasonic backfat was achieved in 14 years. This represents a High flock response 89% above the Control and a Low flock response 36% below the Control. The percentage response in the Low line at Invermay was less than in the High line, mainly because of declining variability as the mean was reduced. The regression of carcass C fat depth over the longissimus muscle at the twelfth/thirteenth rib was 0.48 ± 0.09 mm carcass fat per mm increase in sire's ultrasonic backfat, whilst carcass meat weight increased by 0.20 ± 0.07 kg per mm reduction in sire's ultrasonic fat depth (C fat was the fat depth over the longissimus dorsi). Thus, selection on ultrasonic data was indeed successful in changing muscle and fat composition of the carcass.

In pigs, two of the early experiments in outbred populations were conducted in Maryland (Hetzer and Harvey, 1967) and in North Carolina (Berruecos, Dillard and Robison, 1970). In Maryland, divergent selection in Durocs for 10 generations led to a 26 mm difference in backfat (68% of the original mean) and in Yorkshires for eight generations it led to a 14 mm difference (44% of the original mean). Corresponding estimates of divergence in standard units were $4.7\sigma_p$ in the Durocs and $4.6\sigma_p$ in the Yorkshires. Greater selection differentials (in absolute units) were applied in the Durocs than in the Yorkshires. Realized heritabilities also tended to be higher in the Durocs. In North Carolina, selection was for leaner animals only, alongside a control line for five generations. At the end of the experiment, the selected line was 3.8 mm or 12.8% leaner than the control, or $0.63\sigma_p$. Considering that the Durocs at Beltsville were selected in two directions for twice as long, with a realized heritability which was 76% greater, the results from the two experiments are consistent. Overall, genetic selection can make major changes to characteristics such as backfat depth, which has a reasonably high heritability in each of the farm animal species.

Several recent publications present estimates of relative economic values for the weight of dissected lean and fat in New Zealand lamb carcasses (Waldron, Clarke and Rae, 1991, 1992b; Binnie and Clarke, 1992). Results have indicated that in the late 1980s processing companies were paying producers more for each kilogram of carcass lean, but approximately an equal amount less for each associated kilogram of extra fat, i.e. they were rewarding extra lean providing that it was in excess of the amount of extra fat in the carcasses of heavier lambs. This is an important quantitative market message for sheep breeders.

The expected consequences to ram breeders of the economic antagonism of lean and fat variations have been studied by Simm and Dingwall (1989) and Clarke, Waldron and Rae (1991). The latter authors predicted genetic responses that could be expected from selection for alternative breeding objectives based on a wide range of economic values for lean relative to fat. Their results showed that the same multiple trait indicators, weighted together to derive selection indices giving different levels of emphasis to each trait, were appropriate to a wide variety of breeding objectives. It was concluded that it was possible to make small compromises in maximizing lean growth rate in order to reduce the associated increase in fat weight as promoted by Bennett and Clarke (1984) and Bennett (1990). Furthermore, these compromises were less severe for indices of high discriminating power based on several indicator traits, because such indices are more robust to variations in the relative economic values of lean relative to fat than are indices based on fewer or less accurate indicators. Other more sophisticated tools for measurements of high accuracy as indicators of carcass composition have been discussed by McEwan *et al.* (1995).

For sheep, evidence from these economic studies and the response projections suggest that about half of the selection pressure that is available to improve the growth rate should, in males, be sacrificed in order to maximize economic response through reduced carcass fat content (Clarke, Waldron and Rae, 1991). Owing largely to raw product procurement competition among meat processors to maintain plant efficiencies, which is causing some confused market signals, few New Zealand sheep breeders are making such extreme changes to their immediate selection goals, although some are keeping their options open by incorporating leaner fast-growing breeds and strains into their ram-breeding flocks (e.g. Landcorp's Lamb Supreme programme; Nicoll, 1995).

The genetic parameters on which these predictions were based came from a large database of carcass dissection data (Waldron *et al.*, 1992a) which included, in addition, linear carcass measurements and the major chemical components of lamb carcasses. The major focus of most previous work for sheep was on these latter measurements (Bennett, Kirton and Johnson, 1991a, b).

For cattle, multivariate predictions based on the extensive within-breed genetic parameters for the Charolais by Renand *et al.* (1994) showed that selection for improved growth can be expected to increase leanness and to reduce fat. Furthermore, selection for high lean and low fat simultaneously (i.e. for lean growth rate or, strictly, lean % minus fat %), would reduce by about 50% the gains that could otherwise be achieved in carcass weight for age, which is closely in line with evidence from sheep. Renand (1995) concluded that the contrast between the lean and fat contents of beef carcasses was the major determinant of genetic variability in carcass quality.

It is generally accepted that the effects of fatness on the eating quality of trimmed muscles largely depend on lipid in the perimysial connective tissue surrounding the muscle fibre bundles (marbling fat). Effects of intramuscular fat on juiciness, tenderness and flavour could result from influences on saliva flow and from flavours associated with chemical reactions involving lipids (Wood, Wiseman and Cole, 1994). As mentioned earlier and by Shorthouse and Harris (1991), there is some doubt about the associated effects on tenderness, as this could result from a less rigid muscle structure, a direct dilution of the toughness of myofibrillar proteins and a smoother mouth-feel, or indirectly from muscle fibre type associations with intramuscular fat levels. In either case, as a consequence of these associations with eating characteristics, there seems to be a real need to consider marbling fat independently of total carcass fat in breed improvement programmes. The data of Renand *et al.* (1994) indicate that some degree of genetic independence exists in cattle (genetic correlation = 0.66) as well as in pigs (Wood, Wiseman and Cole, 1994). Other similar evidence from the scientific literature has been reviewed by Kemp (1994), who reported an average heritability of 0.4 for marbling score in beef cattle.

Characteristic breed differences in fat partitioning to subcutaneous versus intramuscular depots have been reported for sheep and cattle (Kemp 1994; Renand *et al.* 1994), as well as for pigs. Fast-growing, lean European 'white' pigs show low carcass marbling compared with dark-skinned Durocs and the fatter American breeds, according to Wood, Wiseman and Cole (1994). They cautioned against further selection for leanness in Europe. It is noteworthy that the Duroc's high concentration of marbling relative to backfat is associated with increased tenderness, juiciness and redness and other changes in muscle characteristics, which are suggestive of an increase in the concentration of red oxidative muscle fibres (Wood, Wiseman and Cole, 1994). Such results point to characteristic associations of eating qualities with lean, as well as with fat, and at the level of individual muscles as well as at the carcass level.

Reduction of fat by genetic means is expected to have an influence beyond that of total carcass fat content and to impact on quality features of the lean component of carcass weight as well. Recent multivariate predictions based on within-breed genetic parameters for Charolais cattle (Renand *et al.*, 1994) show that selection for improved growth is expected to reduce the intramuscular fat content of the longissimus muscle, increase its ultimate pH and collagen content, reduce the solubility of the collagen and increase the oxidative activity and diameter of its muscle fibres. Selection for lean growth rate would be associated with similar changes in ultimate pH but reduced collagen and haem pigment contents in the longissimus muscle, increased collagen solubility and fibre size, as well as greatly enhanced responses in carcass lean content. It would also be genetically associated with an increased glycolytic activity of this muscle. The restrictions on carcass growth needed to control concomitant changes in intramuscular fat were minimal and were genetically associated with a metabolic profile towards low lactate dehydrogenase enzyme activities and high isocitrate dehydrogenase activities (i.e. a strong metabolic shift from glycolytic to oxidative fibre types). The proportion of heavy chain myosins would also increase, especially if changes in collagen content were to be avoided as well (Clarke, Morris and Speck, 1996).

Such physiological changes from selection for reduced fat content could also impact on production and processing aspects of meat production. Some of these seem likely to be associated with concomitant changes in protein accretion, which seems more likely to arise from decreased rates of proteolysis than from increased rates of protein synthesis (Klasing, Calvert and Jarrell, 1987; Baldwin, Calvert and Overbauer, 1991; Goldspink, 1991). For sheep selected over several generations and widely divergent in weaning weight, Oddy *et al.* (1995) demonstrated that genotypic differences in the rates of protein gain were almost exclusively driven by the rate of protein degradation. Evidence from the plasma metabolic profiles of sheep selected for and against lean growth rate also point to genetic variations in protein degradation (Cameron and Cienfuegos-Rivas, 1994). Scheurs (1995) reported a markedly decreased proteolytic capacity of both the calpain and cathepsin enzyme systems in chickens selected for fast growth, while chickens with efficient protein metabolism showed intermediate calpain/calpastatin values but increased cathepsin and cystatin activities. Both strains of chickens showed similar rates of pH decline post-mortem in commercial broilers, and all were higher than the protein-inefficient Leghorn chickens used for egg production. Leghorns showed higher myofibrillar fragmentation indices and lower shear force values in breast muscle than the chickens selected for growth rate. The protein efficient line showed lower shear force values than Leghorns and broilers, which may need ageing times longer than 24 h.

7.5.3 Major genes

Although important gene variants have been identified, most of the identified single gene effects on carcass composition for livestock are more directly associated with traits other than fat content. Gene effects on proteolysis and double muscling have, however, shown important associated effects on fat.

(a) The callipyge gene in sheep

The callipyge gene in sheep first became apparent in an American Dorset flock selected with emphasis on leg conformation. Shey *et al.* (1995) have confirmed the location of the gene on chromosome 18 of the sheep in the interval spanned by the markers CSSM18 and TGLA122. Some unusual inheritance features suggesting dominant negative imprinting are evident (Cockett *et al.*, 1995). The gene has pronounced effects on muscle development. Animals display increased muscle size in the loin and leg, leaner carcasses and increased feed efficiency, but tougher meat (Carpenter and Solomon, 1995). Recent evidence confirms earlier reports on the callipyge phenotype (Al-Hassan and Roberts, 1995; Brown, Hinds and Snowder, 1995; Goodson, Miller and Savell, 1995; Hill, Jackson and Kerth, 1995; Hossner *et al.*, 1995; Khosraviani *et al.*, 1995; Koohmaraie, Wheeler and Shackelford, 1995; Meyer *et al.*, 1995): higher eye muscle area (30–69%), similar slaughter and carcass weights, but higher dressing percentages, lower carcass fat thickness (24–45%), internal fat weights and muscle marbling scores. The gene leads to reduced lung and liver weights and small intestine mass, more efficient liveweight gain under rapid growth, higher nitrogen balance and utilization of crude protein. There is also higher serum growth hormone and IGF-1, reduced serum insulin (39%) and IGF-1 under restricted feeding (20%) and increased circulating IGF BP-2. Higher yields of trimmed cuts and high value cuts from the carcass (by 3–8 percentage points) have been found, and higher but variable muscle weight advantages from the leg and loin regions (20–42%), but not for slow-twitch muscles in the shoulder.

Meat from Callipyge animals has shown similar pH and temperature declines on ageing, lower muscle fragmentation indices (27–37%) and higher shear force in longissimus and semimembranosus muscles. Some muscles respond to electrical stimulation, calcium ion infusions and hydrodyne treatment, but the meat tends to be tougher for some muscles and less acceptable in terms of tenderness, flavour and juiciness (Carpenter and Solomon, 1995; Koohmaraie, Wheeler and Shackelford, 1995; Meyer *et al.*, 1995). The callipyge animals had higher calpastatin and m-calpain levels but no difference in μ-calpain, increased satellite cell proliferation and suggestions of higher capacity for protein synthesis. Genotype × muscle interactions were found for sensory panel tenderness and connective tissue ratings.

(b) Double muscling in cattle

A double-muscling gene has been identified in the Belgian Blue breed of cattle (Hanset and Michaux, 1985), and there is the chance it is also segregating in other breeds, such as Charolais, Limousin and Simmental. The gene leads to higher proportions of high priced cuts in the hind-quarters, which require less fat trim. Hanset, Michaux and Stasse (1987) found that double-muscled Belgian Blue bulls at 12 months of age (relative to conventional Belgian Blue bulls) had a 12 percentage point increase in % lean (70.5 *vs* 58.7%) and a 9 percentage point reduction in the fat content (12.8

vs 22.1%) in the seventh rib-cut. A marker linked to the gene has been identified by Georges *et al.* (1990).

Myosin and several other myofibrillar proteins exist in various isoforms depending on fibre type. Xiong, Nicastroi and Blanchard (1995) reported on the protein solubility of myofibrillar proteins isolated from the semi-membranosus muscle of 10, 16 and 22 month old cattle, which ranged from 15 to 19%. It was highest at 10 and 22 months, the ages at which the proportion of red fibres were lowest. In growing pigs, the proportion of red fibres in most skeletal muscles increased up to 16 weeks of age, while white fibre content decreased steadily (Susuki and Cassens, 1980). Such differences are likely to account for variability (e.g. in gelling behaviour and gel characteristics) of processed muscle foods made from different cuts or muscles.

Key enzymes have been identified which affect the cross-linking of collagen molecules characteristic of different collagen types and that vary with animals and are closely associated with the thermal stability of collagen (Horgan *et al.*, 1991). Meat toughness arising from variation in collagen content can be predicted from the activities of these enzymes (Harper *et al.*, 1995), and a number of candidate genes associated with lysyl hydroxylase, lysyl oxidase, protyl hydroxylase and collagenase (as well as fibrinectin and collagen type 2a.1) have been identified.

(c) The acid meat gene in pigs

A major gene (unfavourable dominant), the 'acid meat' (RN) gene in pigs, increases the glycogen level and branching enzyme activity in the myofibres. It gives rise to a lowered protein content and ultrastructural abnormalities and results in decreased technological abilities associated with high lactic acid levels post-mortem (Le Roy *et al.*, 1990, 1994).

In pigs, intense selection since the 1960s for growth efficiency and high meat (and low fat) percentages have had the side-effect of producing 'pale, soft and exudative' (or watery) meat. This syndrome was associated with malignant hyperthermia (Webb and Simpson, 1986), which turned out to be controlled by a single gene (Smith and Bampton, 1977; Archibald and Imlah, 1985). The gene is now mapped, and a marker on the ryanodine receptor gene has been identified (Fujii *et al.*, 1991). These studies have begun to provide scientists with some understanding of the muscle biology associated with high lean meat yields and the associated low fat levels.

7.5.4 Gene mapping and markers

Excess fat in farm animals could be considered analogous to the human condition of obesity. The complexity and diversity of the physiological and environmental pathways leading to excess fat are enormous, but are beginning to be better understood.

Comparative gene mapping to date indicates that mammalian genomes in general are highly conserved and that the developing new maps of mammalian species can draw from one another and especially the prototype maps of the mouse and humans (Womack, 1991). Genetic maps are rapidly being constructed as a basis for the identification of markers associated with Quantitative Trait Loci (QTLs) in farm animals used for meat production. By early 1996, approximately 1000 markers spaced evenly throughout the cattle genome have been identified (similarly, approximately 600 markers have been identified for the construction of the ovine and porcine maps), sequenced and used to trace the inheritance of DNA segments from parent to offspring in cattle families designed for the development of a linkage map (Bishop *et al.*, 1994). Determination of linkage between marker loci and QTL will depend on the presence of linkage disequilibrium between alleles at the marker locus and alleles at the QTL. This disequilibrium will generate marker-associated quantitative effects that can be detected by appropriate statistical analysis. Linkage disequilibrium can be produced by crossing populations that differ in allele frequencies at marker loci and/or QTL. Within a segregating population, linkage disequilibrium will inevitably be found within a given family as a result of the specific coupling relationships between marker alleles and QTL alleles in the parents of the family (Soller, 1991). Linkage disequilibrium can also be present in a segregating population as a result of genetic drift, founder effects or selection (Soller, 1991).

Reverse genetics has been defined as the use of linked markers to describe the location of a gene involved in a phenotype, followed by efforts to identify and isolate the gene using this starting information. Using this approach, a large number of human disease genes have been identified (Gusella *et al.*, 1983; Reeders *et al.*, 1985; Sherrington *et al.*, 1988)

(a) Candidate gene approach

The candidate gene approach has been used to identify genes causing genetic disorders and genes associated with quantitative traits. It is based on associations between the phenotypes of interest and polymorphisms in genes known to be involved in the biochemical process being studied. While the candidate gene approach appears to be the logical approach, in general we do not possess the prior knowledge required to utilize the technology fully. At least in humans, the association between candidate genes and body mass index has been disappointing. No association has been found between this index and molecular markers of the glucose transporter-1 (Weaver, Kopelman and Hitman, 1992b), glucose transporter-4 (Weaver, Kopelman and Hitman, 1992a), insulin (Weaver, Kopelman and Hitman, 1992a), insulin receptor (Weaver, Kopelman and Hitman, 1992a) and glucocorticoid receptor (Weaver, Kopelman and Hitman, 1992b). However, associations between obesity and body fat levels with apolipoprotein B (Rajput-Williams *et al.*, 1988; Saha *et al.*, 1993), red blood cell acid phosphatase (Lucarini *et al.*, 1990), low density lipoprotein receptor (Zee, Griffiths and Morris, 1992), apolipoprotein D (Vijayaraghavan, Hitman and Kopelman, 1994) and dopamine D_2 receptor have been suggested. Bouchard (1995) noted that a *DraI* polymorphic restriction site in the α_2-adrenoreceptor gene was associated and linked with subcutaneous upper body fat in women, independent of the overall level of fatness. Clearly, such a marker in meat animals has enormous potential implications for modifying fat distribution, particularly in the low value shoulder area. The limited number of published molecular marker studies suggests that there are at least several genes associated and/or linked with human obesity, and thus potentially of value to other animal species of economic importance.

(b) Mouse genetic models of adiposity

Mouse genetic models to date appear to be more informative than human examples from an animal production perspective. Also, five mouse mutations causing obesity are encoded on five different mouse chromosomes (Bouchard, 1995). These mouse obesity genes are in coding regions that have human homologous equivalents. The *agouti* locus on mouse chromosome 2 regulates coat colour pigmentation. Several alleles at the *agouti* locus (A^y, A^{vy}, etc.) shift the relative proportions of black and yellow pigment of the hair of the animal (Kwon *et al.*, 1994; Yen *et al.*, 1994). Pleiotropic effects of the dominant *agouti* mutation are associated with obesity and increased susceptibility to diabetes. The *agouti* gene has been cloned and encodes for a 131 amino acid protein. The gene is normally expressed in neonatal skin (Bultman, Michaud and Woychik, 1992). However, the dominant yellow (A^y) mutation is associated with overexpression in a multitude of adult tissues (Bultman, Michaud and Woychik, 1992; Yen *et al.*, 1994). The overexpression of the *agouti* protein interferes with the binding of the α-melanocyte-stimulating hormone to its receptor in the hair follicle, thus blocking the activation of adenyl cyclase (Lu *et al.*, 1994).

Another mouse obesity gene has also been recently cloned, the *ob* gene, which encodes a 4.5 kb adipose tissue mRNA with a 167 amino acid open reading frame. A nonsense mutation in codon 105 of the mouse gene is observed in the severely obese *ob*/*ob* line, and expresses a 20-fold increase in adipose tissue *ob* mRNA, while a second mutant strain does not exhibit any evidence for expression of the gene (Zhang *et al.*, 1994). The gene product is thought to be associated with appetite regulation.

(c) Mouse QTL experiments

Fisher *et al.* (1993) have recently identified new mouse genes for polygenic obesity. This was accomplished by using a back-cross design between the strains *Mus spretus* and C57BL/6J; the resultant strain is termed the BSB mouse. The BSB strain exhibits a wide range of carcass lipid content ranging from 1 to 50%. On the basis of a QTL approach with a large number of markers, a number of informative loci have been located and have been termed *Mob-1* to *4* (Warden *et al.*, 1993; Bouchard, 1995). In another experiment, a mouse polygenic model of differential susceptibility has been developed by crossing the dietary lipid sensitive strain (AKR/J) with a resistant strain (SWR/J). After 12 weeks of feeding a high fat diet, the AKR/J strain had six times more fat than the resistant SWR/J strain. F2 animals and back-cross data were used and to date three QTLs have been identified, termed *Do1, Do2* and *Do3*. All are related to the level of adiposity and in the case of *Do2* it was related to mesenteric fat (West, Goudey-Lefevre and York, 1994a, b).

(d) Meat producing animals

The leads given by the human and mouse genotyping efforts provide the farm animal geneticist with encouraging leads with respect to a number of QTL experiments which have been initiated in a number of species, including bovine, porcine and ovine, in a number of laboratories around the world. Clearly, the enormous collaborative effort in the human genome project has provided the basis for the genetic maps, which have been generated for farm animal species in the last 2–3 years. These comprehensive linkage maps and linkage studies, using a large number of markers covering the entire genome, are likely to be helpful in elucidating the physiological basis and ultimately the regulation of fat deposition in farm animals.

In the recently published gene map for the pig (Andersson, Haley and Ellegren, 1994), a chromosome has been identified carrying genes (or quantitative trait loci) for backfat depth. They were fatness genes rather than leanness genes, because they came from the wild pig used to cross with the Large White in the experiment. Rothschild *et al.* (1995) have also identified markers in pigs associated with backfat depth.

7.5.5 Gene transfer for improved animal growth and carcass characteristics

(a) Transgenic animals

The term 'transgenic' was first used by Gordon and Ruddle (1981) to describe animals harbouring new genes within their genome. It includes genetic material coming from other species.

Rational prediction of the effects of novel genes is dependent upon a good knowledge of the genetic control of physiological processes affecting growth and development (Kirkpatrick and Rutledge, 1991). These reviewers have stressed the value of development of associations coming from lines selected from a common base population, especially in comparison with animal populations of diverse origin, such as different breeds and strains where physiological associations are much more likely to be coincidental rather than causative. Selection for lean growth has generally resulted in higher levels of circulating growth hormone, generally with a response as well in circulating IGF-1 levels, which is a mediator of some growth hormone effects (Kirkpatrick and Rutledge, 1991). These observations correspond with the physiological effects of exogenous growth hormone and IGF-1. This knowledge is especially relevant to the identification of the responsible genes, a task that is now benefiting from new genetic marker and gene mapping techniques (Soller and Beckmann, 1986; Salmon *et al.*, 1988).

(b) Embryonic stem cells

The first animals containing experimentally introduced foreign DNA were obtained by micro-injection of SV40 DNA into the blastocoel of mice (Jaenisch and Minz, 1974). Another potential strategy for embryo gene transfer is via the use of embryonic stem cells (ES cells) (Evans and Kaufman, 1981). It has been observed that ES cells injected into blastocysts can colonize all tissues of the developing animal, including the germline (Robertson *et al.*, 1986). If the ES cells are genetically modified prior to transfer into blastocysts, they can stably transmit a transgene into chimeras and, through offspring, also through several generations (Pursel *et al.*, 1989a). This procedure also allows for the study of insertional mutations in transgenic offspring (Bradley and Robertson, 1986). The most recent advance using this technology is that it is possible to use ES cells for the production of chimeras and offspring and to obtain transgenic mice with ablated genes or homologous recombined gene loci. This is a powerful tool for the study of gene function and examples are abundant in the field of medicine. Unfortunately, a major limitation for this technology in farm animals is the lack of ES cell lines, because the ES cell technology so far only works in mice. Once this limitation is overcome, progress will be made.

The observation that transgenic mice harbouring constructs of the growth hormone gene produced high levels of the foreign growth hormone, and had markedly increased growth rates and altered body composition (Palmiter *et al.*, 1982, 1983), provided the impetus for the insertion of similar genes into livestock species. The effects of human, rat and bovine transgenes associated with growth hormone have had variable effects on growth in mice (Pursel *et al.*, 1989a), but not without impaired fertility and health problems as has been found in transgenic swine and sheep. Nevertheless, beneficial productivity effects have been achieved in swine. Manipulation of hormone receptors and hormone binding proteins could also be used to control the expression of growth hormone and its intermediaries. In this way, the method is a very powerful research tool for examining the role of genes involved in the regulation of growth and development and in understanding these side effects on other aspects of commercial productivity. The need to continue such studies over several generations is clearly supported by early results (Pursel *et al.*, 1989a). As these reviewers conclude, commercial exploitation will require much better control of transgene expression and of the genetic background of carrier animals and the development of specialist husbandry regimens.

(c) Insertion of fusion genes
The insertion of fusion genes (that is, the gene of interest as well as a piece of DNA) which is regulatory in nature, for example the promoter from the mouse metallothionein-1 gene (Palmiter *et al.*, 1982, 1983), resulted in the immuno detection of hGH in transgenic pigs and rabbits and bGH in lambs. The increased growth rate and body size observed in the GH transgenic mouse models clearly raised the expectation that similar fusion genes may produce similar results in livestock species.

Transgenic pigs, which had elevated levels of GH from the transgene construct, were significantly leaner than their litter mates. When transgenic pigs were compared with their control litter mates at 90 kg liveweight, the average backfat thickness at the tenth rib was 7.0 ± 2.1 versus 18.5 ± 1.8 mm (Pursel, Miller and Pinkert, 1987). Ultimately, the continuous elevation of GH in transgenic pigs posed some considerable health problems. As the transgenic pigs approached maturity they exhibited multiple body features common to acromegaly in humans. Other problems were also encountered, ranging from reduced birth weight, which led to higher neonatal mortality, and a tendency for lethargy and reduced appetite (Pursel, Rexroad and Bolt, 1989b). Clearly, the problems encountered by the transgenic experience may be overcome by the discovery of alternative regulatory sequences that can be tightly controlled by external stimuli and at a specific tissue level.

7.6 Research needs

If meat is to retain its major role in the human diet as a source of high quality protein, minerals and vitamins and as a source of some desirable fats, it is clear that the product supplied to the consumer must contain acceptable levels of fat. There will be some variation in the amount of fat that is acceptable to different consumers and in different markets/-countries. Excess fat is clearly unacceptable and does have health implications, as covered in several chapters in this volume.

7.6.1 Market signals and demand

As trimming of excess fat from the carcass or from cuts is a wasteful option in terms of the energy needed to produce the unnecessary body fat and in terms of the cost (labour or mechanical) to remove it, a clear signalling system is needed between meat producers and processors so that the animal producers are aware of the specifications necessary for their animals to meet processors' requirements with minimal trimming. Payment systems must be tied to these specifications to produce the necessary market systems as the leaner animals produced should be those giving the largest financial returns to the producers. Unfortunately, in many countries the market signals coming to producers do not clearly reflect consumer requirements with extended supply chains, competition between processors and internal political factors such as support prices distorting market signals.

In order to produce clear signals between producers and processors there is need for a common language of carcass description, understood by all parties to any transaction, to describe various attributes, including fatness. This is best achieved by objective carcass classification/grading systems, such as those implemented for pig carcass classification in several countries. Despite a large body of research aimed at developing objective systems suitable for use on beef and sheep carcasses, there are no widely accepted and used systems at an acceptable cost that are currently available, although several systems show promise. Subjective systems of carcass classification are still widely used. Further research is needed to develop economic, sufficiently accurate and acceptable (to both producers and processors) systems for carcass classification of beef and sheep carcasses, which can define fatness levels for marketing purposes.

However, such communication may only be possible where the producer retains ownership to the point of slaughter. If livestock are traded via markets, such as at auction, it is very much more difficult to translate the market requirements to the producer, other than through market prices at auction. The producer may be uninformed of the carcass classification system and how this and the fatness requirements relate to the livestock produced and sold at auction. An information system passing fatness information from the slaughter plant through any intervening links to the farmer is required. This will be assisted if/when all meat animals carry individual identification (presumably electronic) related to ownership history. Sufficient scope exists for research into the development of information systems. Currently such information systems are in use in some slaughter plants in New Zealand linking carcass weight, classification and inspection information back to individual farms and another is feeding back a plot of GR (fatness) measurement taken on each lamb carcass against the carcass weight of those carcasses. This provides useful information to the producers about carcass weights (and therefore estimated liveweights) at which fatness problems start to appear and helps in farm management decisions.

One of the best solutions to assist meat processors to obtain carcasses of desirable fatness is for slaughter plants to contract the supply of animals of required fatness with specifications covering breed, sex, live or carcass weight, nutrition, etc., to target the animals most likely to meet the requirements. Research can provide the particulars required, after taking into account differences due to the several factors known to influence fatness.

7.6.2 Consumer research

Continuing consumer research is required in meat-eating countries to establish the changing levels of fatness considered currently acceptable, noting these changes with time. As mentioned in Chapter 4, fatness can also have an important influence on palatability (particularly flavour and juiciness), which is also an important component of consumer preferences.

7.6.3 Searching for new genes

The search for genes which are linked to animal growth and fat deposition and the molecular variants of these genes still remains a major undertaking. The task will be made more complex when researchers examine other issues such as gene–environment, gene–nutrient intake and gene–gene interactions. Many clues are likely to be derived from the human and mouse research fields. Clearly, the next decade will require the total integration of the disciplines of meat science, quantitative genetics and molecular genetics if any progress is to be made in developing farm animals which meet tight market specifications all of the time. The role of transgenic animals, at least in the short to medium term, is unlikely to yield any practical solutions for meat animal production. However, results from transgenic models will play an important role in revealing gene function and regulation of adiposity. The gene/QTL mapping approaches are likely to lead to discoveries of new genes, and advances in molecular biology are progressing at a rapid pace. In addition, the tools for the discovery and analysis of new genes continue to improve. The detailed analysis of gene structure and regulation of genes will provide a new impetus for the use of even the simplest techniques for the production of transgenic livestock. Statistical methods continue to improve, which provides the quantitative geneticists with improved tools for the development of breeding programmes to exploit QTLs, new genes and transgenic technologies.

7.7 Summary

While fat can be removed from carcasses and cuts by trimming, this is an inefficient and expensive option. It is a waste of time on feed as well as feed energy and cost in producing the excess carcass fat in the first place.

Clarke, J.N., Morris, C.A. and Speck, P.A. (1996) Genetic improvement of meat quality in sheep and cattle. *Proc. N.Z. Soc. Anim. Prod.*, **56**, 157.

Cockett, N.E., Logan, S.P., Jackson, S.P. *et al.* (1995) Genomic imprinting: a possible explanation for the non-Mendelian inheritance pattern of the callipyge phenotype (muscle hypertrophy) in sheep. *J. Anim. Sci.*, **73** (Suppl. 1), 7.

Cross, H.R. and Savell, J.W. (1994) What do we need for a value-based beef marketing system? *Meat Sci.*, **36**, 19.

Eldridge, G.A. (1989) Image analysis for meat and carcass description, in *The Automated Measurement of Beef* (eds L.E. Brownlie, W.J.A. Hall and S.U. Fabiansson), Australian Meat and Livestock Corp., Sydney, pp. 179–185.

Evans, M.J. and Kaufman, M.H. (1981) Establishment in culture of pluripotent cells from mouse embryos. *Nature*, **292**, 154.

Fennessy, P.F. and Drew, K.R. (eds) (1985) Biology of deer production. *R. Soc. N.Z. Bull.* No. 22.

Fensvig, A.T. (1995) The Federation of Danish Pig Producers and Slaughterhouses. *Proc. Int. Conf. Meat Sci. Technol.*, **41**, 156.

Ferguson, D., Thompson, J. and Cabassi, P. (1995) Video image analysis, in *Meat 95*, Australian Meat Industry Research Conference, Cannon Hill, Qld, pp. 7A-13–7A-16.

Fisher, A.V. (1990) New approaches to measuring fat in the carcasses of meat animals, in *Reducing Fat in Meat Animals* (eds J.D. Wood and A.V. Fisher). Elsevier Applied Science, London, pp. 255–343.

Fisher, J.S., Warden, C.H., Pace, M.J. *et al.* (1993) A new mouse model of multigenic obesity. *Obesity Res.* **1**, 271.

Forrest, J.C. (1995) New techniques for estimation of carcass composition, in *Quality and Grading of Carcasses of Meat Animals* (ed. S.D. Morgan Jones), CRC Press, Boca Raton, FL, pp. 167–171.

Fox, D.G., Sniffen, C.J. and O'Connor, J.D. (1992) A net carbohydrate and protein system for evaluating cattle diets. 3. Cattle requirements and diet adequacy. *J. Anim. Sci.*, **70**, 3578.

Fredeen, H.T. (1958) Selection and swine improvement. *Anim. Breed. Abstr.*, **26**, 229.

Fujii, J., Otsu, K., Zorzato, F. *et al.* (1991) Identification of a mutation in porcine ryanodine receptor associated with malignant hyperthermia. *Science*, **253**, 448.

Georges, M., Lathrop, M., Hilbert, P. *et al.* (1990) On the use of DNA fingerprints for linkage studies in cattle. *Genomics*, **6**, 461.

Goldspink, G. (1991) Perspectives for the manipulation of muscle growth. *Adv. Meat Res.*, **7**, 557.

Goodson, K.J., Miller, R.K. and Savell, J.W. (1995) Carcass characteristics, muscle components and palatability assessments from Callipyge versus normal lambs. *Proc. Int. Congr. Meat Sci. Technol.*, **41**, 624.

Gordon, J.W. and Ruddle, F.H. (1981) Integration and stable germ line transmission of genes injected into mouse pronuclei. *Science*, **214**, 1244.

Goutefongea, R. and Dumont, J.P. (1990) Developments in low-fat meat and meat products, in *Reducing Fat in Meat Animals* (eds J.D. Wood and A.V. Fisher), Elsevier Applied Science, London, pp. 398–436.

Gregory, K.E., Cundiff, L.V., Koch, R.M. *et al.* (1994) Breed effects, retained heterosis, and estimates of genetic and phenotypic parameters for carcass and meat traits of beef cattle. *J. Anim. Sci.*, **72**, 1174.

Gusella, J.F., Wexler, N.S., Conneally, P.M. *et al.* (1983) A polymorphic DNA marker genetically linked to Huntington's disease. *Nature*, **306**, 234.

Hanset, R. and Michaux, C. (1985) On the genetic determinism of muscular hypertrophy in the Belgian White and Blue cattle breed. 2. Population data. *Genet. Sel. Evol.* **17**, 369.

Hanset, R., Michaux, C. and Stasse, A. (1987) Relationships between growth rate, carcass composition, feed intake, feed conversion ratio and income in four biological types of cattle. *Genet. Sel. Evol.* **19**, 225.

Harper, G.S., Allingham, P.G., Whan, V.A. *et al.* (1995) Can toughness of connective tissue in meat be predicted from enzyme measurements made on the living animal? *Proc. Aust. Meat Ind. Conf.*, Sept. 1995, **IIB**: 9.

Harrington, G. (1994) Consumer demands: Major problems facing industry in a consumer-driven society. *Meat Sci.*, **36**, 5.

Harris, J.J., Savell, J.W., Miller, R.K. *et al.* (1990) A national market basket survey for lamb. *J. Food Qual.*, **13**, 453.

Hearnshaw, H., Shorthose, W.R., Melville, G. *et al.* (1995) Are carcass grades a useful indication of consumer assessment of eating quality of beef?, in *Meat 95*, Australian Meat Industry Research Conference, Cannon Hill, Qld, pp. 7A-5–7A-9.

Herrod, W. and Treadwell, V. (1995) Chiller assessment – its ramifications and future, in *Meat 95*, Australian Meat Industry Research Conference, Cannon Hill, Qld, pp. 7A-11–7A-12.

Hetzer, H.O. and Harvey, W.R. (1967) Selection for high and low fatness in swine. *J. Anim. Sci.*, **26**, 1244.

Hill, D.A., Jackson, S.P. and Kerth, C.R. (1995) Nitrogen retention of normal and callipyge lambs fed two levels of dietary protein. *J. Anim. Sci.*, **73** (Suppl. 1), 7.

Hopkins, D.L. (1988) Consumer attitudes toward lamb and beef cuts of different fatness. *Proc. Aust. Soc. Anim. Prod.*, **17**, 420.

Hopkins, D.L. (1989) An evaluation of the Hennessy Grading Probe for measuring fat depth in beef carcasses. *Aust. J. Exp. Agric.*, **29**, 781.

Horgan, D.J., Jones, P.N., King, N.L. *et al.* (1991) The relationship between age of animal and the thermal stability and cross-link content of collagen from five goat muscles. *Meat Sci.*, **29**, 251.

Hossner, K.L., Yemm, R.S., Bernal-Santos, G. *et al.* (1995) Alteration of the insulin and insulin-like growth factor I system in response to nutrient restriction in callipyge and non-callipyge sheep. *J. Anim. Sci.*, **73** (Suppl. 1), 138.

Huffman, D.L., Egbert, W.R., Chen, C. and Dylewski, D.P. (1991) Technology for low-fat ground beef. *Proc. Recip. Meat Conf.*, **44**, 73.

Jaenisch, R. and Mintz, B. (1974) Simian virus 40 DNA. *Proc. Natl. Acad. Sci. USA*, **71**, 1250.

Jensen, W.K. (1991) Single probes and classification systems. *Proc. Symp. Electronic Evaluation of Meat in Support of Value-based Marketing, March 27–28, 1992, Purdue University, W. Lafayette, IN*, pp. 11–24.

Johnson, D.L., Baker, R.L. and Morris, C.A. (1986) Reciprocal crossbreeding of Angus and Hereford cattle. 2. Steer growth and carcass traits. *N.Z. J. Agric. Res.*, **29**, 433.

Johnson, R.K. (1981) Crossbreeding in swine: experimental results. *J. Anim. Sci.*, **52**, 906.

Keeton, J.T. (1991) Fat substitutes and fat modification in processing. *Proc. Recip. Meat Conf.*, **44**, 79.

Kemp, R.A. (1994) Genetics of meat quality in cattle. *Proc. World Congr. Genetics Applied to Livestock Prod.*, **19**, 439.

Kempster, A.J. (1990) Marketing procedures to change carcass composition, in *Reducing Fat in Meat Animals* (eds J.D. Wood and A.V. Fisher), Elsevier Applied Science, London, p. 437.

Kempster, A.J., Cuthbertson, A. and Harrington, G. (1982) *Carcase Evaluation of Livestock Breeding, Production and Marketing*, Granada, London.

Kirkpatrick, B.W. and Rutledge, J.J. (1991) Genetics and transgenics of growth. *Adv. Meat Res.*, **7**, 47.

Kirton, A.H. (1989) Principles of classification and grading, in *Meat Production and Processing* (eds R.W. Purchas, B.W. Butler-Hogg and A.S. Davies), New Zealand Society for Animal Production Occasional Publication No. 11, Ruakura Agricultural Centre, Hamilton, Chapter 11, pp. 143–157.

Kirton, A.H., Uljee, A.E., Duganzich, D.M. and Feist, C.L. (1991a) Estimation by drafters of rib tissue thickness (GR) and carcass weight of live lambs. *N.Z. J. Agric. Res.*, **34**, 69.

Kirton, A.H., Uljee, A.E. and Mercer, G.J.K. (1991b) A comparison between lamb drafters and the Phillips Fat Displacement Meter (PFDM) for predicting GR, hot carcass weight and class in live lambs. *Proc. N.Z. Soc. Anim. Prod.*, **51**, 309.

Kirton, A.H., Mercer, G.J.K., Duganzich, D.M. and Uljee, A.E. (1995a) Use of electronic probes for classifying lamb carcasses. *Meat Sci.*, **39**, 167.

Kirton, A.H., Carter, A.H., Clarke, J.N. *et al.* (1995b) A comparison between 15 ram breeds for export lamb production. 1. Liveweights, body components, carcass measurements, and composition. *N.Z. J. Agric. Res.*, **38**, 347.

Khosraviani, M., Bahader, A., Meyer, H.H. *et al.* (1995) The somatotropic axis in callipyge lambs. *J. Anim. Sci.*, **73** (Suppl. 1), 138.

Klasing, K.C., Calvert, C.C. and Jarrell, V.L. (1987) Growth characteristics, protein synthesis and protein degradation in muscles from fast- and slow-growing chickens. *Poult. Sci.*, **66**, 1189.

Koohmaraie, M., Wheeler, T.L. and Shackelford, S.D. (1995) Effects of the callipyge gene on muscle growth and meat quality in lambs. *J. Anim. Sci.*, **73** (Suppl. 1), 63.

Kwon, H.Y., Bultman, S.J., Loffler, C. *et al.* (1994) Molecular structure and chromosomal mapping of the human homolog of the *agouti* gene. *Proc. Natl. Acad. Sci. USA*, **91**, 9760.

LeRoy, P., Naveau, J. and Elsen, J.M. and Sellier, P. (1990) Evidence for a new major gene influencing meat quality in pigs. *Genet. Res. (Camb.)*, **55**, 33.

LeRoy, P., Caritez, J.-C., Elsen, J.-M. and Sellier, P. (1994) Pigmeat quality: experimental study on the RN major locus. *Proc. World Congr. Genetics Applied to Livestock Prod.*, **19**, 473.

Leymaster, K.A. and Jenkins, T.G. (1993) Comparison of Texel- and Suffolk-sired crossbred lambs for survival, growth and compositional traits. *J. Anim. Sci.*, **71**, 859.

Lu, D., Willard, D., Patel, I.R. *et al.* (1994) *Agouti* protein is an antagonist of the melanocyte-stimulating-hormone receptor. *Nature*, **371**, 799.

Lucarini, N., Finocchi, G., Gloria-Bottini, F. *et al.* (1990) A possible genetic component of obesity in childhood. Observations on acid phosphatase polymorphism. *Experientia*, **46**, 90.

Mannion, P. (1993) Ultrasonic grading. *Meat Int.*, **3**, 21.

McEwan, J.C., Davis, G.H., Fennessy, P.F. and Clarke, J.N. (1995) Sires for more valuable lambs. *Proc. N.Z. Soc. Anim. Prod.*, **55**, 324.

McNamara, D.J. (1992) Dietary fatty acids, lipoproteins, and cardiovascular disease. *Adv. Food Nutr. Res.*, **36**, 253.

Meyer, H.H., Busboom, J.R., Bahader, A. and Thompson, J.M. (1995) Carcass and cooked meat characteristics of lambs sired by Dorset rams heterozygous for the callipyge gene, Suffolk rams and Texel rams. *J. Anim. Sci.*, **73** (Suppl. 1), 297.

MLC (1986a) The production of lean lambs, in *Sheep Yearbook*, Meat and Livestock Commission, Bletchley, pp. 55–60.

MLC (1986b) Implications of the ban on hormone implants, in *Beef Yearbook*, Meat and Livestock Commission, Bletchley, pp. 69–75.

MLC (1987) Seam cutting of beef, in *Beef Yearbook*, Meat and Livestock Commission, Bletchley, pp. 109–111.

MLC (1989) Lean meat production, in *Sheep Yearbook*, Meat and Livestock Commission, Bletchley, pp. 81–85.

MLC (1994a) Development of sheep carcase classification, in *Sheep Yearbook*, Meat and Livestock Commission, Bletchley, p. 52.

MLC (1994b) Beef carcase classification, in *Beef Yearbook*, Meat and Livestock Commission, Bletchley, pp. 90–97.

Morris, C.A., Baker, R.L., Carter, A.H. and Hickey, S.M. (1990) Evaluation of eleven cattle breeds for crossbred beef production: carcass data from males slaughtered at two ages. *Anim. Prod.*, **50**, 79.

Morris, C.A., McEwan, J.C., Fennessy, P.F. *et al.* (1997) Selection for high or low backfat depth in Coopworth sheep: juvenile traits. *Anim. Sci.*, in press.

Nicholson, M.J. and Little, D.A. (1989) A note on the usefulness of skinfold thickness vs. condition score for estimating body fat content in Boran cattle. *Anim. Prod.*, **48**, 234.

Nicoll, G.B. (1995) Using genetic variation for greater profit. *Proc. N.Z. Soc. Anim. Prod.*, **55**, 328.

NRC (1988) *Designing Foods. Animal Product Options in the Marketplace*, National Research Council, National Academy Press, Washington, DC.

Oddy, V.H., Speck, P.A., Warren, H.M. and Wynn, P.C. (1995) Protein metabolism in lambs from lines divergently selected for weaning weight. *J. Agric. Sci. (Camb.)*, **124**, 129.

Palmiter, R.D., Brinster, R.L., Hammer, R.E. *et al.* (1982) Dramatic growth of mice that develop from eggs microinjected with metallothionein–growth hormone fusion genes. *Nature*, **300**, 611.

Palmiter, R.D., Norstedt, G., Gelinas, R.E. *et al.* (1983) Metallothionein–human growth hormone fusion genes stimulate growth of mice. *Science*, **222**, 809.

Price, M.A. (1995) Development of carcass grading and classification systems, in *Quality and Grading of Carcasses in Meat Animals* (ed. S.D. Morgan Jones), CRC Press, Boca Raton, FL, pp. 173–199.

Purchas, R.W. (1991) Effect of sex and castration on growth and composition. *Adv. Meat Res.*, **7**, 203.

Pursel, V.G., Miller, K.F. and Pinkert, C.A. (1987) Development of 1-cell and 2-cell pig ova after microinjection of genes. *J. Anim. Sci.*, **65** (Suppl. 1), 402 (Abstr.).

Pursel, V.G., Pinkert, C.A., Miller, K.F. *et al.* (1989a) Genetic engineering of livestock. *Science*, **244**, 1281.

Pursel, V.G., Rexroad, C.E. and Bolt, D.J. (1989b) Gene transfer for enhanced growth of livestock, in *Animal Growth Regulation* (eds D.R. Campion, G.J. Hausman and R.J. Martin), Plenum Press, New York, pp. 21–49.

Rajput-Williams, J., Knott, T.J., Wallis, S.C. *et al.* (1988) Variation of apolipoprotein-β gene is associated with obesity, high blood cholesterol levels, and increased risk of coronary heart disease. *Lancet*, **ii**, 1442.

Rasby, R.J., Wettemann, R.P., Harms, P.G. *et al* (1992) GnRH in the infundibular stalk-median eminence related to percentage body fat in carcasses of beef cows. *Domest. Anim. Endocrinol.*, **9**(1), 71.

Reeders, S.T., Breuning, M.H., Davies, K.E. *et al.* (1985) A highly polymorphic DNA marker linked to adult polycystic kidney disease on chromosome 16. *Nature*, **317**, 542.

Reid, J.T., Bensadoun, A., Bull, L.S. *et al.* (1968) Some peculiarities in the body composition of animals, in *Body Composition in Animals and Man*, Publication No. 1598, National Academy of Sciences, Washington, DC, pp. 19–44.

Renand, G. (1995) Actual basis for genetic selection in beef, in *Expression of Tissue Proteinases and Regulation of Protein Degradation as Related to Meat Quality* (eds A. Quali, D.I. Demeyer and F.J.M. Smulders), European Consortium for Continuing Education in Advanced Meat Science and Technology, Utrecht, pp. 413–430.

Renand, G., Berge, P., Picard, B. *et al.* (1994) Genetic parameters of beef production and meat quality traits of young Charolais bulls progeny of divergently selected sires. *Proc. World Congr. Genetics Applied to Livestock Prod.*, **19**, 446.

Robertson, E., Bradley, A., Kuehn, M., and Evans, M. (1986) Germ line transmission of genes introduced into cultured pluripotent cells by retroviral vector. *Nature*, **323**, 445.

Rose, G. (1990) Dietary fat and human health, in *Reducing Fat in Meat Animals*, (eds J.D. Wood and A.V. Fisher), Elsevier Applied Science, London, pp. 48–65.

Rothschild, M.F., Liu, H.C., Tuggle, C.K. *et al.* (1995) Analysis of pig chromosome 7 genetic markers for growth and carcass performance traits. *J. Anim. Breed. Genet.*, **112**, 341.

Saha, N., Tay, J.S.H., Heng, C.K. and Humphries, S.E. (1993) DNA polymorphisms of the apolipoprotein B gene are associated with obesity and serum lipids in healthy Indians in Singapore. *Clin. Genet.*, **44**, 113.

Salmon, R.K., Berg, R.T., Yeh, F.C. and Hodgetts, R.B. (1988) Identification of a variant growth hormone haplotype in mice selected for high body weight. *Genet. Res. (Camb.)*, **52**, 7.

Savell, J.W., Cross, H.R., Francis, J.J. *et al.* (1989) National consumer retail beef study: interaction of trim level, price and grade on consumer acceptance of steaks and roasts. *J. Food Qual.*, **12**, 251.

Savell, J.W., Harris, J.J., Cross, H.R. *et al.* (1991) National beef basket survey. *J. Anim. Sci.*, **69**, 2883.

Scheurs, F.J.G. (1995) Post mortem processes in breast muscle of chickens with different growth rates and protein efficiencies. *Proc. Int. Congr. Meat Sci. Technol.*, **41**, 666.

Shay, T.L., Cockett, N.E., Nielsen, D. *et al.* (1995) Confirmation of the callipyge gene to ovine chromosome 18. *J. Anim. Sci.*, **73** (Suppl. 1), 297.

Sherrington, R., Brynjolfsson, J., Petursson, H. *et al.* (1988) Localization of a susceptibility locus for schizophrenia on chromosome 5. *Nature*, **336**, 164.

Shorthouse, W.R. and Harris, P.V. (1991) Effects of growth and composition on meat quality. *Adv. Meat Res.*, **7**, 515.

Simm, G. (1992) Selection for lean meat production in sheep, in *Progress in Sheep and Goat Research* (ed. A.W. Speedy), Commonwealth Agricultural Bureaux International, Wallingford, Chapter 9, pp. 193–215.

Simm, G. and Dingwall, W.S. (1989) Selection indices for lean meat production in sheep. *Livest. Prod. Sci.*, **21**, 223.

Sinclair, A.J. and O'Dea, K. (1990) Fats in human diets through history: is the Western diet out of step?, in *Reducing Fat in Meat Animals* (eds J.D. Wood and A.V. Fisher), Elsevier Applied Science, London, pp. 1–47.

Smith, C. and Bampton, P. (1977) Inheritance of reaction to halothane anaesthesia in pigs. *Genet. Res. (Camb.)*, **29**, 287.

Solis-Ramirez, J., Blair, H.T. and Purchas, R.W. (1993) Direct and correlated responses to selection for high or low ultrasonic backfat depth in Southdown sheep. *N.Z. J. Agric. Res.*, **36**, 133.

Soller, M. (1991) Mapping quantitative trait loci affecting traits of economic importance in animal populations using molecular markers, in *Gene Mapping Techniques and Applications* (eds L.B. Shook, H.A. Lewin and D.G. McLaren), Marcel Dekker, New York.

Soller, M. and Beckmann, J.S. (1986) Restriction fragment length polymorphism in poultry breeding. *Poult. Sci.*, **65**, 1474.

Susuki, A. and Cassens, R.G. (1980) A histochemical study of myofibre types in muscle of the growing pig. *J. Anim. Sci.*, **51**, 1449.

Swan, J.E. (1992) Animal by-product processing, in *Encyclopedia of Food Science and Technology* (ed. Y.H. Hui), Wiley, New York, pp. 42–49.

Tatum, J.D. (1981) Is tenderness nutritionally controlled? *Proc. Recip. Meats Conf.*, **34**, 65.

Thatcher, L.P. (1988) New lamb products and the implications for management systems. *Food Technol. Aust.*, **40**(2), 57.

Thatcher, L.P. and Couchman, R.C. (1983) Determining consumer requirements for lamb loin chops – a preliminary study. *Rev. Market. Agric. Econ.*, **51**, 167.

Vijayaraghavan, S.H., Hitman, G.A. and Kopelman, P.G. (1994) Apolipoprotein D polymorphism: a genetic marker for obesity and hyperinsulinemia. *J. Clin. Endocrinol. Metab.*, **79**, 568.

Waldron, D.F., Clarke, J.N. and Rae, A.L. (1991) Analysis of lamb schedules and relative economic values of lean and fat. *Proc. N.Z. Soc. Anim. Prod.*, **51**, 405.

Waldron, D.F., Clarke, J.N., Rae, A.L. *et al.* (1992a) Genetic and phenotypic parameter estimates to improve lamb carcass composition. *N.Z. J. Agric. Res.*, **35**, 287.

Waldron, D.F., Clarke, J.N. and Rae, A.L. (1992b) Comparison of relative economic values for lamb carcass traits in different seasons of the year and regions of New Zealand. *Proc. N.Z. Branch Aust. Econ. Soc.*, **71**, 1778.

Warden, C.H., Fisler, J.S., Pace, J.M. *et al.* (1993) Coincidence of genetic loci for plasma cholesterol levels and obesity in a multifactorial mouse model. *J. Clin. Invest.*, **92**, 773.

Weaver, J.U., Kopelman, P.G. and Hitman, G.A. (1992a) Central obesity and hyperinsulinaemia in women associated with polymorphism in the 5′ flanking region of the human insulin gene. *Eur. J. Clin. Invest.*, **22**, 265.

Weaver, J.U., Kopelman, P.G. and Hitman, G.A. (1992b) Glucose transporter (*Glut-1*) as a possible candidate gene for obesity, in *Obesity in Europe 91. Proc. 3rd Eur. Congr. Obesity* (eds G. Ailhaud, B. Guy-Grand, M. Lafontan and C. Ricquier), John Libbey, France, pp. 889–893.

Webb, A.J. and Simpson, S.P. (1986) Performance of British Landrace pigs selected for high or low incidence of halothane sensitivity. 2. Growth and carcass traits. *Anim. Prod.*, **43**, 493.

West, D.B., Goudey-Lefevre, J. and York, B. (1994a) Genetics of dietary obesity in AKR/J X SWR/J mice: segregation of the trait and identification of a linked locus on chromosome 4. *Mammalian Genome*, **5**, 546.

West, D.B., Goudey-Lefevre, J. and York, B. (1994b) Dietary obesity linked to genetic loci on chromosomes 9 and 15 in a polygenic mouse model. *J. Clin. Invest.*, **94**, 1410.

Widdowson, E.M. and Lister, D. (1991) Nutritional control of growth. *Adv. Meat Res.*, **7**, 67.

Womack, J.E. (1991) Strategies and technologies for comparative gene mapping, in *Gene Mapping Techniques and Applications* (eds L.B. Shook, H.A. Lewin and D.G. McLaren), Marcel Dekker, New York, pp. 3–20.

Wood, J.D. and Fisher, A.V. (eds) (1990) *Reducing Fat in Meat Animals*, Elsevier Applied Science, London.

Wood, J.D., Wiseman, J. and Cole, D.J.A. (1994) Control and manipulation of meat quality, in *Principles of Pig Science* (eds D.J.A. Cole, J. Wiseman and M.A. Valey), Nottingham University Press, Loughborough, pp. 433–456.

Woodward, J. and Wheelock, V. (1990) Consumer attitudes to fat in meat, in *Reducing Fat in Farm Animals* (eds J.D. Wood and A.V. Fisher), Elsevier Applied Science, London, pp. 66–100.

Wright, I.A. and Russel, A.J.F. (1984) Partition of fat, body composition and body condition score in mature cows. *Anim. Prod.*, **38**, 23.

Xiong, L., Nicastroi, F. and Blanchard, P.B. (1995) Functionality of myofibrillar proteins from cattle at different maturities. *Proc. Int. Congr. Meat Sci. Technol.*, **41**, 572.

Yen, T.T., Gill, A.M., Frigeri, L.G. *et al.* (1994) Obesity, diabetes, and neoplasia in yellow $A^{VY}/-$ mice: ectopic expression of the *agouti* gene. *FASEB. J.*, **8**, 479.

Zee, R.Y.L., Griffiths, L.R. and Morris, B.J. (1992) Marked association of a RFLP for the low density lipoprotein receptor gene with obesity in essential hypertensives. *Biochem. Biophys. Res. Commun.*, **189**, 965.

Zhang, Y., Proenca, R., Maffei, M. *et al.* (1994). Positional cloning of the mouse *obese* gene and its human homologue. *Nature*, **372**, 425.

8 Reducing the fat content by production practices

M.E. DIKEMAN

8.1 Introduction

The potential for the manipulation of growth and composition of meat animals has never been greater than at present. Producers, processors and scientists involved in beef, lamb, pork, poultry and fish production and processing acknowledge that removable or trimmable fat as well as non-removable fat in trimmed meat needs to be reduced to make production more efficient and to meet consumer demands for meat products with less fat. At the same time, minimum amounts of intramuscular fat may be necessary to provide some assurance of acceptable flavor, juiciness and tenderness of cooked meat products. Furthermore, methods or technologies used to reduce fat must not cause biochemical and/or morphological changes in muscle that decrease its visual quality, palatability, nutritional value or safety.

8.2 Influence of nutrition

Carcass fatness can be manipulated in cattle, sheep, pigs and poultry by nutrition. However, it is easier to manipulate the fatness of pigs and poultry because they are monogastrics having high voluntary feed intakes. Manipulation of carcass fatness is more difficult for ruminants under practical feeding conditions because of the higher fiber content of their diets and the equalizing nature of the fermentation process in the rumen environment (Jagusch and Rattray, 1979; Kirton *et al.*, 1981). Steele, Boyd and Campbell (1994) suggested that the growing ruminant is substantially constrained by its 'digestive architecture' and further research related to nutrient requirements is needed. Another factor that makes manipulation of carcass fatness in ruminants more difficult is the somewhat uncontrolled quantity and composition of dietary input in grazing situations. Bass, Butler-Hogg and Kirton (1990) stated that, in the practical situation, the potential to influence carcass fatness in sheep and cattle through nutrition is limited and the changes in fatness may not be large enough to reward the producers for their efforts.

Nutritional manipulation of carcass fatness in pigs and poultry is determined largely by changing the proportions of protein (or essential amino acids) and energy in the diet. The swine industry has given considerable attention to protein:energy ratios since the late 1980s because of genetic selection for high lean growth and the corresponding increased dietary protein requirements. Some pig producers even separate barrows and gilts for finishing, because gilts have lower voluntary feed intakes and higher lean growth rates (Table 8.1), thus requiring a diet that is somewhat higher in protein.

8.2.1 Full versus restricted feeding

(a) Pigs

Hays and Preston (1994) stated that the effects of energy intake on performance and carcass composition in pigs may be illustrated by varying the total feed intake or by varying the metabolizable energy density of the diet. Lawrence (1977) reported a reduced percentage of dissectible fat in barrows slaughtered at 87 kg live weight when they consumed *ad libitum* a diet containing 11.4 MJ DE kg^{-1} compared to consuming diets containing 12.9, 15.0, and 18.3 MJ DE kg^{-1} at 87.5, 75.0, and 62.5% of *ad libitum* intake, respectively. However, the percentage of fat and growth rate (GR) declined curvilinearly as the energy density of the diet decreased, but at an increased level of feed intake. However, the practicality of feeding a diet high in energy density at only 62.5% of *ad libitum* intake of a diet low in energy density is a questionable practice.

Campbell and King (1982) allowed boars and barrows to eat a diet containing 14.9 MJ DE kg^{-1} *ad libitum* and at medium and low levels of intake to a live weight endpoint of 70 kg. The greatest reduction in percentage of dissected subcutaneous fat for both boars and barrows and the greatest reduction in percentage of total dissected fat in barrows occurred between the *ad libitum* and medium levels of intake. On the other hand, the greatest decrease in GR occurred between the medium and low levels of intake.

For barrows feeding *ad libitum*, or at 92.5 or 85% of *ad libitum* intake, Leymaster and Mersmann (1991) found that fat deposition was decreased with the restricted intakes. On the other hand, protein intake, even on the 85% intake level, was near adequate and the rate of protein deposition was affected minimally. Thus, to limit fat deposition, diets should be formulated to limit only the energy intake, but not to less than that required for maximum protein deposition. Haydon, Tanksley and Knabe (1989) allowed pigs to eat *ad libitum* and at 85 and 70% of *ad libitum* intake. Days to 110 kg were increased by 14 days, and a trend was seen for improved feed conversion efficiency (FCE) for the 85% *ad libitum* treatment. Separable carcass fat was reduced by 13% for the 85% *ad libitum* treatment. Collectively, these studies show that reducing dietary energy intake results in a reduced percentage of carcass fat as well as a reduced

Table 8.1 Rankings[a,b] of sexes at a constant age within species for growth rate, feed conversion efficiency, carcass traits and meat quality traits

Species and sex	Growth rate	Feed conversion efficiency	Feed intake	Fat thickness	Longissimus muscle area	Meat yield (%)	Marbling	Tenderness	Meat nutrient density[c]
Cattle:									
Bulls	1****	1***	1***	1****	1***	1****	3*	3*	1***
Steers	2**	2**	2**	2**	2**	2**	2***	1**	2*
Heifers	3*	3*	3*	3*	3*	3*	1***	2**	3*
Swine:									
Boars	1***	1***	2*	1***	1***	1***	3*	3*	1**
Barrows	2**	3*	1**	3*	3*	3*	1***	1***	3*
Gilts	3*	2**	3*	2**	2**	2**	2**	2**	2**
Sheep:									
Rams	1***	1**	1**	1***	1***	1***	3*	3*	1**
Wethers	2**	2*	2*	2**	2**	2**	2**	2**	2*
Ewes	3*	3*	3*	3*	3*	3*	1**	1**	3*

[a]1 = Ranks highest for the trait, 3 = ranks lowest for the trait.
[b]Author's interpretation of the relative differences among sexes within a species. The magnitude of differences among rankings is indicated by the difference in the number of asterisks. For example: * versus * = a minor or no difference; * versus ** = a distinct difference; * versus *** = a major difference; and * versus **** = a dramatic difference.
[c]Proportion of protein, vitamins and minerals relative to caloric density of boneless meat trimmed free of subcutaneous fat and accessible intermuscular fat.

GR. However, it may be more economical to reduce carcass fatness by slaughtering the pigs eating *ad libitum* at a 15–20% lighter live weight and reduce the number of days to slaughter by about 30%.

(b) Cattle

Fortin *et al.* (1980) showed the effects of allowing Angus and Holstein bulls, steers and heifers to consume at 70% of *ad libitum* intake a diet estimated to contain 8.4 MJ DE kg^{-1} to 183 kg carcass weight. A breed by sex interaction affected the extent of reduction in percentage of carcass fat. In general, the proportion of subcutaneous fat was reduced by 0.5–1.5 percentage points, whereas total dissected fat was reduced by 3.0–6.3 percentage points. As is the case in most other studies, the economics of feeding at a reduced level of intake were not included in this study. When Murray, Tulloh and Winter (1974) allowed Angus steers feed *ad libitum* or at a restricted level of intake, the percentage of total dissected fat decreased from 25.8 to 21.8%. However, GR was reduced from 1.76 to 0.88 kg per day. Doubling the number of days to attain a specified slaughter endpoint by limiting feed intake does not seem very practical when labor costs, depreciation on facilities and interest on investment costs are considered. Furthermore, meat tenderness and other quality traits are reduced by limiting feed intake.

Williams, Bennett and Keele (1995) simulated the effects of nine backgrounding systems before finishing on either a high or low energy diet for steers sired by 16 sire breeds. For steers finished on the high energy diet, nearly all genotypes in all backgrounding systems (0.25, 0.50 or 0.90 kg per day for 0–400 days) had lower percentages of carcass fat than those finished without backgrounding, except for two genotypes on one of the backgrounding systems. Steers finished at the low ADG directly after weaning produced carcasses with less fat than steers finished at the high ADG. These authors concluded that when steers are backgrounded for a fixed time period, carcasses with less fat may be produced with faster gains during the backgrounding phase. On the other hand, when steers are backgrounded to gain at a fixed rate per day, carcasses with less fat may be produced with longer backgrounding periods.

Hays and Preston (1994) cited six published articles concluding that in cattle fed to usual market grades (22–30% fat), plane of nutrition did not affect final carcass fat level if compared at equal body or, preferably, at equal carcass weights. However, agreement is not universal on the influence or lack of influence of plane of nutrition on carcass composition of cattle. De la Torre, Oltjen and Suing (1993) concluded that 'programmed' feeding (backgrounding vs no backgrounding) can be used to alter carcass composition; however, this was at the expense of reduced GR and FCE. Keele *et al.* (1993) reported that carcass weight accounted for much more of the variation in carcass fat than energy content of the diet and GR. Hays and Preston (1994) concluded from the literature that the carcass composition of cattle can be influenced by energy concentration in the diet, but the difference in energy level must be of a magnitude that would affect GR and FCE to such an extent that it might not be economically practical under present (US) grading and pricing conditions.

Steen and Kilpatrick (1995) studied the effects of plane of nutrition and slaughter weight on carcass composition of serially slaughtered bulls, steers and heifers of three breed crosses. After backgrounding on high forage, cattle consumed a diet consisting of a 2:1 ratio of grass silage and concentrate at either *ad libitum* or at 80% of *ad libitum* intake. They reported that reducing feed intake reduced estimated carcass fat content by 2.2 $g\,kg^{-1}$ and increased lean and bone contents by 1.4 and 4 $g\,kg^{-1}$, respectively. Obviously, those eating at 80% of *ad libitum* required more days to reach the targeted slaughter weight. These results indicate that GR will be decreased and days to slaughter will be increased when either restricted feed intake or reduced dietary energy density is used to reduce carcass fatness. This may not be economical unless sufficient reward is received by producers for the leaner carcasses.

(c) Sheep

Soeparno and Lloyd Davies (1982) demonstrated that the percentage of chemical fat in carcasses of wether sheep fed to a constant live weight of 33.4 kg was reduced 5.3% by *ad libitum* consumption of a diet containing 7.4 $MJ\,ME\,kg^{-1}$ compared with a diet containing 11.6 $MJ\,DE\,kg^{-1}$. Growth rate was reduced from 0.38 to 0.23 kg per day and dressing percentage was reduced from 53.9 to 47.6%. When Jones, Burgess and Dupchak (1983) allowed rams and ewes to consume *ad libitum* or at 70% of *ad libitum* intake, the percentage of subcutaneous fat was reduced in rams (10.2 vs 8.3%) but not in ewes (11.6 vs 11.1%). Growth rate was reduced from 0.46 to 0.40 kg per day in rams but only from 0.36 to 0.34 kg per day for ewes when slaughtered at a constant age.

Arnold and Meyer (1988) studied the effects of gender, time of castration, genotype and feeding regimen on lamb GR and carcass composition (CC). Pasture grazing to a weight of 41 kg before finishing in a feedlot to 52 kg reduced carcass fatness compared with finishing in a feedlot only. Dressing percentage and GR also were decreased (1.3 percentage points and 30.6%, respectively) with pasture grazing. Lord, Fennessy and Littlejohn (1988) stated that sire genotype was the only treatment to affect carcass chemical composition in a comparison of genotype and nutritional effects on lamb carcass composition. Murphy *et al.* (1994) allowed lambs weighing 26 kg to consume at 100, 85 and 70% of *ad libitum* intake and slaughtered them at a final weight to yield 24 kg carcasses. Days-on-feed increased linearly and GR decreased linearly by restricted feeding;

however, FCE was unaffected by intake level. The quantity of muscle was increased and the quantity of fat was reduced with restricted feeding. The reduction in quantity of fat occurred in all depot areas. Glimp, Hart and VonTungeln (1989) fed ewe and wether lambs at *ad libitum* of a 90% concentrate diet, or at 92.5 or 85% of *ad libitum* intake. Feed conversion efficiency was improved by 28.8% when intake was restricted to 92.5% of *ad libitum* intake; however, FCE was improved only 9.9% ($P > 0.05$) when feed intake was restricted to 85% of *ad libitum* intake. Differences in carcass traits were small among the intake levels, but fatness was reduced slightly with restricted intake. With the current (and past) pricing structure of the lamb industry in the USA, the extra time and labor required in restricted feeding and the lack of a premium for the leaner carcasses would not be cost effective.

These results suggest that a minor reduction in carcass fatness of lambs can be attained by reducing dietary energy intake. Other production practices would seem to be more effective in reducing carcass fatness than dietary manipulation.

(d) Poultry

Leenstra (1986) stated in a review article on fat deposition in broiler chickens that in the current situation of overfat broilers, the deposition of waste fat needs to be reduced. This implies that the ratios between maintenance, non-fat growth (not excluding physiologically required fat) and desired fat growth on the one hand and growth of waste fat on the other at a certain level of feed intake must change. Leenstra (1986) further stated that of all the factors affecting fat deposition in broilers, the nutritional aspect has been studied most extensively. Restricted feeding reduces fat deposition, but generally increases the rearing period. He also pointed out that restricted feeding requires more feeding equipment and a lower housing density, because all birds should be able to eat at the same time. Thus, production costs are higher.

Dietary composition can have an effect on feed intake and, thus, on the ratio between non-fat growth and fat deposition. Leenstra (1986) stated that this effect is independent of feed intake. Changes in fat deposition by dietary means will be adopted by the poultry industry only if the value of broiler carcasses with less fat is increased. Also, no method is available for evaluating carcass leanness on a commercial scale, which inhibits progress in reducing fat content.

O'Sullivan, Dunnington and Siegal (1991) evaluated the effects of three feeding regimens on GR and CC of broilers. The regimens were *ad libitum*, alternate day feed restriction from 6 to 27 days and then *ad libitum*, and alternate-day feed restriction from 6 days to the end of the experiment. Broilers reared under the restricted feeding achieved similar body weights at the same age as the broilers fed *ad libitum* and had less carcass fat (13.6 vs 10.1 vs 7.2% at 45 days of age). These relative differences also held up at 64 days. This feeding system appears to be promising for reducing broiler fat content without requiring more days to reach market weight.

8.2.2 Dietary energy and/or protein concentrations and ratios

Bass, Butler-Hogg and Kirton (1990) concluded from the literature that, in general, the greater the energy intake (and hence the faster the growth rate), the fatter the carcass will be at any given weight. However, these authors (citing Sully and Morgan, 1978) indicated that interaction can occur between energy intake and energy density of the diet and may produce different results. Bass, Butler-Hogg and Kirton (1990) further stated that manipulating fatness by changing energy intake is limited in scope (up to 5% units difference in fatness at a given weight) when compared with the changes in composition that can be induced by varying protein intake (up to 15% unit differences) (citing Black, 1974).

(a) Pigs

Campbell and Dunkin (1983) demonstrated in pigs that varying the protein/energy ratio (P/E) can change the chemical composition of live weight gain. Boars and gilts scale fed (based on weight) diets containing 25.2 or 15.1% protein had 15.3 and 20.9% of chemical fat, respectively. Johnston *et al.* (1993) demonstrated that at both 105 and 127 kg slaughter weights, increasing dietary lysine from 0 to 1.4% decreased carcass lipid and increased carcass protein.

Williams *et al.* (1994) fed barrows and gilts 0 or 5% added fat and four energy intake levels (8.3, 8.9, 9.5 and 10.1 Mcal of ME per day for barrows and 7.7, 8.3, 8.9 and 9.5 Mcal of ME per day for gilts). Lysine was fed at 28.5 g per day. Dietary fat improved live weight and lean FCE but not GR or CC. Increasing energy intake resulted in a linear increase in GR (up to 9.5 Mcal for barrows and 8.9 Mcal for gilts). A study by Hansen *et al.* (1994) found that as dietary lysine increased from 0.8, 1.2, 1.6 to 2.0%, barrows showed a linear decrease in GR and daily fed intake and an increase in FCE.

(b) Sheep

Andrews and Orskov (1970) demonstrated that rams and ewes fed diets with only 10% protein were fatter than those fed diets with 20% protein. They found an interaction between protein content in the diet and level of feed intake. Lambs fed on the high (0.916 kg DM per day) and medium (0.818 kg DM per day) intake levels of a diet containing 10% protein had a higher GR and percentage of fat (63.6 vs 60.8% of carcass dry matter) than those fed at the low (0.669 kg DM per day) intake level when slaughtered at 40 kg live weight. On the other hand, lambs fed on the high (0.897 kg DM per day) and medium (0.783 kg DM per day) intake levels of a diet containing 20% protein had less fat (54.0 vs 58.6% of carcass dry matter) than those fed at the low (0.677 kg DM per day) intake level. Orskov *et al.* (1971) compared diets containing 11.0, 15.7 or 19.4% protein consumed *ad libitum* by rams and ewes. As percentage of protein increased, GR increased significantly and percentage of fat decreased significantly, with the greatest increase occurring between the 11.0 and 15.7% protein diets.

(c) Cattle

Gregory *et al.* (1994) conducted an extensive study involving 1599 steers from nine parental breeds and three composite breeds either fed a diet consisting of 3.07 Mcal of ME per kg of dry matter and 11.5% crude protein or a diet consisting of 2.82 Mcal of ME per kg.

Steers were slaughtered serially at four endpoints about 21 days apart with an overall mean slaughter age of 438 days. The lower level of energy density resulted in significantly less twelfth rib fat thickness (0.56 vs 0.75 cm), a lower percentage of intramuscular fat (3.8 vs 4.1%), a lower percentage of fat in the ninth to eleventh rib section (25.4 vs 28.7%), higher Warner–Bratzler shear force values (5.2 vs 5.0 kg) and lower sensory panel flavor scores (4.8 vs 4.95). Most of the effect on carcass fatness was due to a 22 kg reduction in carcass weight and a lower dressing percentage of steers fed the low-energy diet. A similar percentage of carcass fat likely could have been attained for the two dietary energy levels by slaughtering steers on the high energy diet at a lighter weight (and age). For example, a similar reduction in percentage of ninth to eleventh rib fat that was attained by feeding the lower energy diet could have been obtained by slaughtering steers 44 days earlier or at 47 kg lighter live weight (slaughter group 1 vs 3) with no loss in sensory panel tenderness, juiciness or flavor in the longissimus muscle, although the marbling score would have been reduced.

(d) Poultry

Fisher (1984) fed male broilers diets containing either 13.4 or 10.1 MJ DE kg^{-1} DM and reported a decrease in GR from 0.04 to 0.03 kg per day and a decrease in percentage of total body chemical fat from 12.7 to 5.7% for the low energy diet when birds were slaughtered at 8 weeks of age. Jackson, Summers and Leeson (1982) fed male and female broilers together diets ranging from 15.1 to 10.9 MJ DE kg^{-1} DM and observed a decrease in percentage of total body chemical fat ranging from 14.5 to 9.8%, although this decrease was not linear.

Numerous studies on protein and energy manipulation have been conducted with turkeys and broilers. Sell *et al.* (1989) studied the effects of different protein and dietary metabolizable energy levels in turkeys from 1 to 140 days of age. As dietary metabolizable energy density increased from 99 to 111% of 1984 NRC recommendations, the proportions of protein decreased and fat increased linearly, whereas dietary protein density had little effect on carcass composition. Ferket and Sell (1990) evaluated the effects on carcass composition of early protein and energy restriction of large turkey toms fed high or low fat realimentation diets. Percentages of carcass fat of 6-week-old toms were increased by low protein (70% of 1984 NRC recommendations) and decreased by low energy (90% of 1984 NRC recommendations). No carryover effects on body composition occurred at 12 or 20 weeks of age. In agreement with those results, Sell *et al.* (1989) showed that added dietary fat significantly increased GR and body fat after 6 weeks of age, irrespective of early nutrition. Halvorson *et al.* (1991) also reported that supplemental dietary fat increased percentage of carcass fat in tom turkeys.

Castaldo, Jones and Maurice (1990) reported that GR, FCE and CC of turkeys were improved significantly as dietary protein was increased from 14.0 to 16.5 to 19.5%. Summers, Spratt and Atkinson (1992) suggested that level and balance of essential amino acids can have a significant effect on feed intake, thereby influencing GR and CC. Sibbald and Wolynetz (1986) evaluated the effects of five levels of lysine, with each of the five diets diluted with six levels of cellulose to ensure a range of intakes under *ad libitum* feeding. The lysine level required for maximum GR was about 9.6 g per kg of diet, whereas protein accretion continued to increase at greater lysine concentrations. The gain in carcass fat was lowest at the two highest lysine levels. Jackson, Summers and Leeson (1982) evaluated the effects of dietary protein (16, 20, 24, 28, 32 and 36%) and energy (2600, 2800, 3000, 3200, 3400 and 3600 kcal ME kg^{-1}) on broiler composition. Maximum absolute protein deposition was attained at the 20% protein level (Table 8.2), although the percentage of protein continued to slightly increase above this level. The absolute fat deposition increased as energy density increased. Producers would need to receive a premium price for broilers to compensate for the increased dietary cost of higher protein density.

(e) Fish

Much less research has been conducted on manipulating diets for commercial catfish production, partly because it is more difficult to control and measure feed intake and weight gain. Robinson and Robinette (1993) stated that fat content in catfish has been identified as a major problem in the commercial fish industry, partly because of an increase in digestible energy and a decrease in protein level in catfish feeds over the last few years, effectively increasing the digestible energy-to-protein ratio. However, they concluded that increasing dietary protein or restricting feed

Table 8.2 Effect of dietary protein and energy on absolute quantities of carcass composition in broilers

Variable	Carcass fat[1] (g)	Carcass protein[2] (g)	Carcass moisture[3] (g)	Abdominal fat (g)
Dietary protein level (%):	**	**	**	**
16	252[d]	202[a]	833[a]	51[e]
20	237[c]	227[b]	917[b]	44[d]
24	210[b]	233[b]	936[b]	34[c]
28	189[a]	233[b]	982[c]	29[a,b]
32	185[a]	233[b]	965[c]	30[b,c]
36	179[a]	234[b]	967[c]	26[a]
Dietary energy level (kcal ME kg^{-1}):	**	NS[4]	*	**
2600	161[a]	221	909[a]	24[a]
2800	178[b]	225	920[a,b]	29[b]
3000	208[c]	229	934[a,b]	35[c]
3200	211[c]	230	949[b]	37[c]
3400	236[d]	229	943[b]	42[d]
3600	258[e]	229	945[b]	48[e]
Sex:	NS	**	**	**
Male	206	242[b]	990[b]	33[a]
Female	211	213[a]	876[a]	39[b]
Protein × energy	NS	*	*	NS
Protein × sex	NS	NS	NS	NS
Energy × sex	NS	NS	NS	NS
Protein × energy × sex	NS	NS	NS	NS
Standard deviation	31	16	59	7

[a–e]Means followed by like superscripts do not differ significantly ($P > 0.05$).
[1]Determined from proportion of carcass composed of fat, i.e. carcass weight (g) × carcass fat (%)/carcass dry matter (%).
[2]Determined from proportion of carcass composed of protein, i.e. carcass weight (g) × carcass protein (%)/carcass dry matter (%).
[3]Determined from proportion of carcass composed of water, i.e. carcass weight (g) × carcass moisture (%).
[4]NS, not significant.
*$P < 0.05$.
**$P < 0.01$.
Source: Jackson, Summers and Leeson (1982).

intake appeared to have only marginal effects on the fat content of channel catfish.

8.2.3 Antibiotics, probiotics and ionophores

Beermann (1994) concluded that, although ionophores and antibiotics or probiotics are used commercially to enhance FCE and/or GR in meat animals, they do not alter CC. Hays and Preston (1994) stated that, on average, growth-promoting antibacterials result in slightly higher levels of the depot fats. Usually, total feed intake is increased, resulting in increased energy intake and hence slightly fatter animals. Bass, Butler-Hogg and Kirton (1990) and Owens, Zorilla-Rios and Dubeski (1991) stated that antibiotics and ionophores have no significant effect on CC, but do improve FCE and GR.

8.3 Effects of sex condition on composition

8.3.1 Cattle

Purchas (1991) reviewed the literature and reported that intact males of cattle, sheep, pigs and poultry have a faster GR, improved FCE and superior carcass leanness compared with castrates at common weights or ages. Dikeman (1991) also concluded from the literature that intact males of cattle, sheep and pigs have a faster GR, a more desirable FCE and superior carcass muscle to fat ratios than castrates at a constant weight or age (Table 8.1). However, when bulls are fed to the same fatness as steers, the advantage in FCE is lost, because they have to be fed to heavier weights, thus having higher maintenance requirements. Many countries castrate bulls for ease of management and to improve meat quality. The aggressive, agonistic behavior of bulls makes it very difficult to place them on pasture for grazing like steers or heifers, and they can be very destructive of fences, waterers and feedbunks in a confined production system. Bulls are more susceptible to emotional stress, particularly when mixed prior to slaughter, and have a significantly higher incidence of dark, firm and dry meat than steers (Tarrant, 1981). They also generally have less marbling and have less tender meat than steers. Production of bulls for meat in the USA, Australia and New Zealand is very limited because of management problems, meat processor resistance and somewhat lower meat quality. On the other hand, most European countries leave bulls intact because of single ownership from birth to slaughter, a confinement production system, shorter transportation distances, less chance of emotional stress before slaughter and greater meat processor and consumer acceptance. Dikeman *et al.* (1985) summarized research showing that implantation of bulls with zeranol from birth to slaughter and slaughtering them by 15 months of age dramatically reduced masculinity development and agonistic behavior, tended to improve meat quality and did not adversely affect GR. However, implanting tended to increase fatness.

In the comparison of steers and heifers, steers have a faster GR, more desirable FCE and more desirable carcass composition than heifers at a constant age or weight. If a 15% culling rate is assumed in a commercial beef cow herd in the USA, 30% of the heifers raised each year would be kept for replacement. That would result in the calf crop fed for slaughter consisting of 65% steers and 35% heifers. Thus, the only alternatives for dealing with the greater fatness of heifers than steers is to slaughter them at lighter weights, which is commonly done, and/or to implant them with a potent anabolic implant. As stated in the discussion on nutrition, diet manipulation to reduce carcass fatness has limited effects. In the USA, heifers generally are fed separately from steers and marketed at a lighter weight and/or shorter time-on-feed.

8.3.2 *Sheep*

Reasons for castrating rams are less justified than reasons for castrating bulls, because management of rams and dark, firm and dry meat are not significant problems. However, pelt removal can be more difficult for rams than for wethers. Rams have 2–4% less carcass fat than wethers. Thornton and Tume (1988) stated that an increasing proportion of New Zealand farmers are producing ram lambs for meat. With improved pastures, gains of 300–400 g per day and slaughter ages of 3–4 months have been attained. However, lambs are marketed at lighter weights in New Zealand than in the USA. Meat from older rams can be less tender than that from equivalent aged ewes and wethers. Tatum (1981) and Dransfield, Nute and Francombe (1984), however, suggested that this may be due to faster post-mortem chilling rates and cold shortening in the ram lamb carcasses with their thin subcutaneous fat cover. Although a few US lamb processing plants will discriminate against ram lambs, when they are fed a moderately high energy diet from weaning to slaughter, there is little justification for castration. Butler-Hogg, Francombe and Dransfield (1984) suggested that ram lambs should not be castrated in the UK to improve the efficiency of lean meat production, and that recommendation is still justified (Beermann, Robinson and Hogue, 1995).

For lambs expressing the callipyge gene, rams did not differ from wethers in any carcass traits, possibly because wethers already were extremely lean and heavy muscled (Koohmaraie, Shackelford and Wheeler, 1996).

8.3.3 *Pigs*

Most countries castrate boars because 'boar odor' (also called 'boar taint' or 'sex odor') can be detected in the fat of some boars (Seideman *et al.*, 1982; Pearson *et al.*, 1995). Boar taint may result from high levels of androsterone secretion and/or high levels of skatole production (Patterson, 1968; Hansson *et al.*, 1980; Lundstrom *et al.*, 1980; Mottram, Wood and Patterson, 1982; Mortensen, Bejerholm and Pedersen, 1986; Gray, Pearson and Monahan, 1994; Pearson *et al.*, 1995). The latter occurs in some boars because of degradation of typtophan to 3-methylindole (skatole) and absorption in the large intestine into the circulatory system and accumulation in fat. Claus, Weiler and Herzog (1994) stated that, in contrast to androstenone, skatole formation can easily be suppressed by dietary means. Higher concentrations of androsterone and/or skatole have not always led to lower scores for odor or flavor from taste panelists (Malmfors and Lundstrom, 1983). Nearly all male pigs are left intact in the UK, and Wood *et al.* (1986) indicated that taste panelists and consumers consistently have been unable to distinguish between meat from boars and barrows. In the UK, however, boars are slaughtered at about 85 kg live weight before reaching the onset of puberty.

Nearly 100% of boars are castrated in the USA, because market pigs generally have been slaughtered at weights in excess of 110 kg and meat processors recently have been paying premiums for pigs weighing 115–125 kg. At these weights, most boars have reached the onset of puberty and boar odor can be a significant problem. Therefore, it is unlikely that feeding boars for meat production will occur in the USA, even though they have distinct advantages over barrows in FCE and carcass leanness. Boars have a greater demand for protein and amino acids (especially lysine) and, because of their decreased appetites relative to barrows, the dietary concentration of protein or amino acids must be increased (Yen, Cole and Lewis, 1986; Hansen and Lewis, 1993). On the other hand, Fuller (1985) stressed that boars should not be thought of as having higher protein requirements than the other classes, but rather that they have a greater capacity to respond to higher protein levels. Leaving boars intact is a very economical and efficient way to reduce carcass fatness, provided that they are slaughtered at weights not exceeding 100 kg, that meat processors do not discriminate against them and that consumers find meat from young boars to be acceptable.

Barrows have greater feed intakes and higher GR than gilts; however, gilts have a slight advantage in FCE and a significant advantage in carcass leanness (Table 8.1) (Purchas, 1991). For these reasons, gilts require diets with a higher protein concentration. Some US producers are now separating barrows and gilts for finishing, so that diets can be formulated more specifically for their nutrient requirements and, therefore, utilize feed resources more efficiently.

8.3.4 Poultry

Leenstra (1986) suggested that the difference between male and female poultry in GR and fat deposition is worth considering. If broilers are sexed at hatching and reared separately, the females could be used for the production of whole birds and, thus, be slaughtered at earlier ages. The males could be slaughtered at higher weights to produce broiler parts. Females reach maximum non-fat growth at lower dietary protein levels than males (Holsheimer, 1975). Separate rearing may not reduce total broiler fat deposition, but more likely reduce feed costs.

Jones and Wiseman (1985) reported that male broilers have a faster

GR, greater feed intake, greater protein content, less fat and a heavier fat-free carcass weight than females. Mohan *et al.* (1987) reported that male broilers had higher muscle-to-bone ratios than females. Summers *et al.* (1985) reported that tom turkeys actually had lower percentages of carcass fat at 20 weeks of age than females had at 16 weeks of age. These reports suggest potential benefits for segregating male and female broilers and turkeys and feeding and slaughtering them separately.

8.4 Stage of growth curve

Beermann (1994) stated that growth from birth to market weight is characterized by nearly linear increases in skeletal muscle mass and carcass protein content. In contrast, lipid accumulation rate increases dramatically during this period and exceeds protein accretion rates by nearly 200% during the finishing phases of growth. Therefore, as an animal approaches market weight, an increasing percentage of nutrients consumed is used for lipid synthesis, and proportionately less is utilized for protein synthesis. This suggests that reduced carcass fatness and greater FCE will result from slaughtering animals at an earlier stage of finishing or at lighter slaughter weights.

8.4.1 Pigs

Cisneros *et al.* (1996) evaluated two genotypes of two sexes of pigs at slaughter weights of 100, 115, 130, 145 and 160 kg. Increases in slaughter weight were associated with increases in feed intake, backfat depth and longissimus muscle area, whereas percentage of trimmed boneless cuts decreased. Changes in GR or FCE were minimal. These authors concluded that the breeding-company hybrid and the Hampshire × Yorkshire × Duroc genotypes they studied could be slaughtered at live weights up to 160 kg with limited impact on growth performance and commercial meat yields. These results generally contradict the results of Gu, Schinckel and Martin (1992), who showed a curvilinear increase in carcass fat weight relative to slaughter weight, suggesting an increase in carcass fat percentage with an increase in slaughter weight.

8.4.2 Cattle

Gregory *et al.* (1994) conducted an extensive study involving 1599 steers from nine parental breeds and three composite breeds. Steers were slaughtered serially at four age endpoints about 21 days apart with a mean slaughter age of 438 days. Dressing percentage increased from 59.56 to 61.52% and live weight increased from 512 to 584 kg from the first to last slaughter age (about 63 days apart). Percentage of carcass lean decreased from 60.0 to 57.6% from the first to the last slaughter age, primarily because twelfth rib adjusted fat thickness increased an average of 0.10 cm with each additional 21 days on feed. However, when lean was expressed as a percentage of live weight instead of carcass weight, the percentage of lean ranged only from 35.83 to 35.39%. Therefore, when the increased dressing percentage that resulted from the greater slaughter age (and weight) was considered, the proportion of live weight that was lean remained relatively constant over slaughter ages.

The results of Toelle *et al.* (1986) are in general agreement with those of Gregory *et al.* (1994). They found that the proportion of fat in gains of 409 bulls of various breed types slaughtered at 320, 440 or 560 kg increased with slaughter weight for all breeds, but the differences were small. However, they did speculate that steers or heifers may increase in fat percentage more rapidly with increasing weight than bulls. These results suggest that cattle should be slaughtered when dressing percentage and carcass lean percentage are optimized, and that lean meat yield should be expressed as a proportion of live weight.

8.4.3 Sheep

Snowder, Glimp and Field (1994) stated that optimal slaughter weights to produce lean lamb carcasses for different breeds have not been identified conclusively for the industry. Optimal leanness is only of minor concern to lamb slaughterers in the USA, because lambs weighing in excess of 65 kg often bring a top market price, even though they may be yield grade 4 or 5 (>0.90 cm fat thickness). Per capita lamb meat consumption in the USA is less than 1 kg, and lamb meat is consumed by only a small percentage of consumers.

Figure 8.1 shows the change in percentage of chemical fat in the seven-rib racks of lamb carcasses of different breed types as reported by Snowder, Glimp and Field (1994). The data clearly show that slaughtering Rambouillet, Polypay and Targhee wethers at different weights can have dramatic effects on fat content. For example, rack chemical fat percentage increased from approximately 37 and 32.5% for Rambouillet and Polypay, and Targhee, respectively, at 40.5 kg slaughter weight to 47.5 and 50%, respectively, at 63 kg slaughter weight. Even the larger framed, faster growing Columbia wethers showed an increase from approximately 28 to 44.5% fat at 40.5 and 63 kg slaughter weights, respectively. These data clearly show that the fat content of lamb carcasses could be reduced significantly by slaughtering lambs at 45–50 kg rather than at 60–65 kg.

Arnold and Meyer (1988) demonstrated that delaying castration of ram lambs until weaning (23 kg) or until 39 kg decreased fat thickness. Delayed

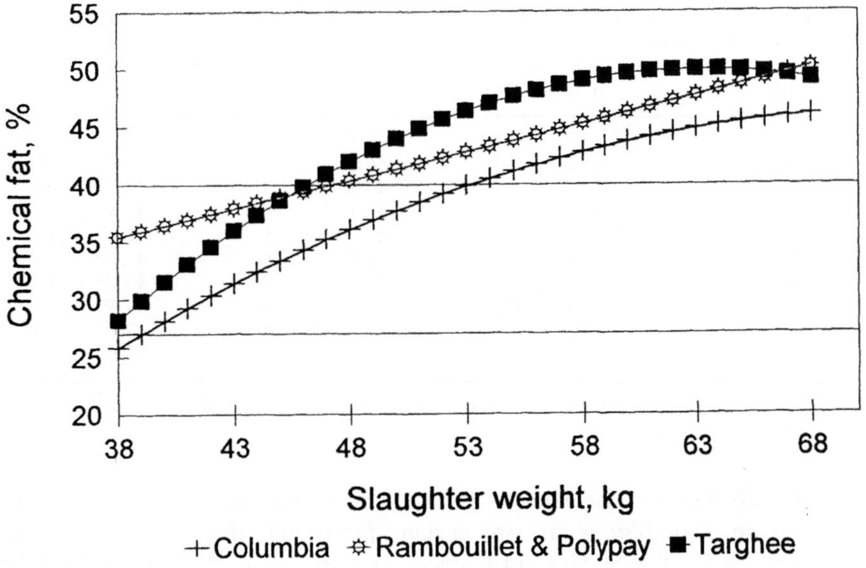

Figure 8.1 Breed-specific regression lines predicting chemical fat of lamb racks over a range of live slaughter weights of wether lambs. (Source: Snowder, Glimp and Field, 1994.)

castration also improved GR and FCE in whiteface lambs but not in blackface lambs with greater inherent growth potential.

8.4.4 Poultry

Table 8.3 contains data showing the effects of slaughter age on carcass composition of broilers in a study by Perreault and Leeson (1992). The data illustrate that the percentage of protein decreases and that of fat increases with slaughter age. Therefore, selecting the appropriate slaughter age (or weight) is very important.

8.5 Biological type of animal

8.5.1 Cattle

Although genetic selection against fatness is discussed in Chapter 7, it is appropriate to discuss briefly how differences in biological types of livestock may be used to reduce fat content. Gregory *et al.* (1994) evaluated nine sire breeds of cattle for carcass and meat traits when slaughtered at the same age. Limousin steers dressed out 63.43% and had 64.6% carcass lean compared with a 60.3% dress and 53.8% carcass lean for Herefords. Thus, the Limousin steers produced 202.4 kg of lean compared with

Table 8.3 Mean (±SE) composition of eviscerated broiler carcasses (protein and fat expressed on dry matter basis)

Age (days)	Moisture (%)	Protein (%)	Fat (%)
21	67.3[a] ± 0.3	48.9[a] ± 0.7	44.7[e] ± 0.5
28	65.6[b] ± 0.3	46.8[bc] ± 0.6	45.0[e] ± 0.5
35	64.7[c] ± 0.3	47.2[b] ± 0.7	45.8[de] ± 0.8
39	64.1[c] ± 0.2	45.7[bcd] ± 0.4	47.4[cd] ± 0.4
43	63.9[c] ± 0.3	45.7[bcd] ± 0.5	47.7[cd] ± 0.7
47	63.9[c] ± 0.3	44.7[d] ± 0.6	47.4[cd] ± 0.5
50	64.4[c] ± 0.3	45.5[cd] ± 0.4	45.1[e] ± 0.7
55	62.8[d] ± 0.3	42.7[e] ± 0.4	49.5[b] ± 0.5
60	62.8[d] ± 0.3	42.8[e] ± 0.4	47.8[bc] ± 0.4
70	60.4[e] ± 0.3	39.3[f] ± 0.7	52.2[a] ± 0.9

[a–e]Values followed by the same letter in a column are not significantly different at $P < 0.05$.
Source: Perreault and Leeson (1992).

156.5 kg of lean for the Herefords. This occurred because the Limousin steers were slaughtered at an earlier point on their growth curve. However, 60% of the Hereford steers graded ≥Choice compared with only 14% for the Limousin steers. Wheeler *et al.* (1996) reported even greater differences in percentage of retail product trimmed free of surface fat between 50% Piedmontese × 50% Hereford or Angus steers and Angus × Hereford reciprocal crossbred steers. These results clearly show that certain biological types of cattle can be chosen to reduce fat content if slaughtered at comparable weights. Because intramuscular fat is the most important factor in determining USDA quality grade and because quality grades have a distinct effect on value of carcasses, producers must breed and feed cattle to attain a certain level of intramuscular fat in order to receive an equitable market price. Therefore, crosses of Continental × British breeds generally optimize carcass leanness and meat quality under US conditions.

8.5.2 Sheep

Snowder, Glimp and Field (1994) demonstrated that Columbia wethers had a lower percentage of chemical fat in the rack than Rambouillet, Polypay and Targhee wethers, especially at weights of 45–50 kg (Figure 8.1). This illustrates the effect of biological type on CC.

8.6 Growth promotants

8.6.1 Anabolic steroids

Implants containing androgens and/or estrogens have been used widely in cattle for several decades to improve GR, FCE and CC. These commer-

Table 8.4 Anabolic steroid implant effect based on the amount of growth promotant and release time[a]

Implant	Estrogenic effect (mg)	Androgenic effect (mg)	Release time (days)
Synovex-C	7.2	0	120
Calf-oid	7.2	0	120 (est.)
Ralgro	11–13	0	70
Compudose	24	0	168
Synovex-S	14.4	0	120
Implus-S	14.4	0	120 (est.)
Synovex-H	14.4	200 testosterone	120
Implus-H	14.4	200 testosterone	120 (est.)
Finaplix-S	0	140 TBA	105
Finaplix-H	0	200 TBA	105
Revalor-S	24	120 TBA	120

[a]When calculating the steroid effect, 1 mg of trenbolone acetate (TBA) has 8–10 times the growth promoting effect of 1 mg of testosterone.

cial anabolic agents can be divided into two basic types, natural hormones and xenobiotics, although some implants contain combinations of these two types. Commercial anabolic implants for cattle are listed in Table 8.4. These implants are easy to use, are cost effective, result in increased lean production and are unquestionably safe.

(a) Cattle

Heifers and steers show the greatest response to steroid implants, whereas bulls do not show much response. In fact, bulls may deposit more, rather than less, fat when implanted. Implanted steers often achieve similar GR and FCE as bulls (Fisher, Wood and Whelehan, 1986). Bouffautt and Willemart (1983) indicated that GR has been increased as much as 70% in heifers implanted with trenbolone acetate. In general, estrogenic implants are more effective in steers and androgenic implants are more effective in heifers. Combination implants generally produce an additive effect compared with implants containing only an estrogenic or androgenic compound.

A review of many research reports shows that anabolic steroid implants increase GR by an average of 10–20%. In general, breeds or types of cattle that have the greatest potential for muscle growth show the greatest response to implants. Keane and Drennan (1987) observed that steers compensating for earlier growth restriction had a 50% increase in GR after being implanted. Greater responses often are observed during the first few weeks after implantation, suggesting a peak and then a decline in circulating concentrations of the hormones (Hayden, Bergen and Merkel, 1992).

The improvement in FCE in response to implanting cattle with anabolic implants ranges from 5 to 14%. The improvement results from proportional increases in muscle growth relative to fat deposition (Keane and

Drennan, 1987) and increases in feed intake (Beermann, 1994), which reduces the proportion of feed utilized for maintenance. Beermann (1994) stated that implanting cattle increases their potential mature size. Consequently, implanting increases the quantity of muscle growth and tends to delay fat deposition. At the same live weight, implanted cattle will have a higher percentage of muscle and a lower percentage of fat. At the same age or time-on-feed, implanted cattle will have a greater weight of muscle and an equal or lower weight of fat. Dressing percentage will be increased slightly at a constant time endpoint.

Heifers implanted with 36 mg of zeranol at birth and at 100 and 200 days of age were fed a finishing diet for 78 days to slaughter at 21–22 months by King *et al.* (1992). Rate of fat deposition and marbling score were distinctly less and carcass weight, ribeye area and lean-meat yield were noticeably greater for implanted than for nonimplanted heifers. Keane and Drennan (1987) implanted steers and heifers with trenbolone acetate and resorcylic acid lactone to slaughter ages of 491, 612 or 731 days and reported 23.1 kg more muscle, 2.6 kg less fat, 3.2 kg more bone and 23.6 kg heavier side weights than the nonimplanted controls. The cattle implanted the longest showed the greatest increase in muscle. The increase in weight of muscle and minor changes in fat and bone resulted in a significantly higher percentage of muscle in both steers and heifers at heavier weights.

Because of the emphasis on quality grade (marbling) in the USA, beef processors have been critical of implanting cattle with trenbolone acetate alone or in combination with an estrogenic agent within 60–70 days of slaughter. They believe that marbling is depressed, which results in a lower proportion of cattle grading USDA Choice. Perry *et al.* (1991) evaluated the effects of trenbolone acetate–estradiol implants on GR and carcass composition of Holstein, Angus and Simmental crossbred steers. Because of the differences in frame size and propensity for fat deposition, the authors attempted to slaughter the cattle at a constant marbling endpoint. Empty body protein GR was increased more than 25% in all three breed groups. The live weight required to attain the small degree of marbling necessary for the Choice quality grade was 25–45 kg heavier for the implanted than for the nonimplanted cattle. Quality grade and rib section composition, however, were not different between the two groups. The results demonstrate that one of the most potent anabolic steroid implants can be used to increase GR and protein accretion without negative effects on USDA quality grade, but the cattle need to be fed to heavier weights to attain the same degree of marbling.

Samber *et al.* (1996) evaluated the effects of several implants in various combinations at several time sequences during the feeding period. They found that trenbolone acetate–estradiol (Revalor) implants resulted in a decreased percentage of Choice carcasses when cattle were implanted three times with the last implant 60 days before slaughter and fed a diet containing 12.5% protein. When cattle receiving the same implants in the same sequence were fed a diet containing 14.5% protein, however, the percentage of carcasses grading Choice was increased from 55 to 73%. They did not, however, reach the same percentage of Choice carcasses as the nonimplanted controls, or those implanted with estradiol and progesterone 180 days before slaughter and reimplanted with Revalor 80 days before slaughter. As in most other studies, the implanted cattle were significantly heavier than the nonimplanted controls. Warner–Bratzler shear values were increased with the Revalor implants, although the mean shear values were still acceptable.

Implanting is a practical technology to utilize, is cost effective, increases GR and FCE and improves CC. There is no withdrawal time and residues in edible tissues are almost immeasurable or undetectable. No human health problem has been traced to the approved anabolic steroid implants used properly. The Food and Drug Administration has rigid requirements that must be met before any implant is approved for use. The approval process requires much research, takes several years and is an extremely costly procedure (Collins *et al.*, 1989). Bass, Butler-Hogg and Kirton (1990) stated that the 1986 EEC decision to ban all growth promotants appears to be related to political pressure rather than scientific evidence.

In a study on the comparative effects of somatotropin and anabolic steroids in feedlot cattle, Preston *et al.* (1995) reported that steers implanted with estradiol benzoate + progesterone followed by trenbolone acetate on day 63 had faster GR than steers implanted with 80 or 160 mg of bovine somatotropin (bST) per week. Both treatments improved FCE. These authors further reported that the effects of bST and anabolic steroid implants were additive. The percentage of USDA Choice carcasses was reduced by both treatments because of reduced marbling. The bST implants reduced fat thickness, whereas the anabolic steroid implants only slightly reduced fat thickness; however, the carcasses of anabolic steroid-implanted steers were 16 kg heavier.

(b) Poultry

Maurice *et al.* (1985) and Castaldo, Jones and Maurice (1990) demonstrated that GR and FCE were improved significantly with trenbolone acetate or trenbolone acetate + zeranol implants in turkeys. This response was enhanced as dietary protein density increased. Carcass fat was not affected. Anabolic steroid implants have not, however, been approved in the USA for use in poultry.

(c) Sheep and pigs

Only one implant (zeranol) is approved for lambs and it is not used widely because it is not effective. Hancock, Wagner and Anderson (1991) stated that, under the appropriate conditions, anabolic steroids can be very efficacious in swine. Historically, however, estrogens and androgens have not been considered to be particularly effective growth enhancers in pigs, according to Roche and Quirk (1986). Anabolic steroids are not approved for growth regulation in pigs in the USA.

8.6.2 Somatotropin

Somatotropin (ST) is secreted endogenously by the anterior pituitary gland in response to positive stimulation by a peptide called somatotropin-releasing hormone or growth-hormone-releasing factor (Beermann, 1994). Normal ST secretions are essential for normal growth, and higher than normal secretions are associated with more rapid growth of muscle, bone and some organs. Beermann (1994) reviewed the literature on the effects of exogenous administration of ST on GR and CC and stated that nutrients are partitioned toward greater rates of protein synthesis and deposition and toward much reduced rates of lipogenesis and lipid accumulation in growing pigs, sheep and cattle.

Since recombinant DNA technology began to be used in the early 1980s to produce somatotropin efficiently, many research studies have been conducted on its growth-promoting and fat-depressing effects in cattle, pigs and sheep and on its milk-secretion effects in dairy cattle. At the time of writing this chapter, only bovine ST has been approved for injection of dairy cattle to increase milk production. Although there has been much controversy among dairy producers and among segments of society regarding whether or not it should be used, it is safe and it improves efficiency. Until recently, there has not been much promise for use of ST in meat animals other than by daily injection. Recently, however, several reports have demonstrated potential for an implant allowing the release of ST over one to several weeks (Knight *et al.*, 1991; Klindt, Buonomo and Yen, 1992, 1995; Hacker *et al.*, 1993). This would make the application of ST much more practical and feasible.

(a) Pigs

Clearly and unequivocally, exogenous porcine somatotropin (pST) increases GR, FCE and CC of pigs when daily doses of 50–150 μg are utilized (Table 8.5). Only minimal negative, if any, effects on meat visual quality (color, firmness and marbling) and on cooked meat palatability (tenderness, flavor and juiciness) have been observed (McKeith, Lan and Beermann, 1994). Numerous studies with pigs have shown that pST increases GR by an average of 20% and improves FCE by 20–35%, depending on the dosage, genotype, dietary protein or other variables (Table 8.5). Unlike the improvement in FCE in cattle from anabolic steroid implants by stimulating greater feed intake, pST depresses feed intake. The dra-

Table 8.5 Effects of porcine somatotropin (pST) dose on growth and carcass composition (reproduced by permission of Academic Press)[a]

Item	pST dose ($\mu g\ kg^{-1}$ BW per day)				
	0	50	100	150	200
Number of pigs	10	10	10	10	10
Initial weight (kg)	30.8	31.2	31.3	30.7	30.8
Average daily gain (g)	890^{b}	990^{c}	1030^{c}	1000^{c}	1040^{c}
Daily feed intake (kg)	2.86^{b}	2.51^{c}	2.35^{c}	$2.20^{c,d}$	2.15^{d}
Feed : gain	3.23^{b}	2.53^{c}	2.30^{d}	$2.20^{d,e}$	2.09^{e}
Carcass composition:					
Chilled side weight (kg)	33.7^{b}	34.0^{b}	33.4^{b}	32.5^{c}	32.3^{c}
Muscle (kg)	15.7^{b}	19.6^{c}	19.9^{c}	$20.5^{c,d}$	21.4^{d}
Adipose (kg)	11.6^{b}	7.2^{c}	5.0^{d}	$3.8^{d,e}$	3.0^{e}
Bone (kg)	3.7^{b}	4.0^{c}	4.1^{c}	4.2^{c}	4.3^{c}
Skin (kg)	1.8^{b}	2.1^{b}	2.5^{c}	2.5^{c}	2.5^{c}
Composition of carcass gain (g per day):					
Protein	93^{b}	138^{c}	150^{d}	152^{d}	158^{d}
Lipid	264^{b}	144^{c}	104^{d}	59^{e}	30^{f}
Ash	24^{b}	31^{c}	34^{d}	34^{d}	34^{d}

[a]Pigs received daily injections of excipient or the specified dose of recombinant pST. Data are least-squares means. Data are summarized from those published from a larger experiment by Krick *et al.* (1992) and Theil *et al.* (1993).
[b–f]Means within a row with different superscripts differ ($P < 0.05$).
Source: Beermann (1994).

matic improvement in FCE results from the reduction in feed intake and the highly significant increase in carcass muscling. Steele *et al.* (1994) stated that pST caused reductions of *ad libitum* intake by pigs of 8% during the growing phase and 28% during the finishing phase. Beermann (1994) summarized recent research reports and concluded that pST increases total carcass skeletal muscle by 28–38% and reduces total adipose tissue mass by 36–74% at a 90 kg final weight. In addition, pST increases bone by 10–17% and skin by 15–38%. Evock *et al.* (1988) reported that organ weights have been increased by 25–40%. Beermann (1994) further concluded that protein accretion shows a diminishing response to pST dosages $>100\,\mu g$ per kg body weight, but the reduction in lipid accretion rate is linear from 50 to 200 μg pST per kg body weight.

The relative effects of exogenous pST administration appear to be greatest in genotypes that have inferior protein accretion rates. Furthermore, reduction of lipid accretion rate appears to be greatest during the later phases of growth, when animals are approaching normal market weights (Beermann, 1994, citing McNamara *et al.*, 1991; Moseley *et al.*, 1992). Salbutamol, a β-agonist (BA), fed at 2.75 ppm of the diet increased fat depth in barrows, but 4 mg per day of pST caused a dramatic reduction and negated the effects of salbutamol (Hansen *et al.*, 1994).

Steele, Boyd and Campbell (1994) stated that, in general, the effect of pST on protein accretion in pigs cannot be exploited fully without adequate protein intake. In general, 30 g per day intake of lysine or 1.1–1.2% dietary lysine concentration is required for pST-treated pigs when consuming feed *ad libitum* during the finishing phase. This represents a >50% increase in the daily requirement for lysine. Johnston *et al.* (1993) demonstrated that pST-treated pigs showed a greater response to increased lysine from 59 to either 105 or 127 kg live weight. Several studies suggest that pST increases the partial efficiency of amino acid utilization for protein accretion by 25%.

Knight *et al.* (1991), Klindt, Buonomo and Yen (1992; 1995) and Hacker *et al.* (1993) demonstrated that weekly or biweekly implants of pST produced improvements in GR, FCE and CC in growing pigs, but not to the same extent as daily injections. The only probable way in which pST technology would be adopted widely by the swine industry, if approved, is if implants are proved to be effective and reasonably priced. If pST is approved for commercial use, is effective in an implant form and is adopted, the swine industry would make a giant step towards having a competitive edge over most other segments of the meat industry.

(b) Sheep

Growing sheep also respond to exogenous bST and ovine somatotropin (oST), but the responses generally are not as great as those observed in pigs. Beermann *et al.* (1990) demonstrated a 12–19% increase in GR and a 20–22% response in FCE with oST treatment. However, the improvement in FCE did not result from a decrease in feed intake as generally has been observed in pigs. These authors reported a 36% increase in carcass protein content and a 33% decrease in lipid accretion rates. Beermann *et al.* (1990) found that ewe lambs exhibited greater reductions in fat accretion and greater responses in GR than wethers when oST was administered over an 8-week period prior to slaughter. Ewe lambs generally have slightly slower GR and deposit more fat at the same weight or age as wethers.

Johnsson *et al.* (1987) studied the effects of injecting or infusing 0.025, 0.1 and 0.25 mg kg^{-1} per day of bST into ewe and wether lambs between 10 and 22 weeks of age. Lambs injected with bST tended to have lighter carcasses and heavier non-carcass components. The 0.1 and 0.25 mg kg^{-1} per day doses reduced fatness of the neck plus shoulder joint.

The quantity or composition of absorbed amino acids, or both, limit nitrogen retention by growing lambs, according to Beermann *et al.* (1991). They used casein infusion of control lambs and lambs treated with bST (100 μg per kg body weight for 15 days). Casein infusion (4–5 g nitrogen per day) increased the nitrogen balance by 43.4%, whereas oST increased it by 33.5%, without interaction. These authors concluded that oST increases the efficiency of utilization of absorbed amino acids for protein.

Prolonged release delivery of bST in an oil-based formulation (formulated sometribove) was injected at 50, 100 or 150 mg as a single treatment once every 2 weeks or as two equal treatments once per week by McLaughlin *et al.* (1994). Feed conversion efficiency was improved by 9 and 10% in lambs injected once and twice per 2 weeks, respectively. Fat thickness was reduced by 17, 30 and 42% with increasing dosages. Percentage of muscle increased and weight of fat decreased.

Clearly, oST would have to be effective in an implant or prolonged release form before the sheep industry would adopt this technology. Sheep often graze or run in larger lots than pigs and their quick, flighty nature would make it extremely impractical to inject them daily with oST. An implant release form of oST might be accepted readily by the sheep industry, because no really effective anabolic steroid implant currently is available for use in sheep.

(c) Cattle

Effects similar to those observed with oST injection in sheep have been observed with cattle, although the responses have been more variable. Injection of 33, 100 or 300 μg per kg body weight of bST (experiment 1) or 8.25, 16.5, 33 or 66 μg per kg body weight (experiment 2) in steers decreased GR (linearly), increased FCE (quadratically) and decreased dry matter intake (linearly) as the dosage increased, according to Moseley *et al.* (1992). The 33 $\mu g\,kg^{-1}$ bST treatment improved GR and FCE, 100 $\mu g\,kg^{-1}$ improved FCE and 200 $\mu g\,kg^{-1}$ decreased GR and FCE. The optimum dosage for GR and FCE appeared to be between 16.5 and 33 $\mu g\,kg^{-1}$; however, the dose response for changes in carcass chemical composition had not attained a plateau even at 300 $\mu g\,kg^{-1}$.

Implants delivering 40, 80 or 160 mg per week of bST were administered weekly to cattle subcutaneously in the tailhead by Dalke *et al.* (1992). Steers were fed a high concentrate diet. Dry matter intake was decreased and FCE increased in a dose-dependent manner. Fat thickness, marbling score and percentage of fat in the ninth to eleventh rib section all decreased in a dose-dependent manner, but dressing percentage was not affected. Implanting feedlot steers weekly with bST appears to be an efficacious method of improving feedlot performance and partitioning nutrient deposition. Preston *et al.* (1995) also reported increased GR and FCE and reduced fat content using a sustained-release bST implant treatment. Furthermore, the effects of bST were additive with anabolic steroid treatment. The negative effect of bST was a reduction in the percentage of carcasses grading Choice.

Beermann (1994) summarized the limited research that has been conducted with growth-hormone-releasing factor. The duration of elevated ST concentration is short (30–45 min), and returns to near basal levels within 60–90 min. This is much shorter than the 8–10 h of elevated ST concentrations achieved with direct administration of ST. Therefore, to obtain a chronic elevation of ST concentration in the blood, intermittent administration of growth-hormone-releasing factor or continuous release in the form of an implant would be necessary. Bass, Butler-Hogg and Kirton (1990) stated that the lengthy and costly process of producing the growth-hormone-releasing factor limits its usefulness as an anabolic agent. These commercial limitations may be overcome by developing patent synthetic analogues and slow-release formulations for use as implants.

(d) Poultry
In general, ST administration does not significantly alter growth or composition in avian species (Beermann, 1994).

(e) Fish
Recombinant bST was injected intraperitoneally in 2-year-old trout at 10, 20 and 30 μg per g body weight in a single dose by Garber *et al.* (1995), and fish were weighed at 0, 14, 28 and 56 days. Recombinant bST reduced feed intake by 17.6% and improved GR by 44.8% by day 14. Treated fish had improved FCE from day 0 to 28 and reduced carcass fat (linear). These authors concluded that this treatment may be an efficacious method of modulating growth in rainbow trout. However, the practicality of this treatment has not been demonstrated.

8.6.3 β-Adrenergic agonists

According to Beermann (1994), the β-adrenergic agonists (or β-agonists, BAs) exhibit profound effects on growth and metabolism of skeletal muscle and adipose tissue in growing animals. Some β-agonists also cause negative effects (real or perceptual) on meat quality. The BAs commonly studied are clenbuterol, cimaterol, L-644-969, salbutamol and ractopamine. The great advantage of BAs over ST is that they can be incorporated in diets, which eliminates animal handling for injections. These compounds increase skeletal muscle mass and reduce fat deposition in a dose-dependent manner, and the results can be dramatic. The repartitioning effect is expressed to a greater extent in ruminants than in non-ruminants. In contrast to the effects of ST, BAs do not increase organ or bone weights and, consequently, dressing percentage is increased. They improve the efficiency of production of muscle in cattle, sheep, pigs and poultry.

(a) Sheep
Response to dietary administration of BAs in ruminants generally has been greater than that in pigs and poultry (Moloney *et al.*, 1991). Beer-

mann (1993) reviewed the effects of BAs in meat animals and stated that a 10–20% increase in GR and a 15–20% improvement in FCE occurred in lambs fed 1–10 ppm cimaterol, L-644-969 or L-655-871 in typical mixed concentrate diets consumed *ad libitum*. Carcass dissectible adipose tissue decreased by 15–30% and dissectible muscle increased by 10–30%. Table 8.6 shows some of the responses that different researchers have found in lambs fed BAs. Beermann (1994) stated that young animals still nursing, reared on milk replacer or recently weaned exhibit little improvement in GR or CC when fed BAs.

Galbraith *et al.* (1990) evaluated the effects of cimaterol and estradiol-17β alone or combined on growth and body composition of wether lambs. Cimaterol increased GR, empty body weight, dressing percentage and weight of protein. Estradiol-17β also increased GR, empty body weight, carcass protein and dry matter intake. The effects of cimaterol were slightly greater than those of estradiol-17β; however, no additive effects of cimaterol and estradiol-17β were observed. Coleman, Ekeren and Smith (1988) fed clenbuterol at 2 ppm to wether lambs and studied lipid synthesis and adipocyte growth. They concluded from *in vitro* data that clenbuterol did not decrease subcutaneous fat accretion in sheep by inhibiting lipogenesis but by increasing triacylglycerol turnover. Pringle *et al.* (1993) concluded that BAs may act on the calpain–calpastatin system to stimulate muscle growth through a decrease in protein degradation, which has been suggested as the most efficient means of increasing muscle protein accretion. Table 8.7 shows that shear values for cooked meat are increased significantly in lambs fed BAs.

Koohmaraie, Shackelford and Wheeler (1996) studied the effects of L-644-969 and male sex condition on muscle growth of callipyge lambs. They concluded that muscle GR was near maximum in lambs expressing the callipyge gene, regardless of male sex condition or L-644-969 treatment. Tenderness at 20 days post-mortem was decreased significantly by feeding L-644-969, and was associated with increased calpastatin activity.

(b) Cattle

Studies with cattle have produced responses similar to those observed in lambs fed BAs. Effects on GR have been variable, ranging from a 20% decrease to a 30.4% increase. Effects on FCE have been more consistent, ranging from 8 to 31% improvement. Table 8.8 contains some of the CC data from studies in which cattle were fed BAs. The effects of feeding BAs on CC are a dramatic reduction in carcass lipid, an increase in carcass lean and an increase in dressing percentage.

Feeding 12 pairs of monozygotic twin Friesian bulls cimaterol at 0.05 mg per kg body weight for 90 days to ending liveweights of 162, 299 or 407 kg resulted in dramatic effects on GR, CC and meat tenderness in a

Table 8.6 Effects of β-agonists on selected slaughter traits and carcass composition of sheep

β-Agonist and composition procedure[a]	Dose (ppm)	% of untreated control						
		Carcass weight	Dressing weight	Kidney pelvic fat (%)	Fat (%)	Lean (%)	L.D. area	Fat depth
1a	1	106**	106*	86	80*	110*	128*	N.R.
	10	107*	107*	79	73*	112*	134*	N.R.
	100	102	102	65*	77*	110*	126*	N.R.
	0.5	107*	107*	75	77*	112*	129*	N.R.
	2	110*	110*	93	81*	108*	145*	N.R.
	10	107*	107*	67*	77*	110*	133*	N.R.
	2	105*	105*	63*	N.R.	N.R.	142*	63*
1b	50 μg per day	100	N.R.	N.R.	73*	114	N.R.	N.R.
	100 μg per day	99	N.R.	N.R.	71*	116	N.R.	N.R.
1	2	107	104	124	N.R.	N.R.	132*	63*
2	2	N.R.	N.R.	80	N.R.	N.R.	115*	176*
	10	N.R.	N.R.	65*	N.R.	N.R.	125*	79
	6	N.R.	N.R.	73*	N.R.	N.R.	141*	79*
	0.01 mg per day	N.R.	N.R.	78	N.R.	N.R.	115*	83
	0.05 mg per day	N.R.	N.R.	70*	N.R.	N.R.	131*	75*
	0.25 mg per day	N.R.	N.R.	62	N.R.	N.R.	140*	64*
2b	0–11.42	111^{L*Q*}	106	61^{L*Q*}	63^{L*}	110^{L*}	124^{L*}	N.R.
	0.1 mg kg^{-1} per day	108	106*	81*	73*	109*	124*	N.R.
2	1.8	113*	107*	61*	N.R.	N.R.	110*	86
2	10	105	107*	61*	N.R.	N.R.	129*	62*
	10	107	107*	75*	N.R.	N.R.	129*	80*
2c	10	121*	110*	72	95	102	139*	77
2	2	111*	N.R.	67*	59*	110*	N.R.	N.R.
3d	0.25	106	N.R.	87	101	102	101	86
	1	109	N.R.	82	91*	110*	107	80
	4	109	N.R.	77*	88*	111*	113*	57*

[a]1, 2 and 3 are clenbuterol, cimaterol and L-644-969, respectively. Procedures used to determine carcass composition; a, b, c and d are chemical analysis of the hindquarter, half carcass dissection, carcass density and chemical analysis of half carcass, respectively.
[b]$P < 0.05$; L and Q are linear and quadratic effect of β-agonist dose, respectively.
[c]N.R. = not reported.
Source: Moloney (1988). Data from several sources.

Table 8.7 Effect of β-adrenergic agonists on the sensory properties of the longissimus (reproduced by permission of Academic Press)

Reference	Compound	Trait	Response
Pork:			
Jones, Burgess and Dupchak (1985)	Cimaterol	Shear force	Up to 0.5 kg↑
Warriss *et al.* (1990a)	Salbutamol	Shear force	0.9 kg↑
Warriss *et al.* (1990b)	Salbutamol	Shear force	0.8 kg↑
Merkel (1988)	Ractopamine	Shear force	ND[a]
Stites *et al.* (1994)	Ractopamine	Tenderness	ND
		Juiciness	ND
		Shear force	ND
Uttaro *et al.* (1993)	Ractopamine	Shear force	0.5 kg↑
Beef:			
Miller *et al.* (1988)	Clenbuterol	Shear force	0.7 kg↑
Chikhou *et al.* (1993)	Cimaterol	Shear force	Up to 62 N↑[b]
Lamb:			
Koohmaraie *et al.* (1991)	L644,969	Shear force	Up to 3.5 kg↑
Koohmaraie and Shackelford (1991)	L644,969	Shear force	Up to 2.2 kg↑
Pringle *et al.* (1993)	L644,969	Shear force	3.5 kg↑
Merkel (1988)	Cimaterol	Shear force	Up to 2.0 kg↑
Poultry:			
Morgan, Jones and Calkins (1989)	Cimaterol	Shear force	Up to 1.3 kg↑
Gwartney, Calkins and Jones (1991)	Cimaterol	Shear force	Up to 1.2 kg↑

[a]ND = no difference.
[b]Value cited in newtons in the original reference.
Source: McKeith, Lan and Beermann (1994).

Table 8.8 Effects of β-agonists on selected slaughter traits and carcass composition of cattle

		% of untreated control						
β-Agonist and composition procedure[a]	Dose (mg per day)	Carcass weight	Dressing weight	Kidney pelvic fat (%)	Fat (%)	Lean (%)	L.D. area	Fat depth
1a	10	102	101	77*	80*	113*	111*	64*
	500	97	100	67*	71*	114*	116*	58*
1b	10	99	103	62*	52	N.R.	118*	60*
1c	0.1 ppm	105	105	79	88*	109	N.R.	N.R.
	1.0 ppm	109	107	60*	74*	117	N.R.	N.R.
2d	4 ppm	106*	106*	N.R.	66*	110*	N.R.	N.R.
2c	0–66	112^{L*Q*}	109^{L*Q*}	56^{L*Q*}	60^{L*Q*}	117^{L*Q*}	142^{L*Q*}	N.R.
2c	4 ppm	103	106*	52*	54*	115*	136*	N.R.
3c	0–4 ppm	109^{L*Q*}	108^{L*Q*}	58^{L*Q*}	64^{L*Q*}	116^{L*Q*}	137^{L*Q*}	52^{L*Q*}

[a]1, 2 and 3 are clenbuterol, cimaterol and L-644-969, respectively. Procedures used to determine carcass composition: a, b, c and d are chemical analysis of the hindquarter, half carcass dissection, carcass density and chemical analysis of half carcass, respectively.
*$P < 0.05$; L and Q are linear and quadratic effects of β-agonist dose, respectively.
[c]N.R. = not reported.
Source: Moloney (1988). Data from several sources.

Table 8.9 Effects of β-agonists on growth and carcass composition of growing-finishing swine (reproduced by permission of Academic Press)

		Control values and proportional responses (%)				
				Carcass composition		
Treatment dose (ppm)	Treatment period (days)	ADG[a] (g per day)	Feed:gain[b]	Protein	Lipid	Reference
Cimaterol:						
0	51	760	3.86	–	2.58 cm[c]	Jones *et al.* (1985)
0.25		+5.3	–9.7*	+8.2[d]	+10.5[c]	
0.5		+1.3	–8.4*	+11.9	–12.0	
1.0		+3.9	–12.1*	+11.2	–17.4	
Ractopamine:						
0	45	818	3.74	50.9%[e]	29.9%[e]	Watkins *et al.* (1990)
5		+7.2*	+8.0*	+3.1	–1.6	
10		+7.8*	–10.2*	+7.9*	–9.0*	
20		+9.0	–12.0*	+12.0*	–14.0*	
0	40–45	890	3.25	40.4 kg[f]	19.7 kg[f]	Bark *et al.* (1992)
20		+10.0%	–12.8%**	+20.6%**	–26.9%**	
Ro 17-8714:						
0	40	1016	3.41	13.4%	29.7%	Bracher-Jakob *et al.* (1990)
20		+0.7	–4.7	+2.6	–4.4	
60		+6.1	–10.9*	+3.0	–11.8	
180		+8.8	–12.6*	+4.5	–10.1	

[a]Average daily live-weight gain.
[b]Kilograms feed per kilogram live-weight gain.
[c]Data are expressed on the basis of tenth rib fat thickness.
[d]Data are expressed on the basis of semimembranosus and biceps femoris muscle weights.
[e]Data are percentage physically dissected lean from one carcass side.
[f]Total carcass muscle and adipose mass.
*$P < 0.05$.
**$P < 0.01$.
Source: Beermann *et al.* (1994).

study of Vestergaard, Sejrsen and Klastrup (1994). Cimaterol increased estimated gain and weight of prime rib by 55 and 20%, respectively, and lean and protein content of the longissimus dorsi muscle by 26 and 25% and concomitantly reduced fat content by 16%. Cimaterol reduced total collagen content but not collagen solubility. However, Volodkevich shear values were increased by 290%!

(c) Pigs

Feeding BAs to pigs also improves GR, FCE and CC (Table 8.9), although not as much as in lambs and cattle. In addition, the negative effect on tenderness generally is not as great as in lambs and cattle. Van der Wal (1990) compiled data from 12 swine experiments involving four different BAs that showed an average of 10% increase in muscle mass.

Watkins *et al.* (1990) conducted six trials involving 888 pigs (study 1) and three trials involving 360 pigs (study 2) at various geographical locations to evaluate the effects of ractopamine hydrochloride on the performance and carcass characteristics of finishing pigs. All ractopamine levels (2.5, 5, 10, 15, 20 or 30 ppm) improved GR and FCE when compared with control pigs. Levels of 10–30 ppm improved dressing percentage, and all levels except 2.5 ppm increased dissectable lean mass. When gilts and barrows were fed 20 ppm of ractopamine, it increased GR and FCE and reduced fat depth, according to Uttaro *et al.* (1993). They also observed heavier trimmed loins and bellies from the pigs fed ractopamine. Greater reductions in backfat and fat in primal cuts were observed for gilts compared with barrows than for ractopamine-fed pigs (20 ppm) compared with controls; however, these effects were additive. Gilts and ractopamine-fed pigs had higher WBS values than barrows and control pigs, respectively. Table 8.7 contains tenderness data from BAs studies.

Stites *et al.* (1991) fed ractopamine at 5, 10 or 20 ppm to pigs for 48 days and observed increased slaughter weight, GR and FCE compared with the controls. Dressing percentage increased linearly with ractopamine dose. However, no differences in percentage of trimmed primal cuts were observed, although trimmed ham and loin percentage increased at the 20 ppm ractopamine level. No adverse effects on the processing characteristics of bacon occurred. In contrast to the results of Stites *et al.* (1991), Aalhus *et al.* (1990) found no increase in GR and FCE when ractopamine was fed at 10, 15 or 20 ppm. However, dissected lean increased by 2.6, 3.3 and 4.4 percentage points, respectively. Surprisingly, marbling increased, but so did the WBS value (decreased tenderness). Williams, Bennett and Keele (1995) fed ractopamine to barrows and gilts at 44.7 mg per day and found that GR was increased at the onset of feeding until a plateau was reached at about 22 days, after which a linear decline in GR response occurred. When energy density of the diet was increased, fat mass increased. Crome *et al.* (1996) found that ractopamine fed at 10 or 20 ppm had positive effects on GR, CC and carcass cutting yields.

Salbutamol fed at 2.75 ppm increased GR and FCE, but also increased fat depth, whereas pST caused dramatic reductions and negated the effects of salbutamol on fat depth, according to Hansen *et al.* (1994). Treatments were additive in their effects on increased percentage of muscle.

(d) Poultry

Several studies have established that cimaterol decreases fat accumulation and increases protein accretion in the muscle of poultry (Dalrymple *et al.*, 1984; Buttery and Dawson, 1987; Morgan, Jones and Calkins, 1989). Takahashi, Akiba and Horiguchi (1993) reported that feeding clenbuterol at 0.33 mg kg^{-1} for 14 days to 7-day-old female broilers improved FCE and carcass protein percentage, but the effects were not very large.

Morgan, Jones and Calkins (1989) reported increases of 6.0 and 5.6% in percentage of carcass protein and reductions of 1.0 and 19.9% in percentage of carcass lipid when cimaterol was fed for 38 or 56 days to broilers, respectively. Rickes *et al.* (1987) reported a 5.1 and 6.0% improvement in percentage of carcass protein when feeding L-644-969 at 0.25 and 1 ppm, respectively. Buyse *et al.* (1991) noted that the action of BAs may be modulated by factors such as type of BA, species, breed or sex, time of initiation and duration of supplementation. Gwartney, Calkins and Jones (1991) studied the effects of cimaterol and its withdrawal in 21-day-old broilers. They found that cimaterol increased muscle weight and percentage of protein, decreased percentage of fat and reduced tenderness. Withdrawal for 7 days before killing them eliminated the differences. Morgan, Jones and Calkins (1989) and Gwartney, Calkins and Jones (1991) also reported higher shear-force values in cooked broilers when supplementing diets with cimaterol. A summary of the research on supplementing broiler diets with BAs indicates that the response in either GR or CC is not very large and that reduced tenderness occurs. Hence the overall benefit of feeding BAs to broilers seems questionable.

(e) Effects of BAs on meat quality

The effects of BAs on visual quality of meat are rather minor (McKeith, Lan and Beermann, 1994). However, most studies show that Warner–Bratzler shear force is increased, in some cases up to 3.5 kg (Table 8.7). One of the explanations is that BAs decrease μ-calpain activity and increase calpastatin activity, which decreases the rate and extent of postmortem tenderization (Kretchmar *et al.*, 1990; Koohmaraie *et al.*, 1991; Pringle *et al.*, 1993).

Cimaterol and clenbuterol are not approved for use in the USA and in several other countries. Considerable negative publicity occurred in the USA from 1994 to 1996 concerning the illegal use of these two BAs in market cattle, lambs and pigs exhibited at livestock shows. Most livestock-show authorities have imposed very strict regulations and penalties for violations. Meat processing plants have threatened to not slaughter market animals from livestock shows.

Moloney *et al.* (1991) concluded from the literature that a critical factor in the usefulness of BAs is likely to be the degree to which their effects are retained after withdrawal, since a recommended (if not statutory) period is likely. Van der Wal (1990) stated that these compounds involve a more complicated risk assessment than ST. There have been reports of acute poisoning in humans in Spain following the consumption of liver containing high residues of clenbuterol (Guyer and Miller, 1994). Another report implicated clenbuterol residues in an acute poisoning episode in France following the consumption of contaminated beef liver (Guyer and Miller, 1994). In both cases, the animals were fed large quantities of clenbuterol just before slaughter. In response to these reports, the Center for Veterinary Medicine in the USA has requested additional acute toxicity information for all BAs being investigated for use in food-producing animals. Moloney *et al.* (1991) (citing David, 1989) stated that drastic measures have been taken by the Federal German authorities when clenbuterol use was discovered in calf fattening rations. Malucelli, Ellendorff and Meyer (1994) measured residues after withdrawal of various BAs in 160 broilers. After feeding for 35 days, BAs were withdrawn and tissues collected on days 0, 1, 2, 3, 7, 14 and 43. A withdrawal period greater than 2 weeks was required for residues in edible tissues to decline below detectable levels.

Because of the beneficial effects of ST on GR, FCE and CC, the lack of significant negative effects on meat tenderness and the significantly fewer public and regulatory agency concerns, ST use by the livestock industry seems to have a greater chance of being implemented than BA use.

8.7 Research needs

A balance of fundamental (basic) and applied research needs to be conducted in the future in order for animal agriculture to become more efficient. Fundamental or basic research is needed to provide new information and the knowledge base for applied research. Improvements in disease resistance and ability to synthesize essential amino acids, and reductions in energy expenditures for maintenance and growth would be very beneficial. Applied research should include an economic focus, but be scientifically and statistically valid.

To the author's knowledge, there has been no research on feeding β-agonists to fish and this needs to be done to determine effects on growth rate, feed conversion efficiency, carcass composition and meat palatability. For sheep, research needs to be conducted on the combination of anabolic steroid and somatotropin implants to determine if they have an additive or synergistic effect on performance and carcass composition. This combination of implants should be studied in lambs on diets of different energy and/or protein densities. In addition, more fundamental research needs to be conducted on defining amino acid and/or protein requirements for growing lambs.

For pigs, more fundamental research needs to be done on boar taint and methods for controlling it. In addition, more fundamental and applied research needs to be conducted on slow-release forms for somatotropin. Somatotropin injection is extremely effective in increasing growth rate, improving feed conversion efficiency and decreasing carcass fatness, but an implant or other practical slow-release form needs to be developed so that it could be used widely by the pig industry. Research must be continued on optimum energy density and amino acid balance in diets of growing pigs. Further research needs to be conducted on the role and importance of feed additives such as carnitine and chromium. More fundamental and innovative research on the potential for manipulating prenatal muscle development by altering feed intake or amino acid density or by somatotropin injection of pregnant sows is suggested.

For cattle, more research needs to be conducted on the combination of somatotropin and anabolic steroid implants to determine if they have an additive or synergistic effect on performance and carcass composition. The effects of this combination of growth promotants during the growing phase when cattle are fed high roughage diets and in heifers or early maturing types of cattle need to be evaluated. Fundamental and applied research to establish amino acid requirements for growing cattle certainly needs to be conducted. Even in intensive management systems for cattle, rumen microbiota impose a constraint on the availabilities of amino acids and energy to cattle. Therefore, amino acid requirements need to be established in order to improve amino acid balance and improve efficiency of protein use. Both fundamental and applied research with an economic emphasis need to be conducted.

Poultry researchers should give attention to developing methods for measuring or predicting carcass composition. A value-based marketing system will need to be implemented to provide incentive to reduce the fat content of poultry.

8.8 Summary

The extent to which fat content of meat animals can be reduced and the methods to accomplish the reduction are dependent upon the species. All species currently produce waste or unwanted fat, but some produce more than others. The least potential to reduce fat may be in fish, mainly because some species already are relatively low in fat, but also because the options or technologies are fewer and handling is impractical. Trout have responded to a single injection of bovine somatotropin for a 56-day period with an increased growth rate, improved feed conversion efficiency and reduced fat. Manipulating the dietary energy to protein ratio may be the most practical option for reducing fat in fish, but the reduction may be limited.

The sheep industry in the USA is small and lamb slaughterers are few in number. Processors generally place little emphasis on USDA yield grade differences among carcasses. The most economical and a very effective way to reduce the fatness of males is to not castrate them. This would work especially well in farm flocks on an accelerated production system. For whiteface breeds and/or range flocks, castration after weaning will result in some reduction in fatness and should be practiced. For ewe lambs, the most critical control point is to slaughter them at the right stage on their growth curve because within a 15 kg live weight change, carcass fatness can increase dramatically. Restricting feed intake and/or feeding a growing diet up to 40 kg before finishing will result in some reduction in fatness. However, more time and labor will be required to reach market, and it may not be economical. Daily somatotropin injection is very effective for improving growth rate, feed conversion efficiency and carcass leanness of sheep with only minor effects on meat tenderness and overall palatability. However, daily injection of somatotropin is not practical in sheep production. Supplementing diets with β-agonists would be a very practical and an extremely effective way of reducing fatness. However, the two most effective β-agonists for reducing fatness also increase meat toughness significantly and are not approved for use in meat animals in the USA. Furthermore, these compounds need to be proved safe for humans consuming meat from animals to which they have been fed.

Options for reducing fatness in pigs are very promising. Increasing dietary protein intake, and/or improving amino acid balance and/or reducing caloric density have been demonstrated to be effective in reducing carcass fatness with negligible effects on growth rate. Another option is to restrict the feed intake of a high energy diet. The disadvantages of this approach are the increased days to reach market weight and the increased feeding space required so all pigs can eat at the same time. Coupled with changing dietary energy to protein ratios, producers can segregate barrows and gilts and formulate diets more specifically for their respective growth requirements. Feeding boars does not seem feasible in the USA because pork processors pay premiums for heavy (112–125 kg) slaughter weights, which would allow many boars to reach puberty and cause boar odor problems in pork. Recombinant DNA-synthesized somatotropin significantly increases growth rate and feed conversion efficiency and dramatically reduces carcass fatness. Daily injections may be feasible for pigs, because they usually are reared in confinement facilities. Research on implants to deliver somatotropin look very promising. This technology would probably be adopted widely and would put the pig industry in an extremely competitive position relative to all other segments of the meat industry. Negative effects on meat quality are negligible. The β-agonist ractopamine hydrochloride has been found to improve efficiency and carcass composition only slightly. Supplementing diets with ractopamine is very practical, but decreased meat tenderness would be expected as a consequence of using it or other β-agonists. Improving dietary protein intake and amino acid balance coupled with the use of somatotropin appears to offer the most promise for reducing fatness in pork carcasses.

Several options exist for reducing fatness in cattle. The opportunities for slaughtering at earlier points on the growth curve, leaving bulls intact and effective use of anabolic steroid implants probably are greater for cattle than for the other species. Life-long implantation with appropriate implants is a very practical, cost-effective method for increasing growth rate, feed conversion efficiency and quantity of lean. Reduced dietary-energy density can be additive with the effects of implanting and slaughtering at an earlier point on the growth curve. Dietary manipulation will add increased time to reach market weight, but is not necessarily a disadvantage when cattle are harvesting roughages. Production of bulls requires more specialized management, but is used in many countries to produce lean carcasses efficiently. Somatotropin injection in cattle can improve feed conversion efficiency and reduce fat. However, this delivery system would probably not be adopted by the beef cattle industry. Delivery of somatotropin via implants shows promise and would probably be adopted if proved effective. Limited research shows that the effects of anabolic steroid and bST implants are additive in increasing growth rate and feed conversion efficiency. The combination will probably decrease marbling. The effects of β-agonists on growth rate in cattle have been variable, but they consistently improve feed conversion efficiency. As in the other species, they would be very practical in increasing dressing percentage and reducing fatness dramatically. However, they also may increase meat toughness significantly. Increased meat toughness coupled with the fact that some β-agonists have not been proved safe and are not approved for use in the USA, the EEC and other countries for meat animals raises questions about their worth or overall benefit.

Opportunities for reducing fat in poultry are moderately promising. As with all species, but more specifically in broilers and turkeys, a value-based marketing system will be necessary to provide the proper incentives for reducing fatness. Male and female broilers can be segregated so that each sex can be fed optimally, the females can be slaughtered at lighter weights for whole carcass merchandizing and males can be fed to heavier weights and marketed as broiler parts. Dietary protein concentration can be increased by up to 20% to reduce fatness, but diet costs will increase. Alternate-day feeding has been shown to reduce fatness without a reduction in weight. However, feeding space will be needed for all broilers to eat at the same time. Feeding β-agonists to poultry would be expected to improve feed conversion efficiency slightly, but significantly reduce carcass fatness significantly and increase carcass protein slightly. Negative effects on meat quality and tenderness and their safety raise questions about their potential.

References

Aalhus, J.L., Jones, S.D.M., Schaefer, A.L. *et al.* (1990) The effect of ractopamine on performance, carcass composition and meat quality of finishing pigs. *Can. J. Anim. Sci.*, **70**, 943.

Andrews, R.P. and Orskov, E.R. (1970) The nutrition of the early weaned lamb. II. The effect of dietary protein concentration, feeding level and sex on body composition at two liveweights. *J. Agric. Sci. (Camb.)*, **75**, 19.

Arnold, A.M. and Meyer, H.H. (1988) Effects of gender, time of castration, genotype and feeding regimen on lamb growth and carcass fatness. *J. Anim. Sci.*, **66**, 2468.

Bass, J.J., Butler-Hogg, B.W. and Kirton, A.H. (1990) Practical methods of controlling fatness in farm animals, in *Reducing Fat in Meat Animals* (eds J.D. Wood and A.V. Fisher), Elsevier Applied Science, London, pp. 145–200.

Beermann, D.H. (1993) Beta-adrenergic agonists and growth, in *The Endocrinology of Growth, Development, and Metabolism in Vertebrates* (eds M.P. Schreeibman, C.G. Scanes and P.K.T. Pang), Academic Press, San Diego, pp. 345–366.

Beermann, D.H. (1994) Carcass composition of animals given partitioning agents, in *Low-Fat Meats: Design Strategies and Human Implications* (eds H.D. Hafs and R.G. Zimbelman), Academic Press, San Diego, pp. 203–232.

Beermann, D.H., Hogue, D.E., Fishell, V.K. *et al.* (1990) Exogenous human growth hormone releasing factor and ovine somatotropin improve growth performance and composition of gain in lambs. *J. Anim. Sci.*, **68**, 4122.

Beermann, D.H., Robinson, I.F., Byrem, I.M. *et al.* (1991) Abomasal casein infusion and exogenous somatotropin enhance nitrogen utilization by growing lambs. *J. Nutr.*, **121**, 2020.

Beermann, D.H., Robinson, T.F. and Hogue, D.E. (1995) Impact of composition manipulation on lean lamb production in the United States. *J. Anim. Sci*, **73**, 2493.

Black, J.L. (1974) Manipulation of body composition through nutrition. *Proc. Aust. Soc. Anim. Prod.*, **10**, 211.

Bouffault, J.C. and Willemart, J.P. (1983) Anabolic activity of trenbolone acetate alone or in association with oestrogens, in *Anabolics in Animal Production* (ed. E. Meissonier), Office International des Epizooties, Paris, pp. 155–179.

Butler-Hogg, B.W., Francombe, M.A. and Dransfield, E. (1984) Carcass and meat quality of ram and ewe lambs. *Anim. Prod.*, **39**, 107.

Buttery, P.J. and Dawson, J.M. (1987) The mode of action of beta-agonists as manipulators of carcass composition, in *Beta-Agonists and Their Effects on Animal Growth and Carcass Quality* (ed. J.P. Hanrahan), Elsevier Applied Science, London, pp. 29–43.

Buyse, J., Decuypere, E., Huyghebaert, G. and Herremans, M. (1991) The effect of clenbuterol supplementation on growth performance and plasma hormone and metabolite levels of broilers. *Poult. Sci.*, **70**, 993.

Campbell, R.G. and Dunkin, A.C. (1983) The influence of nutrition in early life on growth and development of pigs. *Anim. Prod.*, **36**, 415.

Campbell, R.G. and King, R.H. (1982) The influence of dietary protein and level of feeding on the growth performance and carcass characteristics of entire and castrated male pigs. *Anim. Prod.*, **35**, 177.

Castaldo, D.J., Jones, J.E. and Maurice, D.V. (1990) Growth and carcass composition of female turkeys implanted with anabolic agents and fed high-protein and low-protein diets. *Arch. Anim. Nutr.*, **40**, 703.

Cisneros, F., Ellis, M., McKeith, F.K. *et al.* (1996) Influence of slaughter weight on growth and carcass characteristics, commercial cutting and curing yields, and meat quality of barrows and gilts from two genotypes. *J. Anim. Sci.*, **74**, 925.

Claus, R., Weiler, U. and Herzog, A. (1994) Physiological aspects of androsterone and skatole formation in the boar – a review with experimental data. *Meat Sci.*, **38**, 289.

Coleman, M.E., Ekeren, P.A. and Smith, S.B. (1988) Lipid synthesis and adipocyte growth in adipose tissue from sheep chronically fed a beta-adrenergic agent. *J. Anim. Sci.*, **66**, 372.

Collins, S.S., Belk, K.E., Cross, H.R. and Smith, G.C. (1989) The EEC ban against growth-promoting hormones. *Nutr. Rev.*, **47**, 238.

Crome, P.K., McKeith, F.K., Carr, T.R. *et al.* (1996) Effect of ractopamine on growth performance, carcass composition, and cutting yields of pigs slaughtered at 107 and 125 kilograms. *J. Anim. Sci.*, **74**, 709.

Dalke, B.S., Roeder, R.A., Kasser, T.R. *et al.* (1992) Dose–response effects of recombinant bovine somatotropin implants on feedlot performance in steers. *J. Anim. Sci.*, **70**, 2130.

Dalrymple, R.H., Baker, P.K., Gingher, P.E. *et al.* (1984) A repartitioning agent to improve performance and carcass composition of broilers. *Poult. Sci.*, **63**, 2376.

David, H. (1989) The hormone case in North Rhine Westphalia. *Fleischwirtschaft*, **69**, 1802.
de la Torre, F., Oltjen, J.W. and Suing, R.D. (1993) Effects of limit feeding and ration energy level on feed intake, efficiency and carcass traits in feedlot steers. *J. Anim. Sci.*, **71** (Suppl. 1), 259.
Dikeman, M.E. (1991) Growth, carcass characteristics and meat quality, in *37th International Congress of Meat Science and Technology*, Kulmbach, Germany, Vol. 1, p.1.
Dikeman, M.E., Cross, H.R., Crouse, J.D. *et al.* (1985) Recommendations for the production of young bulls for meat. *Kan. Agric. Exp. Sta. Bull.*, 648.
Dransfield, E., Nute, G.R. and Francombe, M.A. (1984) Comparison of eating quality of bull and steer beef. *Anim. Prod.*, **39**, 37.
Evock, C.M., Etherton, T.D., Chung, C.S. and Ivy, R.E. (1988) Pituitary porcine growth hormone (ppGH) and a recombinant pGH analog stimulate pig growth performance in a similar manner. *J. Anim. Sci.*, **66**, 1928.
Ferket, P.R. and Sell, J.L. (1990) Effects of early protein and energy restriction of large turkey toms fed high-fat or low-fat realimentation diets. 2. Carcass characteristics. *Poult. Sci.*, **69**, 1982.
Fisher, A.V., Wood, J.D. and Whelehan, O.P. (1986) The effects of a combined androgenic–oestrogenic anabolic agent in steers and bulls. 1. Growth and carcass composition. *Anim. Prod.*, **42**, 203.
Fisher, C. (1984) Fat deposition in broilers, in *Fats in Animal Nutrition* (ed. J. Wiseman), Butterworths, London, pp. 437–470.
Fortin, A., Reid, J.T., Maiga, A.M. *et al.* (1980) Effect of energy intake level and influence of beef and sex on the physical composition of the carcass of cattle. *J. Anim. Sci.*, **51**, 331.
Fuller, M.F. (1985) Sex differences in the nutrition and growth of pigs, in *Recent Developments in Pig Nutrition* (eds J.A. Cole and W. Haresign), Butterworths, London, pp. 177–189.
Galbraith, H., Hatendi, P.R., Alderson, E.M. and Scaife, J.R. (1990) The effect of cimaterol and oestradiol-17β alone or combined on growth and body composition of wether lambs. *Anim. Prod.*, **51**, 311.
Garber, M.J., DeYonge, K.G., Byatt, J.C. *et al.* (1995) Dose–response effects of recombinant bovine somatotropin (Posilac™) on growth performance and body composition of two-year-old rainbow trout (*Oncorhynchus mykiss*). *J. Anim. Sci.*, **73**, 3216.
Glimp, H.A., Hart, S.P. and VonTungeln, D. (1989) Effect of altering nutrient density (concentrate to roughage ratio) and restricting energy intake on rate, efficiency and composition of growing lambs. *J. Anim. Sci.*, **67**, 865.
Gray, J.I., Pearson, A.M. and Monahan, F.J. (1980) Flavor and aroma problems and their measurement in meat, poultry and fish products. *Adv. Meat Res.*, **9**, 250.
Gregory, K.E., Cundiff, L.V., Koch, R.M. *et al.* (1994) Breed effects and retained heterosis for growth, carcass, and meat traits in advanced generations of composite populations of beef cattle. *J. Anim. Sci.*, **72**, 833.
Gu, Y., Schinckel, A.P. and Martin, T.G. (1992) Growth, development, and carcass composition in five genotypes of swine. *J. Anim. Sci.*, **70**, 1719.
Guyer, C.G. and Miller, M.A. (1994) Human food safety evaluation of repartitioning agents, in *Low-Fat Meats: Design Strategies and Human Implications* (eds H.D. Hafs and R.G. Zimbelman), Academic Press, San Diego, pp. 53–63.
Gwartney, B.L., Calkins, C.R. and Jones, S.J. (1991) The effect of cimaterol and its withdrawal on carcass composition and meat tenderness of broiler chickens. *J. Anim. Sci.*, **69** 1551.
Hacker, R.R., Deschutter, A., Adeola, O. and Kasser, T.R. (1993) Evaluation of long-term somatotropin implants in finishing pigs. *J. Anim. Sci.*, **71**, 564.
Halvorson, J.C., Waibel, P.E., Oju, E.M. *et al.* (1991) Effect of diet and population density on male turkeys under various environmental conditions. 2. Body composition and meat yield. *Poult. Sci.*, **70**, 935.
Hancock, D.L., Wagner, J.F. and Anderson, D.B. (1991) Effects of estrogens and androgens on animal growth. *Adv. Meat Res.*, **7**, 255.
Hansen, B.C. and Lewis, A.J. (1993) Effects of dietary protein concentration (corn:soybean meal ratio) on the performance and carcass characteristics of growing boars, barrows and gilts: mathematical description. *J. Anim. Sci.*, **71**, 2122.

Hansen, J.A., Nelssen, J.L., Goodband, R.D. and Laurin, J.L. (1994) Interactive effects among porcine somatotropin, the beta-adrenergic agonist salbutamol, and dietary lysine on growth performance and nitrogen balance of finishing swine. *J. Anim. Sci.*, **72**, 1540.

Hansson, K.E., Lundstrom, K., Fjelkner-Modig, S. and Persson, J. (1980) The importance of androstenone and skatole for boar taint. *Swed. J. Agric. Res.*, **10**, 167.

Hayden, J.M., Bergen, W.G. and Merkel, R.A. (1992) Skeletal muscle protein metabolism and serum growth hormone, insulin, and cortisol concentrations in growing steers implanted with estradiol-17β, trenbolone acetate, or estradiol-17β plus trenbolone acetate. *J. Anim. Sci.*, **70**, 2109.

Haydon, K.D., Tanksley, T.D., Jr and Knabe, D.A. (1989) Performance and carcass composition of limit-fed growing-finishing swine. *J. Anim. Sci.*, **67**, 1916.

Hays, V.W. and Preston, R.L. (1994) Nutrition and feeding management to alter carcass composition of pigs and cattle, in *Low-Fat Meats: Design Strategies and Human Implications* (eds H.D. Hafs and R.G. Zimbelman), Academic Press, San Diego, pp. 13–33.

Holsheimer, J.P. (1975) The effect of changing energy–protein ratios on carcase composition of broilers, in *The Quality of Poultry Meat: Proc. 2nd European Symposium on Poultry and Poultry Meat Quality* (ed. B. Erdtsieck), Spelderholt Institute for Poultry Research, Beekbergen, p.1.

Jackson, S., Summers, J.D. and Leeson, S. (1982) Effect of dietary protein and energy on broiler carcass composition and efficiency of nutrient utilization. *Poult. Sci.*, **61**, 2224.

Jagusch, K.T. and Rattray, P.V. (1979) Carcass production for the consumer. Nutritional manipulation of carcass composition of lambs grown in New Zealand. *Proc. Lincoln Coll. Farmers Conf.*, 20.

Johnsson, I.D., Hathorn, D.J., Wilde, R.M. *et al.* (1987) The effects of dose and method of administration of biosynthetic bovine somatotropin on live weight gain, carcass composition and wool growth in young lambs. *Anim. Prod.*, **44**, 405.

Johnston, M.E., Nelssen, J.L., Goodband, R.D. *et al.* (1993) The effects of porcine somatotropin and dietary lysine on growth performance and carcass characteristics of finishing swine fed to 105 or 127 kilograms. *J. Anim. Sci.*, **71**, 2986.

Jones, R.L. and Wiseman, J. (1985) Effect of nutrition on broiler carcase composition: Influence of dietary energy content in the starter and finisher phases. *Br. Poult. Sci.*, **26**, 381.

Jones, S.D.M., Burgess, T.D. and Dupchak, K. (1983) Effects of dietary energy intake and sex on carcass tissue and offal growth in sheep. *Can. J. Anim. Sci.*, **63**, 303.

Keane, M.G. and Drennan, M.J. (1987) Lifetime growth and carcass composition of heifers and steers non-implanted or sequentially implanted with anabolic agents. *Anim. Prod.*, **45**, 359.

Keele, J.W., Ferrell, C.L., Arnold, R.N. *et al.* (1993) Influence of controlled energy intake on body composition of steers. *Beef Res. Prog. Rep.*, **4** (ARS-71), 151.

King, B.D., Bergen, R.D., McKinnon, J.J. *et al.* (1992) Effect of zeranol on fat deposition and carcass traits in beef heifers. *Can. J. Anim. Sci.*, **72**, 965.

Kirton, A.H., Sinclair, D.P., Chrystall, B.B. *et al.* (1981) Effect of plane of nutrition on carcass composition and the palatability of pasture-fed lamb. *J. Anim. Sci.*, **52**, 285.

Klindt, J., Buonomo, F.C. and Yen, J.T. (1992) Administration of porcine somatotropin by sustained-release implant: Growth and endocrine responses in genetically lean and obese barrows and gilts. *J. Anim. Sci.*, **70**, 3721.

Klindt, J., Buonomo, F.C. and Yen, J.T. (1995) Administration of porcine somatotropin by sustained-release implant: Growth, carcass, and sensory responses in crossbred white and genetically lean and obese boars and gilts. *J. Anim. Sci.*, **73**, 1327.

Knight, C.D., Kasser, T.R., Swenson, G.H. *et al.* (1991) The performance and carcass composition responses of finishing swine to a range of porcine somatotropin doses in a 1-week delivery system. *J. Anim. Sci.*, **69**, 4678.

Koohmaraie, M., Shackelford, S.D., Muggli-Cockett, N.E. and Stone, R.T. (1991) Effect of the β-adrenergic agonist L-644,969 on muscle growth, endogenous proteinase activities, and postmortem proteolysis in wether lambs. *J. Anim. Sci.*, **69**, 4823.

Koohmaraie, M., Shackelford, S.D. and Wheeler, T.L. (1996) Effects of β-adrenergic agonist (L-644,969) and male sex condition on muscle growth and meat quality of callipyge lambs. *J. Anim. Sci.*, **74**, 70.

Kretchmar, D.H., Hathaway, M.R., Epley, R.J. and Dayton, W.R. (1990) Alterations in postmortem degradation of myofibrillar proteins in muscle of lambs fed a β-adrenergic agonist. *J. Anim. Sci.*, **68**, 1760.

Lawrence, T.J.L. (1977) The effect of dietary nutrient density on growth in the pig. *Anim. Prod.*, **25**, 261.

Leenstra, F.R. (1986) Effect of age, sex, genotype and environment on fat deposition in broiler chicken – a review. *World Poult. Sci. J.*, **42** (1), 12.

Leymaster, K.A. and Mersmann, H.J. (1991) Effect of limited feed intake on growth of subcutaneous adipose tissue layers and on carcass composition in swine. *J. Anim. Sci.*, **69**, 2837.

Lord, E.A., Fennessy, P.F. and Littlejohn, R.P. (1988) Comparison of genotype and nutritional effects on body and carcass characteristics of lambs. *N. Z. J. Agric. Res.*, **31**, 13.

Lundstrom, R., Hansson, K.E., Fjelkner-Modig, S. and Persson, J. (1980) Skatole – another contributor to boar taint. *Proc. Eur. Mtg. Meat Res. Workers*, **26**, 300.

Malmfors, B. and Lundstrom, K. (1983). Consumer reactions to boar meat – a review. *Livestock Prod. Sci.*, **10**, 187.

Malucelli, A., Ellendorff, F. and Meyer, H.H.D. (1994) Tissue distribution and residues of clenbuterol, salbutamol, and terbutaline in tissues of treated broiler chickens. *J. Anim. Sci.*, **72**, 1555.

Maurice, D.V., Jones, J.E., Whisenhunt, J.E. and Castaldo, D.J. (1985) Response of turkeys to the anabolic agent trenbolone acetate. *Nutr. Rep. Int.*, **31**, 59.

McKeith, F.K., Lan, Y.H. and Beermann, D.H. (1994) Sensory characteristics of meat from animals given partitioning agents, in *Low-Fat Meats: Design Strategies and Human Implications* (eds H.D. Hafs and R.G. Zimbelman), Academic Press, San Diego, pp. 233–252.

McLaughlin, C.L., Hedrick, H.B., Veenhuizen, J.J. *et al.* (1994) Performance, clinical chemistry, and carcass responses of finishing lambs to formulated sometribove (methionyl bovine somatotropin). *J. Anim. Sci.*, **72**, 2544.

McNamara, J.P., Brekke, C.J., Jones, R.W. and Dalrymple, R.H. (1991) Recombinant porcine somatotropin alters performance and carcass characteristics of heavyweight swine fed alternative feedstuffs. *J. Anim. Sci.*, **69**, 2273.

Mohan, B., Narahari, D., Venkatesan, E.S. *et al.* (1987) Studies on the influence of age and sex on the meat characteristics of broiler chicken. *Cheiron*, **16**(3), 119.

Moloney, A. (1988) Application of beta agonists in ruminants explored. *Feedstuffs*, **11**, July, 13.

Moloney, A., Allen, P., Joseph, R. and Tarrant, V. (1991) Influence of beta-adrenergic agonists and similar compounds on growth. *Adv. Meat Res.*, **7**, 455.

Morgan, J.B., Jones, S.J. and Calkins, C.R. (1989) Muscle protein turnover and tenderness in broiler chickens fed cimaterol. *J. Anim. Sci.*, **67**, 2646.

Mortensen, A.B., Bejerholm, C. and Pedersen, J.K. (1986) Consumer test of meat from entire males, in relation to skatole in backfat. *Proc. 32nd Eur. Mtg. Meat Res. Workers*,**1**, 23.

Moseley, W.M., Paulissen, J.B., Goodwin, M.C. *et al.* (1992) Recombinant bovine somatotropin improves growth performance in finishing beef steers. *J. Anim. Sci.*, **70**, 412.

Mottram, D.S., Wood, J.D. and Patterson, R.L.S. (1982) Comparison of boars and castrates for bacon production. 3. Composition and eating quality of bacon. *Anim. Prod.*, **35**, 75.

Murphy, T.A., Loerch, S.C., McClure, K.E. and Solomon, M.B. (1994) Effects of restricted feeding on growth performance and carcass composition of lambs. *J. Anim. Sci.*, **72**, 3131.

Murray, D.M., Tulloh, N.M. and Winter, W.H. (1974) Effect of three different growth rates on empty body weight, carcass weight and dissected carcass composition of cattle. *J. Agric. Sci. (Camb.)*, **82**, 535.

Orskov, E.R., McDonald, I., Fraser, C. and Corse, E.L. (1971) The nutrition of the early weaned lamb. III. The effect of *ad libitum* intake of diets varying in protein concentration on performance and on body composition at different liveweights. *J. Agric. Sci. (Camb.)*, **77**, 351.

O'Sullivan, N.P., Dunnington, E.A. and Siegal, P.B. (1991) Growth and carcass characteristics of early- and late-feathering broilers reared under different feeding regimens. *Poult. Sci.*, **70**, 1323.

Owens, F.N. Zorilla-Rios, J. and Dubeski, P. (1991) Effects of ionophores on metabolism, growth, body composition and meat quality. *Adv. Meat Res.*, **7**, 321.

Patterson, R.L.S. (1968) 5α-Androst-16-ene-3-one: compound responsible for taint in boar meat. *J. Sci. Food Agric.*, **19**, 31.

Pearson, A.M., Dickson, R.L., Hill, T.W. and Holtan, D.W. (1995) Boar taint: androstenone or skatole? A review. *Proc. Int. Congr. Meat Sci. Technol.*, **41**(II), 78.

Perreault, N. and Leeson, S. (1992) Age-related carcass composition changes in male broiler chickens. *Can. J. Anim. Sci.*, **72**, 919.

Perry, T.C., Fox, D.G. and Beermann, D.H. (1991) Effect of an implant of trenbolone acetate and estradiol on growth, feed efficiency, and carcass composition of Holstein and beef steers. *J. Anim. Sci.*, **59**, 4696.

Preston, R.L., Bartle, S.J., Kasser, T.R. *et al.* (1995) Comparative effectiveness of somatotropin and anabolic steroids in feedlot steers. *J. Anim. Sci.*, **73**, 1038.

Pringle, T.D., Calkins, C.R., Koohmaraie, M. and Jones, S.J. (1993) Effects over time of feeding a β-adrenergic agonist to wether lambs on animal performance, muscle growth, endogenous muscle proteinase activities, and meat tenderness. *J. Anim. Sci.*, **71**, 636.

Purchas, R.W. (1991) Effect of sex and castration on growth and composition. *Adv. Meat Res.*, **7**, 203.

Rickes, E.L., Olson, G., Duquette, P.F. and Convey, E.M. (1987) Effect of β-adrenergic agonist L-644,969 on growth and carcass composition of broiler chickens. *Poult. Sci.*, **66** (Suppl. 1), 166.

Robinson, E.H. and Robinette, H.R. (1993) Effects of dietary protein level and feeding regimen on growth and on fattiness of channel catfish, *Ictalurus punctatus*. *J. Appl. Aquacult.*, **3**(1/2), 67.

Roche, J.F. and Quirke, J.F. (1986) The effects of steroid hormones and xenobiotics on growth of farm animals, in *Control and Manipulation of Animal Growth* (eds P.J. Buttery, D.B. Lindsay and N.B. Haynes), Butterworths, London, p. 39.

Samber, J.A., Tatum, J.D., Morgan, J.B. *et al.* (1996) Implant program effects on performance and carcass quality of steer calves fed 212 days. *J. Anim. Sci.*, **75**, 1470.

Seideman, S.C., Cross, H.R., Oltjen, R.R. and Schanbacher, B.D. (1982) Utilization of the intact male for red meat production: a review. *J. Anim. Sci.*, **55**, 826.

Sell, J.L., Ferket, P.R., Angel, C.R. *et al.* (1989) Performance and carcass characteristics of turkey toms as influenced by dietary protein and metabolizable energy. *Nutr. Rep. Int.*, **40**, 979.

Sibbald, I.R. and Wolynetz, M.S. (1986) Effects of dietary lysine and feed intake on energy utilization and tissue synthesis by broiler chicks. *Poult. Sci.*, **65**, 98.

Snowder, G.D., Glimp, H.A. and Field, R.A. (1994) Carcass characteristics and optimal slaughter weights in four breeds of sheep. *J. Anim. Sci.*, **72**, 932.

Soeparno, A.B. and Lloyd Davies, H. (1982) The effect of dietary energy concentration of growth and carcass in Daldale wether sheep. *Proc. Aust. Soc. Anim. Prod.*, **14**, 503.

Steele, N.C., Boyd, R.D. and Campbell, R.G. (1994) Growth, metabolic modifiers, and nutrient considerations, in *Low-Fat Meats: Design Strategies and Human Implications* (eds H.D. Hafs and R.G. Zimbelman), Academic Press, San Diego, pp. 167–189.

Steen, R.W.J. and Kilpatrick, D.J. (1995) Effects of plane of nutrition and slaughter weight on the carcass composition of serially slaughtered bulls, steers and heifers of three breed crosses. *Livestock Prod. Sci.*, **43**, 205.

Stites, C.R., McKeith, F.K., Singh, S.D. *et al.* (1991) The effect of ractopamine hydrochloride on the carcass cutting yields of finishing swine. *J. Anim. Sci.*, **69**, 3094.

Sully, R.J. and Morgan, J.H.L. (1978) The influence of growth rate and diet on beef carcasses. *Proc. Aust. Soc. Anim. Prod.*, **12**, 234.

Summers, J.D., Spratt, D. and Atkinson, J.L. (1992) Broiler weight gain and carcass composition when fed diets varying in amino acid balance, dietary energy, and protein level. *Poult. Sci.*, **71**, 263.

Summers, J.D., Leeson, S., Bedford, M. and Spratt, D. (1985) Influence of dietary protein and energy on performance and carcass composition of heavy turkeys. *Poult, Sci.*, **64**, 1921.

Takahashi, K., Akiba, Y. and Horiguchi, M. (1993) Effects of beta-adrenergic agonist (clenbuterol) on performance, carcase composition, hepatic microsomal mixed function oxidase and antibody production in female broilers treated with or without corticosterone. *Br. Poult. Sci.*, **34**, 167.

Tarrant, P.V. (1981) The occurrence, causes and economic consequences of dark-cutting in beef – a survey of current information, in *The Problem of Dark-Cutting in Beef. Current Topics of Veterinary Medicine and Animal Science* (eds D.E. Hood and P.V. Tarrant), Vol. 10, Martinus Nijhoff, The Hague, pp. 3–36.

Tatum, J.D. (1981) Is tenderness nutritionally controlled? *Proc. Recip. Meat Conf.*, **34**, 65.

Thornton, R.F. and Tume, R.K. (1988) Manipulation of growth in domestic animals. *Proc. Int. Congr. Meat Sci. Technol*, **2**, 6.

Toelle, V.D., Tess, M.W., Johnson, T. and Anderson, B.B. (1986) Lean and fat growth patterns of serially slaughtered beef bulls fed different energy levels. *J. Anim. Sci.*, **63**, 1347.

Uttaro, B.E., Ball, R.O., Dick, P. *et al.* (1993) Effect of ractopamine and sex on growth, carcass characteristics, processing yield, and meat quality characteristics of crossbred swine. *J. Anim. Sci.*, **71**, 2439.

Van der Wal, P. (1990) Wide world of new technologies, in *Proceedings of Pork Strategies 2000 Conference., Des Moines, Iowa*, p. 9.

Vestergaad, M., Sejrsen, K. and Klastrup, S. (1994) Growth, composition and eating quality of longissimus dorsi from young bulls fed the β-agonist cimaterol at consecutive developmental stages. *Meat Sci.*, **38**, 55.

Watkins, L.E., Jones, D.J., Mowrey, D.H. *et al.* (1990) The effect of various levels of ractopamine hydrochloride on the performance and carcass characteristics of finishing swine. *J. Anim. Sci.*, **68**, 3588.

Wheeler, T.L., Cundiff, L.V., Koch, R.M. and Crouse, J.D. (1996) Characterization of biological types of cattle (Cycle IV): carcass traits and longissimus palatability. *J. Anim. Sci.*, **74**, 1023.

Williams, C.B., Bennett, G.L. and Keele, J.W. (1995) Simulated influence of postweaning production system on performance of different biological types of cattle: II. Carcass composition, retail product, and quality, *J. Anim. Sci.*, **73**, 674.

Williams, N.H., Cline, T.R., Schinckel, A.P. and Jones, D.J. (1994) The impact of ractopamine, energy intake, and dietary fat on finisher pig growth performance and carcass merit. *J. Anim. Sci.*, **72**, 3152.

Wood, J.D., Jones, R.C.D., Francombe, M.A. and Whelehan, O.P. (1986) The effects of fat thickness and sex on pig meat quality with special reference to the problems associated with overleanness. 2. Laboratory and trained taste panel results. *Anim. Prod.*, **43**, 535.

Yen, H.T., Cole, D.J.A. and Lewis, D.J. (1986) Amino acid requirements of growing pigs: 8. The response of pigs from 50 to 90 kg live weight to dietary ideal protein. *Anim. Prod.*, **43**, 155.

9 Mimetic and synthetic fat replacers for the meat industry

P.J. SHAND

9.1 Introduction

Continuing consumer interest in reduced fat foods has created a growing need for reduced fat and low fat meat products in the marketplace. Almost half of the new red meat product introductions in the USA in 1995 contained a reduced fat claim (Krizner, 1995). Developing lean or extra lean products, while assuring the necessary palatability demanded by consumers, is not as simple as just removing fat (Troutt *et al.*, 1992). The active approach to fat replacement is to add fat mimetic ingredients, such as carbohydrates or proteins, either to replace fat or to modify the interactions of the remaining compounds. Numerous nonmeat ingredients are available today as texture-modifying and water binding agents. While not always marketed as such, they act as fat replacers, modifying texture and mouth-feel and binding water to retain juiciness.

There have been a number of excellent general reviews on fat replacers for the food industry (Setser and Racette, 1992; Lucca and Tepper, 1994; Jones, 1995; Calorie Control Council, 1996). This chapter provides a general overview of fat replacers that are being used or are proposed for the meat industry. Some of the ingredients described are approved for use in meat systems, while others are pending approval or are still in the experimental stage. Specific applications of fat replacers in meat systems have been reviewed by Keeton (1994) and are also discussed in Chapters 4, 10 and 12 of this volume.

9.2 Functional properties of fat

Successful fat reduction in meat and/or poultry products requires a good understanding of the functions of fat in meat products and how those functions can be replicated with nonfat ingredients. A successful fat replacement product must replicate the texture, mouth-feel and flavor of the original fat (Drewnowski, 1992). The functional and organoleptic properties of fat as they relate to specific types of meat products are complex and somewhat product dependent. From a nutritional point of view, fat contributes 9 $kcal\,g^{-1}$ of energy and supplies essential fatty acids and fat soluble vitamins to the diet.

9.2.1 *Effects of fat in meat products*

Key properties of meat products that are influenced by fat include appearance, mouth-feel, texture, juiciness, flavor and storage stability. Appearance is the first thing that will influence a consumer's quality perception. Fat gives products a certain sheen, provides opacity and contributes to browning. The mouth-feel provided by fat is difficult to describe. Fatty or oily mouth-feel is considered to be a combination of several basic parameters, including viscosity (thickness, body, fullness), lubricity (creaminess, smoothness), absorption/adsorption and others, which are not yet clearly defined, such as cohesiveness and mouth-coating (Glicksman, 1991). Fat affects the texture, tenderness and bite of meat products. It also contributes flavor and serves as a carrier for fat soluble flavors. In meat products, juiciness and moisture release are often affected by changes in the fat level.

9.2.2 *Functional properties of fat*

Fat provides meat products with many functional properties during processing which will change when the fat level is altered. For example, fat is responsible for heat distribution during cooking. Fat also serves to dilute protein content and delay protein extraction by physically separating muscle proteins, therefore allowing more tolerance in mixing procedures.

Fats may modify the time–intensity profile of flavors. Oil coating of the tongue tends to slow or reduce flavor perception. On replacing fat with water, the increased moisture levels within a food system can create an unbalanced flavor as the perceptions of saltiness, acidity, sweetness and bitterness are often accentuated (Yackel and Cox, 1992).

Keeton (1994) indicated that processed meat products have the greatest opportunity for fat reduction or modification through the use of fat replacement technologies. However, the eating quality of comminuted meats is highly dependent on the fat content, because fat contributes flavor, texture and juiciness to such products. With marked fat reduction, these products can be hard, rubbery or mealy (Keeton, 1994). The meat industry has found that reducing fat on inherently lean cuts of meat (such as ham) has been easier than removing fat from sausages, which start out with a much higher fat content (Vafiadis, 1996).

9.3 Fat replacer definitions and classification

9.3.1 Fat replacers

The terms fat replacers, fat substitutes and fat mimetics are usually used interchangeably. However, some authors give different descriptions for each. Fat replacer is a general term, which encompasses fat mimetics and fat substitutes (Jones, 1995). Fat replacers are ingredients that contribute fewer or no calories to formulated foods without altering flavor, mouth-feel, viscosity or other organoleptic properties (Keeton, 1994). Most fat replacers are used for partial rather than total replacement of fat.

9.3.2 Fat mimetics

Fat mimetics are defined as a partial replacement for fat by mimicking or imitating a particular function, but not all functions of fat in a food (Calorie Control Council, 1996). Fat mimetics are compounds that help replace the mouth-feel of fats but cannot substitute for fat on an equivalent weight basis. Fat mimetics can be further classified as protein based or carbohydrate based. Generally, fat mimetics are not useful as a frying medium. Because they mimic fat by entrapping water, protein and carbohydrate mimetics are best used when the application has sufficient moisture.

9.3.3 Fat substitutes and analogs

Fat substitutes and analogs are molecules whose physical and thermal properties resemble fat but have either fewer calories than fat or no calories. Fat analogs or synthetic fat substitutes can replace fat in a food by direct substitution and many can be used for frying.

9.3.4 Fat barrier compounds

While not technically fat replacers, fat barriers are ingredient systems which reduce the fat content of a food by retarding fat absorption into the product during frying (Calorie Control Council, 1996).

9.3.5 Strategies for fat reduction

Strategies for fat reduction or replacement in meat products include substitution of leaner meats, replacement of fat with added water, addition of carbohydrate or protein based fat mimetics and use of other fats or synthetic fat substitutes (Keeton, 1994). Substitution of leaner meats is discussed in Chapter 7. Water replacement alone makes low fat meat products too soft and may result in additional shrink during cooking and excess accumulation of purge in packages. The most popular fat replacer strategy today is the binding of added water with a carbohydrate or protein based fat replacer.

9.3.6 *Selection of fat replacers*

Some factors to consider when selecting a fat replacer include ingredient composition and economics, solubility in water and salt solutions, hydration properties, solution viscosity, gelation conditions, gel characteristics, such as texture, setting temperature, gel strength, incompatibilities, caloric contribution, degree of heating needed for optimum functionality, effects on flavor and texture, and labeling requirements. Processing procedures may also be different than that traditionally used by the meat industry. Some of the fat mimetics should be dry blended with other ingredients, others may need to be prehydrated or added as a prepared 'creme'. Order of addition to meat systems may be important.

9.3.7 *Mechanism of action of fat replacers*

We still do not fully understand the specific mechanism of action of each fat replacer. For example, fat mimetics entrap water in different ways. Some mimetics form a rigid, organized matrix of long polymeric molecules into a gel network, some form particle/aggregate gels, while others act as non-interacting particles to break up the meat protein gel matrix (Murray, 1995). The effects of each ingredient type on meat protein gelation and functionality has not been fully explored. Research on mixed or multicomponent gels such as filled gels, complex gels and interpenetrating polymer networks (Oakenfull, 1987; Ziegler and Foegeding, 1990) could aid in our understanding of these systems.

It should be noted that adding other foods, such as rice or cherries to meat products, is considered by USDA to be a meat or poultry extender and not a fat replacer (USDA, 1995).

9.4 Protein-based fat replacers

Protein-based fat replacers are fat mimetics since they cannot fully replace the functionality of fat in foods. Most contribute $4\,\mathrm{kcal\,g^{-1}}$ on a dry weight basis, but when hydrated with 3 or more parts of water would provide about $1\,\mathrm{kcal\,g^{-1}}$ on a wet weight basis.

9.4.1 *Traditional plant and animal proteins*

Traditional use of plant and animal proteins in meat systems has been as fillers, binders and extenders to improve the flavor, texture, appearance, yield and nutritional value of comminuted meats (Keeton, 1994). Soy protein concentrates and isolates, milk proteins and other nonmeat proteins can serve as fat replacers owing to their ability to bind water and form gels. Protein hydrolysates may also be added to low-fat meat formulations to achieve a brothy or meat flavor note (Shand and Schmidt, 1990).

Use of proteins in low fat meat products is an ongoing area of research and the subject of numerous reviews (Van den Hoven, 1987; Mandigo and Eilert, 1993; Keeton, 1994). Chapter 10 also outlines applications of plant and animal proteins in low fat meat systems.

9.4.2 Microparticulated proteins

Microparticulated proteins can mimic certain properties of fat (Singer and Moser, 1993). While the specific procedures and raw materials may vary, the process of microparticulation basically involves the simultaneous application of heat and high shear to proteins, such as whey protein concentrate, or egg albumin and skim milk. The coagulating proteins are reduced to spherical microparticles of 0.1–2 μm in diameter. The round shape and uniform size of the particles allow them to roll easily over one another. Particles smaller than approximately 3 μm are not detectable as individual particles, but are perceived as a creamy fluid with the smoothness and richness typically associated with fat (Singer and Dunn, 1990). Figure 9.1 provides a description of how texture perception may be influenced by the size of particles.

Microparticulated proteins are also thought to coat the mouth and taste buds in a similar way to fat. This coating action not only allows flavors to reach the receptors slowly and gradually, but also helps mask some of the off-flavors that seem more pronounced in low fat, high added water systems (Lucca and Tepper, 1994).

Protein microparticles occur naturally in foods such as milk, egg, grains and legumes (Kalab, 1990), but are not found in sufficient abundance or in the necessary size or shape to function as fat replacers. It has been suggested that microparticles from vegetable proteins, such as soy are too

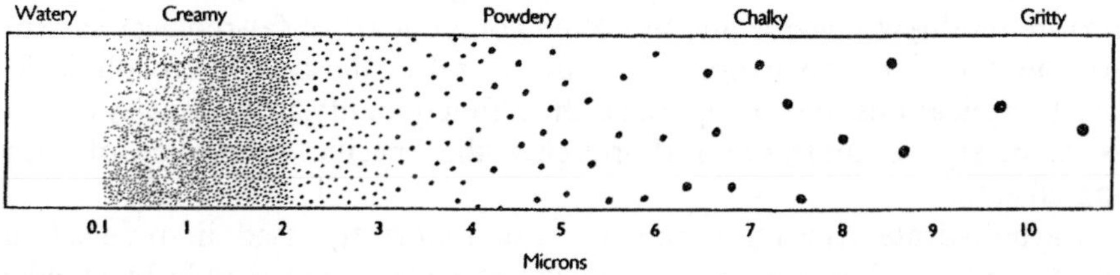

Figure 9.1 Size of protein particles and texture. (Reprinted with permission of The NutraSweet Kelco Company.)

large (up to 80 μm) and casein micelles too small (< 1 μm) to simulate the creamy mouth-feel of fat (Singer and Moser, 1993).

Microparticulated protein fat mimetics are considered to be physically modified proteins and are therefore classified as GRAS by the FDA. Nutritionally, they are digested and absorbed as protein and would contribute 4 cal g^{-1} (dry state).

(a) Examples of protein-based microparticulates

Simplesse from NutraSweet Kelco is by far the most widely known protein-based fat mimetic (Jones, 1995). Originally, Simplesse was sold as a wet ingredient, with about 20–40% solids and a short shelf-life. A dry form is now available (Simplesse Dry 100), as also are several Simplesse-based blends for bakery applications. While some food manufacturers may include the brand name Simplesse on the label, food ingredient statements normally would read 'whey protein (concentrate)'.

Other microparticulated protein fat mimetics include Trailblazer (Kraft General Foods) and Finess (Reach Associates) (Anon., 1990a). Some whey-protein fat substitutes such as Dairylight (Ault Foods) and Dairy-Lo (Cultor Food Science) have been treated with heat (but are not necessarily microparticulated).

Simplesse and other microparticulated proteins cannot be used as a frying medium and cannot be used in some fried foods as excessive heat would cause these ingredients to coagulate and lose their fat-like mouthfeel (Anon., 1990a). However, Simplesse would probably withstand the heating associated with sausage production or normal cooking of meat products, but this has not been verified experimentally. Research on Simplesse in low fat hot dogs shows some promise (Ruby, 1994). It is not known if moisture limitations in some meat applications may impair functionality and sensory mimetic properties of microparticulated proteins. Also, cost considerations may limit their widespread use.

9.5 Carbohydrate-based fat replacers

Starches and polysaccharide gums are very effective water binding and texture modifying agents and are often components of fat-replacer systems for low fat meat applications. In order to select the best hydrocolloid for meat applications and to optimize their functionality in a meat system, it is necessary to understand thoroughly the specific properties of these ingredients.

Carbohydrate based fat mimetics are frequently used in reduced fat meat products. These mimetics achieve fat replacement mainly by stabilizing the added water in a gel-like matrix, resulting in lubricity and moisture release similar to higher fat products (Glicksman, 1991; Yackel and

Cox, 1992). Pre-formed carbohydrate gels such as konjac flour can also mimic particle definition (visible fat) in sausage.

9.5.1 *Starches and derivatives*

Starches are carbohydrate polymers of glucose and usually contain linear, straight-chain amylose and branched amylopectin components. Granule size, shape and size distribution and the amylose to amylopectin ratio of a starch vary with botanical source (potato, corn, wheat, tapioca, rice) (Oakenfull, 1987). In addition to starch source and composition, starch gelatinization temperatures can be influenced by water to starch ratios, lipids, sucrose, salt content and processing conditions (Lund, 1984).

Several modified starches and starch derivatives are being promoted for their fat mimetic properties (Glicksman, 1991). Maltodextrins and dextrins are nonsweet starch hydrolysates that have a dextrose equivalent (DE) of less than 20 (starch DE = 0, dextrose DE = 100). In the process of reducing the molecular weight, the pasting and gelling properties of these ingredients are altered. Figure 9.2 illustrates the properties of starch, maltodextrins and corn syrup solids. Corn syrup solids and maltodextrins act similar to starch, but result in products with a cleaner flavor and mouth-feel (Shand and Schmidt, 1990). Low DE maltodextrins are thought to function more effectively as fat replacers than high DE maltodextrins (Setser and Racette, 1992). When used at high concentrations, dextrins and maltodextrins contribute significantly to mouth-feel, body and viscos-

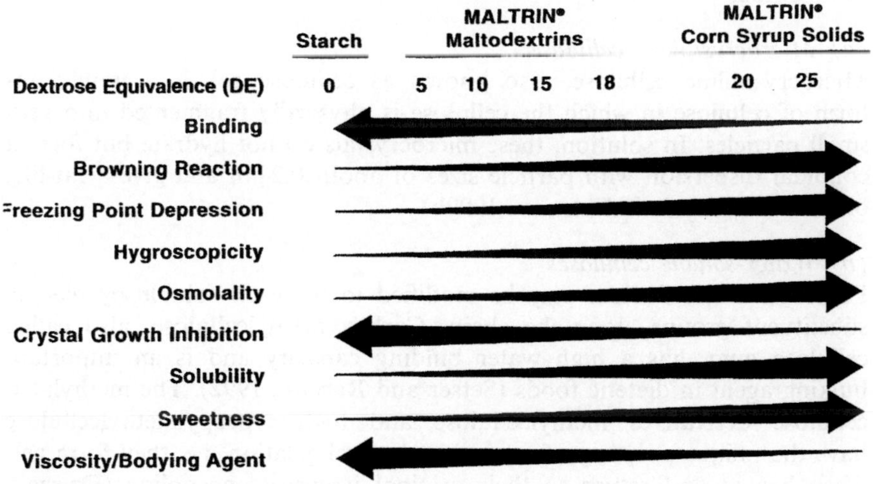

Figure 9.2 Properties of starch, maltodextrins and corn syrup solids. (Reprinted with permission of the Grain Processing Corporation, Muscatine, IA.)

ity (Lucca and Tepper, 1994). Some starches are also modified by physical and chemical means to maintain small, swollen granules similar in size to lipid micelles (approximately 2 μm) to simulate the texture and mouth-feel of fat (Glueck *et al.*, 1994; Lucca and Tepper, 1994).

Modified starches offer the opportunity to tailor a starch to a meat processor's exact requirements. For burger applications, it is important to use a starch that will swell in cold water to allow thickening before any heat is applied. It also needs to have freeze/thaw stability. For ease of handling, some starch blends for low fat meat applications include an instant or pregelatinized starch to increase the viscosity of the raw batter.

Benefits to the use of starches and derivatives are reasonable costs, wide availability and conventional storage and handling procedures. However, depending on the type of starch and the level used, some flavor masking and/or flavor contribution can be expected (Lucca and Tepper, 1994). Proper selection of a modified starch or maltodextrin will help to ensure that it will be fully functional during standard cook cycles for poultry, ham or cooked sausage.

9.5.2 Celluloses and derivatives

Cellulose is found in almost all land plants and is the structural material of the cell wall. Chemically, cellulose differs from starch simply by having β-1,4-linkages rather than α-1,4-linked glucose units (Sanderson, 1981). Cellulose cannot be metabolized by humans and provides insoluble fiber or roughage to the diet. On the other hand, other plant-based materials such as hemicelluloses, β-glucans, pectins and lignins are sources of soluble fiber.

(a) Microcrystalline cellulose

Microcrystalline cellulose, also known as cellulose gel, is a nonfibrous form of cellulose in which the cellulose is physically fragmented into very small particles. In solution, these microcrystals do not hydrate but form a colloidal dispersion with particle sizes of about 0.2 μm and give a fat-like oil in water mouth-feel (Anon., 1990b).

(b) Water-soluble celluloses

For food use, celluloses may be modified to be water-soluble by placing substituent groups along the chain. Carboxymethylcellulose, also called cellulose gum, has a high water binding capacity and is an important bulking agent in dietetic foods (Setser and Racette, 1992). The methylated cellulose derivatives methylcellulose and hydroxypropylmethylcellulose have the unique property of reversible thermal gelation, i.e. they form gels when heated and return to their original viscosity on cooling (Dziezak, 1991). These products, although not true fat mimetics, are useful as fat barriers in fried foods as they may reduce oil absorption into the batter or breading and may also improve adhesion of the batter and coating to the food.

(c) Oatrim

Fat replacement in low fat meats can also be achieved through the use of vegetable fibers, which contain cellulosic and hemicellulosic materials. Oatrim is a cold water dispersible amylodextrin derived from whole oat or debranned whole oat flour and is rich in β-glucans (Setser and Racette, 1992). The amylodextrin and the β-glucan contribute body, stability, creaminess, mouth-feel and the glossy appearance of fat. Inglett at the National Center for Agricultural Utilization Research developed the patented process for producing oatrim and licensed its commercialization to ConAgra, Quaker Oats and Rhone-Poulenc (Haumann, 1993). Oatrim has been used in a variety of food products ranging from muffins to meat products. Under specific processing conditions, thermo-irreversible gels can be produced from oatrim in combination with carrageenan or xanthan and locust bean gum (Jenkins and Wild, 1994). The ground gel can then be used to simulate fat particles in low-fat sausage applications.

(d) Oat bran

Oat bran has also been used in low-fat meat products for textural enhancement (Giese, 1992). Leanmaker is an oat bran-based fat replacer system (which also includes flavorings and seasonings) developed by Webb Technical Group and licensed to Heller Seasonings (Pszczola, 1991). Currently, Leanmaker is being used mainly in ground beef and fresh sausage. Poor product bind, discoloration, mealy texture, off-flavors and shortened microbiological shelf-life can sometimes result with the overuse or misuse of oat bran or oat fiber products in meat patties (Keeton, 1994).

9.5.3 Gums

Gums are hydrocolloids that provide viscosity or thickening and in some cases gel formation. The importance of gums is based upon their ability to immobilize water and contribute viscosity. They generally contribute no calories and also have potential health benefits as dietary fibers (Setser and Racette, 1992). There are a number of gums available to the food industry (Dziezak, 1991). Some have applications as fat replacers as they can provide a slippery, creamy, lubricious mouth-feel simulating the sensory properties of fats and oils (Glicksman, 1991). Gums, such as xanthan, guar and locust bean, are frequently included in reduced fat salad dressing formulations to help stabilize the system.

(a) Seaweed gums

Carrageenan is finding widespread application in low fat meats and further processed poultry (Murphy, 1995). The term carrageenan describes

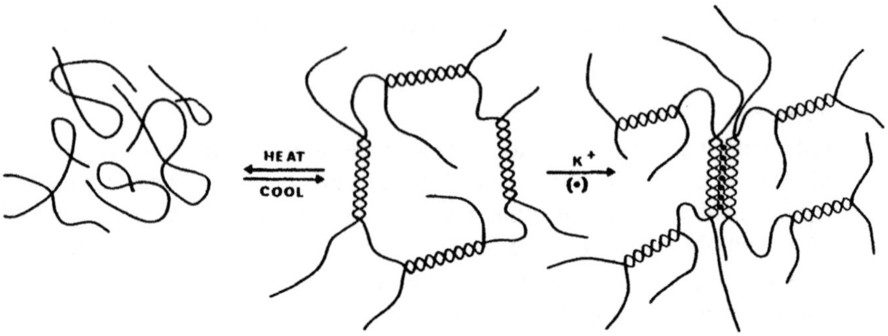

Figure 9.3 The 'domain model' for carrageenan gelation proposed by Rees and co-workers. Potassium or other gel-promoting cations are indicated by • [Reprinted with permission from Oakenfull (1987). Copyright CRC Press, Boca Raton, Florida © 1987.]

a group of sulfated polysaccharides extracted from various red seaweeds (Sanderson, 1981). The three main fractions are kappa, iota and lambda. Kappa- and iota-carrageenans form thermoreversible gels, i.e. gel on cooling and 'remelt' on heating. Lambda-carrageenan does not gel but can serve as a thickener. Iota-carrageenan is preferred for meat patty applications (Egbert *et al.*, 1991), while the kappa type may be more functional in whole muscle items, such as hams, poultry rolls or restructured roasts (Shand, Sofos and Schmidt, 1994). A proposed mechanism for carrageenan gelation is illustrated in Figure 9.3.

Semi-refined or 'high fiber' carrageenan (Philippine Natural Grade or PNG grade) can be used and labeled as carrageenan in all regulated uses. PNG carrageenan is less refined and contains cellulose fiber along with the carrageenan polysaccharides. Murphy (1995) described some of the benefits of PNG carrageenan in meat applications, including cost and opacity. Possible disadvantages include lack of standardization, presence of impurities and less effective gelling ability.

An interesting fat replacer made from carrageenan is called CarraFat (Carrageenan Company, no date). It can be incorporated into various meat products as a dry ingredient or as a 'creme' by shearing with 12 parts of water.

Alginates, extracted from various brown seaweeds, are widely used in the food industry. Alginates are linear polymers composed of D-mannuronic acid and L-guluronic acid. The most useful and unique property of alginates is their ability to form chemical rather than thermally induced gels. Divalent cations are required for gelation and, unlike carrageenan, the gels are not thermo-reversible (Oakenfull, 1987). An alginate based meat binding system has been developed (Means and Schmidt, 1986) which can be used to restructure lean meats into steakettes or other forms.

(b) Plant extracts
Like starch and cellulose, pectin is a structural carbohydrate present in all plants. Pectins are straight chain polymers of D-galacturonic acid and D-galacturonic acid methyl esters (Thestrup, 1993). High (methyl) ester pectins, used mainly for jams and jellies, require sugar and acid conditions to gel. Low methyl pectins can form gels in the presence of calcium ions (Oakenfull, 1987).

Pectin functions as a gelling agent, thickener and in some cases as a fat mimetic. A pectin based fat replacer has recently been introduced. Slendid (Hercules) is a non-amidated low methyl pectin derived from citrus peel, which is being promoted as a fat replacer for a wide range of products, e.g. salad dressings, ice cream, meat and bakery products (Thestrup, 1993).

(c) Seed gums
Guar gum and locust bean gum are plant seed derivatives consisting of straight chains of D-mannose units with side branches of single units of galactose (Sanderson, 1981). They give viscous solutions at low concentrations and are useful as thickening, stabilizing and water binding agents. In meat systems, they would act to soften the texture and provide a smooth mouth-feel. Dziezak (1991) indicated that locust bean gum has been used in specialty meat products, such as bologna and sausage, to facilitate extrusion and stuffing.

(d) Microbial gums
Xanthan gum is a high molecular weight polysaccharide produced in pure culture fermentation of *Xanthonas capestris* (Dziezak, 1991). It is an effective thickening agent at very low concentrations. Xanthan solutions are highly pseudoplastic (i.e. thin rapidly with shear and recover original viscosity when shearing stops), which may account for the desirable flavor-release properties and non-gummy mouth-feel. Xanthan gum was useful as a marinate component for lean pork (Shand and McCurdy, 1995).

Gellan gum, produced by culturing *Pseudomonas elodea*, has a linear backbone consisting of four saccharide units: glucose, glucuronic acid, glucose and rhamnose (Dziezak, 1991). Gellan gum has similar properties to alginates in that gelation is induced chemically, rather than thermally. However, gellan gum needs to be heated to dissolve and requires cations to bring about gelation. Gellan may have potential as a partial replacement for other gums, such as carrageenan, and in applications where clarity is important (Sanderson and Clark, 1983). Gellan will gel at concentrations as low as 0.05%. Laaman and Tye (1991) suggested that the micelle structure of a thermally stable gellan dispersion would make it an ideal fat mimic for meat systems. More research is needed on meat applications for gellan gum and xanthan gum.

9.5.4 *Other polysaccharides*

(a) Konjac flour
Konjac flour, extracted from the konjac root, contains a high molecular weight glucomannan consisting of 3:2 mannose to glucose with β-1,4-linkages (Tye, 1991). Konjac flour can produce meltable or heat stable gels, which can be modified with the addition of starch or various gums. Konjac gels can be used to simulate the sensory properties and appearance of hardened fat and connective tissue in the mouth (Tye, 1991). When ground into fat-like particles, these gels can be used to replace visible fat in meat products such as pepperoni or fresh sausage (Osburn and Keeton, 1994).

(b) Polydextrose
Polydextrose is a low calorie bulking agent that can replace some of the functions of fat or sucrose in food products (Frye and Setser, 1993). It is formed by the random polymerization of glucose with small amounts of sorbitol and citric acid. It provides only $1\,kcal\,g^{-1}$ as it shows poor gastrointestinal absorption and high resistance to microbial degradation in the colon (Haumann, 1986). Owing to a possible laxative effect at high concentrations, usage at more than 15 g per serving must carry specific labeling details. Polydextrose can also provide bulk solids content to reduced fat foods. Troutt *et al.* (1992) found that polydextrose in combination with other fat replacers was beneficial in low fat patties.

9.6 Fat-based fat replacers

9.6.1 *Alternative fats and oils*

Rather than replacing fat in meat products with something else, the future of reduced fat products may lie in modifying fats so that they are healthier (Kevin, 1995). Fatty acid modification of animal fats by genetic (Chapter 7) or dietary (Chapter 8) means has already been described, and could be used to produce more healthful processed meat products. Fatty acid profiles of meat products can also be altered by substituting fat with various vegetable oils (Giese, 1992; Keeton, 1994). However, since many consumers want 'all or none', the use of alternative fats may not have as much consumer appeal as reduction of total fat (and calories).

9.6.2 *Structured triglycerides*

Structured lipids are created by esterification or interesterification of glycerol with various fatty acids. While the current generation of these modified fats is being produced synthetically, genetically engineered crops may supply future needs (Kevin, 1995). Caprocaprylobehenin (caprenin) is a triglyceride made from two short-chain saturated fatty acids, capylic (C_8) and capric (C_{10}), and one long-chain one, behenic acid (C_{22}). Because behenic acid is poorly absorbed, this triglyceride is estimated to contribute $5\,kcal\,g^{-1}$. It is considered to be GRAS and is promoted mainly as a cocoa butter substitute (Haumann, 1986).

Salatrim is another structured triacylglycerol which provides about $5\,kcal\,g^{-1}$. Functionality can be manipulated by varying fatty acid composition. Salatrim was developed by the Nabisco Food Group and is presently being commercialized (Benefat brand) by Cultor Food Science under license (Kosmark, 1996).

Medium chain triglycerides (MCT) have been used for many years in special dietetic products (LaBarge, 1988). MCT deliver around 90% ($8.3\,kcal\,g^{-1}$) of the energy of normal fats, thus their potential contribution to reduced fat meat products appears limited.

9.6.3 *Emulsifiers*

Emulsifiers are fat or fatty acid derived compounds that have the ability to modify the surface properties of solids or liquids due to their hydrophilic and lipophilic properties (Giese, 1996b). They have become important ingredients in many low fat food products, e.g. salad dressings and bakery products (Jones, 1995). Emulsifiers are not fat replacers in a traditional sense but can function as part of a fat replacer system. Examples include lecithin, mono- and diglycerides, polyglycerol esters, polysorbates and sodium stearolyl lactylate (Glicksman, 1991). Some emulsifier properties of potential interest to meat processors include emulsification, lubrication, starch complexing, protein complexing, aeration and freeze/thaw stability. Emulsifiers generally have an oily or fatty consistency and may be liquid, semisolid or solid in appearance. They are used at low concentrations ($<0.5\%$) and can exert a 'fat sparing' or 'fat extending' effect on any remaining fat in a reduced fat product.

USDA (1995) has proposed that emulsifiers can be used in combination with fat mimetics in various low fat meat products. However, the efficacy of these ingredients in low fat meat products has not been documented by research.

9.7 Synthetic fat replacers

9.7.1 *Sucrose polyester*

Research to formulate food products with specially engineered fat ingredients has been underway for four decades. The most widely known is olestra, a sucrose polyester (SPE), developed by Procter & Gamble (Cin-

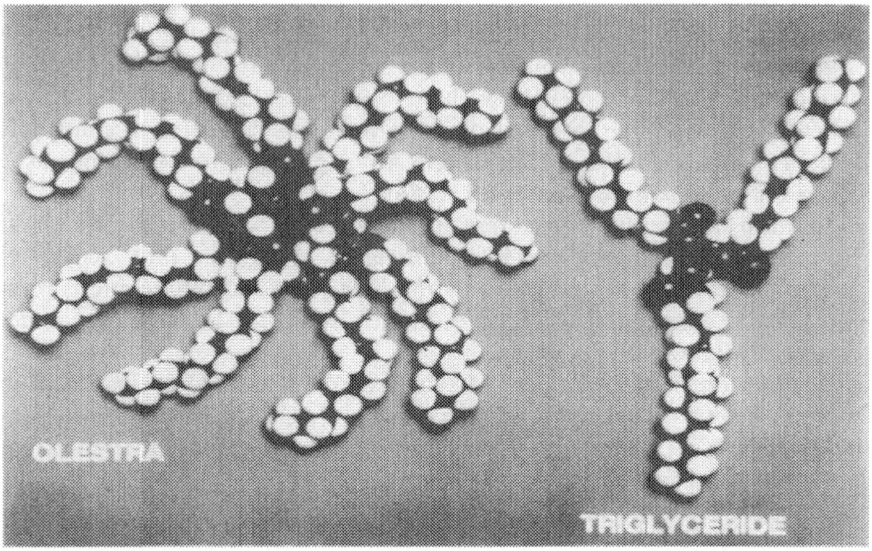

Figure 9.4 Structures of olestra and triglyceride.

cinnati, OH). Olestra was discovered serendipitiously by Fred Mattson and Bob Volpenhiem of Procter & Gamble while looking for a more digestible source of fatty acids for premature infants (Harrigan and Breene, 1993). Owing to their large bulky size, SPE are resistant to hydrolysis by pancreatic and microbial enzymes. Thus, they are not digested or absorbed and do not contribute calories (Haumann, 1986; LaBarge, 1988). Reviews by Harrigan and Breene (1993) and the FDA (1996) provide excellent background information on SPE. Figure 9.4 illustrates the different structures of a triglyceride and of olestra.

Olestra is synthesized by esterifying 6–8 of the hydroxyl groups of sucrose with fatty acids in the presence of catalysts. The fatty acids of chain length C_{12} to C_{20} and higher can be derived from various sources, including soybean, corn, palm, coconut and cottonseed oils. The functionality and physicochemical properties of olestra and other SPE can be altered by varying the esterification level and by changing the constituent fatty acid length and degree of saturation (Harrigan and Breene, 1993). They can, therefore, be formulated for many different applications, such as fried foods, dairy products, cooking oils, spreads and shortenings, and retain all the sensory properties of the triglycerides that they replace, but without the calories (Giese, 1996a).

Olestra is claimed to have excellent sensory characteristics as a fat substitute. It has an appearance, flavor, heat stability, flash point and shelf-life similar to those of its constituent fats (Anon., 1990a). Olestra would probably work well as a zero-calorie frying medium for fried or prebrowned meat items (Aylward, 1996). It is also expected to have similar functionality to conventional fats and oils in comminuted meat products. Although there are no published data on SPE evaluation in meat products, Swanson at Washington State University has observed that pork sausages made with SPE were comparable to products processed with normal lard (B. Swanson, personal communication).

Sucrose polyesters are not a completely new class of fat additive, since the lower esters (mono-, di- and tri-) have been used as emulsifiers for many years (Harrigan and Breene, 1993). However, olestra presented an interesting regulatory dilemma as it was difficult to design appropriate safety testing protocols for olestra as it could be present in the diet at levels far in excess of those of traditional food additives (Bergholz, 1992). Extensive research on olestra has improved its functional properties and minimized problems with 'anal leakage'. Since olestra can affect the absorption of fat soluble vitamins in the digestive tract, vitamins A, D, E and K should be supplemented in olestra-containing foods (Giese, 1996a).

Olestra has been approved as a direct food additive in salted snacks and crackers (FDA, 1996). However, any other uses would require a separate FDA review and approval process (Aylward, 1996). Several other companies hold patents on sucrose polyesters and are probably waiting to see the public's response to Procter & Gamble's olestra (Olean brand). Consumer acceptability, functionality and economics of synthetic lipids such as olestra will play a role in determining the extent of utilization of these ingredients by the meat industry.

9.7.2 Other experimental synthetic fat substitutes

A number of synthetic fat substitutes are still in the developmental stages and may be many years from commercialization. Trialkoxytricarballylate (TATCA) is under development by CPC International (Harrigan and Breene, 1993). In TATCA, fatty acids are esterified on to a polycarboxylic acid backbone. The resulting ester linkages are reversed from the conventional esters present in triglycerides, leading to resistance to hydrolysis by lipase enzymes. Toxicological evaluation of these compounds is required to verify safety (Singhal, Gupta and Kulkarni, 1991).

Frito-Lay (PepsiCo) holds patents on dicarboxylic acid esters (DDM), which are esters of malonic and alkyl malonic acids. DDM is purported to be suitable for high temperature applications such as frying (Anon, 1990a).

Esterified propoxylated glycerols (EPG) are a new family of propylene oxide derivatives developed by ARCO Chem Co. (Singhal, Gupta and Kulkarni, 1991). EPGs are similar in structure to normal triglycerides, except that propylene glycol has been inserted between the glycerol and fatty acid segments (Harrigan and Breene, 1993). Preliminary studies suggest that EPGs are resistant to enzymatic hydrolysis in the gut (Singhal, Gupta and Kulkarni, 1991).

Research by Akoh and Swanson at Washington State University and Curtice–Burns Foods has led to several patents for raffinose polyester, alkyl glycoside fatty acid polyester and methyl glucoside polyester as synthetic fat substitutes (Harrigan and Breene, 1993). Physical properties of these compounds are said to be similar to sucrose polyesters and commercial vegetable oils.

Sorbital fatty acid esters are being studied by Cultor Food Science for possible development for a number of applications including meat analogs (Haumann, 1993). Sorbestrin, a mixture of sorbitol and sorbitan polyesters, is a high temperature fat substitute, which is only partially metabolized ($1.5\,kcal\,g^{-1}$) (Murray, 1995).

Polydimethylsiloxane and other silicone based oils (Dow Corning) show promise as a frying medium (Morehouse and Zabik, 1989).

Most of these experimental fat substitutes are considered to be acaloric. Recent patent activity indicates that research and development of synthetic fat substitutes is continuing. However, none of these are currently approved for food use. Some companies are probably waiting to see how Olean is accepted by consumers before undergoing the rigorous regulatory process.

9.8 Research needs

Replacement of fat in meat products is a difficult technical challenge. We still do not know enough about all the functions of fat in meat products in order to mimic its properties adequately using other ingredients. No one fat replacer is perfect for all applications. Problems with flavor, juiciness, texture and mouth-feel have not been adequately solved by current fat replacers.

Continued research into the effects of fat replacers on sensory properties of meat products is essential. Most information describing advantages of the various fat replacers comes from manufacturers' literature rather than from basic research sources. There is a particular lack of information on the effects of microparticulated proteins and emulsifiers in low fat meats. Further research on synthetic fat replacers should be a long-term goal.

In order to select the most effective and economical fat replacer system, it is necessary to develop a better understanding of the specific mechanism of action of the various fat replacers as well as synergistic and additive effects. How each fat mimetic entraps water differs (Murray, 1995). The effects of these ingredients on meat protein gelation and functionality should also be explored.

9.9 Summary

The ideal fat replacer would be a 'colorless, odorless, liquid that looks, tastes and functions like oil but has no calories and is less expensive than water', according to Glicksman (1991). However this 'Oily Grail' does not exist. There is no simple strategy for selecting the best fat replacer for low fat meat products. In this chapter, a number of current and potential fat replacers were described. Often, the solution may be to use several ingredients in combination as part of a functional blend.

Technically, fat replacement systems for low fat meat products are improving and ingredient suppliers are developing a better understanding of the needs of the meat industry. However, much is still unknown about the interactions of various fat substitutes with classical sausage ingredients, processing procedures, storage conditions and final product preparation (Keeton, 1994). Mandigo and Eilert (1993) observed that the success of lower fat and low fat processed meats will depend on the palatability, economics and health perceptions of these products. The ultimate goal is a great tasting, low fat meat product that consumers do not just try once or twice but enjoy as part of an ongoing healthier diet and lifestyle.

Acknowledgements

The author acknowledges the assistance of the Saskatchewan Food Product Innovation Program and L. Sparrow in the preparation of this chapter.

References

Anon. (1990a) Fat substitute update. *Food Technol.*, **44**(3), 92.

Anon. (1990b) Lose the fat, retain the taste: cellulose gel helps do both. *Prepared Foods*, **159**(5), 157.

Aylward, L. (1996) Ole for olestra? Is it applicable to meat products? *Meat Market. Technol.*, **4**(4), 48.

Bergholz, C.M. (1992) Safety evaluation of olestra, a nonabsorbed, fatlike fat replacement. *Crit. Rev. Food Sci. Nutr.*, **32**(2), 141.

Calorie Control Council (1996) *Fat Reduction in Foods*, Calorie Control Council, Atlanta, GA.

Carrageenan Company (no date) *CarraFat™ Product Specification Data Sheet*, Carrageenan Company, Santa Ana, CA.

Drewnowski, A. (1992) Sensory properties of fat and fat replacements. *Nutr. Rev.* **50**(4), 17.

Dziezak, J.D. (1991) A focus on gums. *Food Technol.*, **45**(3), 116.

Egbert, W.R., Huffman, D.L., Chen, C. and Dylewski, D.P. (1991) Development of low-fat ground beef. *Food Technol.*, **45**(6), 64.

FDA (1996) Food additives permitted for direct addition to food for human consumption: Olestra; final rule. *Fed. Reg.*, **61**, 3117.

Frye, A.M. and Setser, C.S. (1993) Bulking agents and fat substitutes, in *Low Calorie Foods Handbook* (ed. A.M. Altschul), Marcel Dekker, New York, pp. 211–251.

Giese, J. (1992) Developing low-fat meat products. *Food Technol.*, **46**(4), 100.
Giese, J. (1996a) Olestra: properties, regulatory concerns and applications. *Food Technol.*, **50**(3), 130.
Giese, J. (1996b) Fats, oils and fat replacers. *Food Technol.*, **50**(4), 78.
Glicksman, M. (1991) Hydrocolloids and the search for the 'Oily Grail'. *Food Technol.*, **45**(10), 94.
Glueck, C.J., Streicher, P.A. Illig, E.K. and Weber, K.D. (1994) Dietary fat substitutes. *Nutr. Res.*, **14**, 1605.
Harrigan, K.A. and Breene, W.M. (1993) Fat substitutes: sucrose polyesters and other synthetic oils, in *Low Calorie Foods Handbook* (ed. A.M. Altschul), Marcel Dekker, New York, pp. 181–210.
Haumann, B.F. (1986) Getting the fat out – researchers seek substitutes for full-fat fat. *J. Am. Oil Chem. Soc.*, **63**, 278.
Haumann, B.F. (1993) Status of selected fat substitutes *INFORM* **4**(11), 1232.
Jenkins, R.K. and Wild, J.L. (1994) Thermo-irreversible gel particles for use in foods. US Pat. 5 294 457.
Jones, S.A. (1995) Fat replacers: the broad perspective. *World of Ingredients*, May/June, 8.
Kalab, M. (1990) Microparticulate protein in foods. *J. Am. Coll. Nutr.*, **9**, 374.
Keeton, J.T. (1994) Low-fat meat products – technological problems with processing. *Meat Sci.*, **36**, 261.
Kevin, K. (1995) The next generation of fat replacers. *Food Process.*, **56**(7), 63.
Kosmark, R. (1996) Salatrim: properties and applications. *Food Technol.*, **50**(4), 98.
Krizner, K. (1995) Born free. Reduced fat claims spice up red meat product introductions in 1995. *Meat Market. Technol.*, **3**(11), 40.
Laaman, T.R. and Tye, R.J. (1991) Application of gellan gum to meat systems. *Res. Disclosure*, **323**, 212.
LaBarge, R.G. (1988) The search for a low-caloric oil. *Food Technol.*, **42**(1), 84.
Lucca, P.A. and Tepper, B.J. (1994) Fat replacers and the functionality of fat in foods. *Trends Food Sci. Technol.*, **5**, 12.
Lund, D. (1984) Influence of time, temperature, moisture, ingredients, and processing conditions on starch gelatinization. *CRC Crit. Rev. Food Sci. Nutr.*, **20**, 249.
Mandigo, R.W. and Eilert, S.J. (1993) Processing of meat to meet consumer demand: developments in restructured and low-fat processed products. *Proc. Int. Congr. Meat Sci. Technol.*, **39**, 305.
Means, W.J. and Schmidt, G.R. (1986) Algin/calcium gel as a raw and cooked binder in structured beef steaks. *J. Food Sci.*, **51**, 60.
Morehouse, S.E. and Zabik, M.E. (1989) Evaluation of polydimethylsiloxane fluids as noncaloric frying media. *J. Food Sci.*, **54**, 1061.
Murphy, D. (1995) Carrageenan – coming into its own. *Nat. Prov.*, **209**(5), 60.
Murray, P.R. (1995) Exploring the foundation of fat replacement. *Food Technol. Eur.*, **2**(1), 24.
Oakenfull, D. (1987) Gelling agents. *CRC Crit. Rev. Food Sci. Nutr.*, **26**(1), 1.
Osburn, W.N. and Keeton, J.T. (1994) Konjac flour gel as fat substitute in low-fat prerigor fresh pork sausage. *J. Food Sci.*, **59**, 484.
Pszczola, D.E. (1991) Oat-bran-based ingredient blend replaces fat in ground beef and pork sausage. *Food Technol.*, **45**(11), 60.
Ruby, B.F. (1994) Chemical, physical and sensory characteristics of low-fat frankfurters. *MS Thesis*, University of Wyoming, Laramie, WY.
Sanderson, G.R. (1981) Polysaccharides in foods. *Food Technol.*, **35**(7), 50.
Sanderson, G.R. and Clark, R.C. (1983) Gellan gum. *Food Technol.*, **37**(4), 63.
Setser, C.S. and Racette, W.L. (1992) Macromolecule replacers in food products. *CRC Crit. Rev. Food Sci. Nutr.*, **32**, 275.
Shand, P.J. and McCurdy, A.R. (1995) Marination improves palatability of PSE and normal pork loin roasts. *Proc. Int. Congr. Meat Sci. Technol.*, **41** Vol. II, 497.
Shand, P.J. and Schmidt, G.R. (1990) New technology for low-fat meat products. *Proc. Recip. Meat Conf.*, **43**, 37.
Shand, P.J., Sofos, J.N. and Schmidt, G.R. (1994) Kappa-carrageenan, sodium chloride and temperature affect yield and texture of structured beef rolls. *J. Food Sci.*, **59**, 282.

Singhal, R.S., Gupta, A.K. and Kulkarni, P.R. (1991) Low calorie fat substitutes. *Trends Food Sci. Technol.*, **2**, 241.

Singer, N.S. and Dunn, J.M. (1990) Protein microparticulation: the principle and the process. *J. Am. Coll. Nutr.*, **9**(4), 388.

Singer, N.S. and Moser, R.H. (1993) Microparticulated proteins as fat substitutes, in *Low Calorie Food Handbook* (ed. A.M. Altschul), Marcel Dekker, New York, pp. 171–179.

Thestrup, N. (1993) The new potential of pectin. *Food Manuf.*, **68**(9), 55.

Troutt, E.S., Hunt, M.C., Johnson, D.E. *et al.* (1992) Characteristics of low-fat ground beef containing texture-modifying ingredients. *J. Food Sci.*, **57**, 19.

Tye, R.J. (1991) Konjac flour: properties and applications. *Food Technol.*, **45**(3), 87.

USDA (1995) Food Standards: requirements for processed meat and poultry products named by use of an expressed nutrient content claim and a standard term. *Fed. Regis.*, **60**(250), 67474.

Vafiadis, D.K. (1996) Fat war: the struggle continues. *Nat Prov.*, **210**(1), 50.

Van den Hoven, M. (1987) Functionality of dairy ingredients in meat products. *Food Technol.*, **41**(10), 72.

Yackel, W.C. and Cox, C. (1992) Application of starch-based fat replacers. *Food Technol.*, **46**(6), 146.

Ziegler, G.R. and Foegeding, E.A. (1990) The gelation of proteins. *Adv. Food Nutr. Res.*, **34**, 203.

10 Use of additives from plant and animal sources in production of low fat meat and poultry products

S.J. EILERT and R.W. MANDIGO

10.1 Introduction

The production and marketing of processed meat products has undergone a tremendous evolution in the last 10 years. The number and variety of non-meat ingredients utilized in these products has continued to increase. Mandatory nutritional labeling, sublisting of ingredients and safe handling statements have resulted in processed meat labels containing considerable information. If one would have looked at a 1996 version for 97% fat-free frankfurters in 1976, one might have laughed at the possibility of even one package being sold. This is due to the inconceivable notion that consumers would purchase low fat items in the first place and, second, purchase meat items with unfamiliar nonmeat ingredients listed on a lengthy and sometimes complicated package label.

The meat industry has adapted to the need for lower fat formulations. Consumers have expanded their willingness to consume a variety of nonmeat ingredients (Camerer, 1992). Taste, convenience and the demand for lower fat items have minimized the fear of words such as 'carrageenan', 'maltodextrin' and 'sodium caseinate'. Government regulations have been altered to allow processors greater flexibility in formulating low fat items. As stressed in previous articles (Mandigo, 1991; Mandigo and Eilert, 1994), the movement to lower fat formulations is the answer to a niche market demand, that will be met if the lower fat items are of high quality and wholesome. The demand is present and requires attention by the processor.

Mandigo and Eilert (1994) extensively emphasized the point that in order to produce low fat processed meats, one must adequately manage water. The 40% rule (fat + added water $\leqslant$ 40% in sausage products; USDA, 1988) expanded the amount of water allowed in cooked sausage items. One is not obligated to utilize additional nonmeat ingredients to bind the water in low fat/high added water formulations, and extensive research has been devoted to the manipulation of raw meat materials in order to accomplish this goal. However, these strategies often result in more raw material handling and increased time on the production floor.

These measures are not without cost. The need for ingredients to provide additional water binding has prompted the approval of many new nonmeat ingredients, including regulations allowing purge controllers in cured pork items (USDA, 1992) and proposed legislation that will change the standards of identity of many processed meat items and further allow addition of nonmeat ingredients.

There is no one universally proven strategy or ingredient that will allow a low fat, high water added meat item to have the sensory qualities of higher fat items while maintaining the price competitiveness of higher fat products. Replacing fat with water is the underlying strategy outlined in this chapter, which also is discussed in Chapter 4. Our goal is to address ingredients derived from animal and plant sources that will assist in proper management of added water.

10.2 Addition of plant products

There are a variety of plant products that are used in low fat meat products and they are discussed in this section.

10.2.1 Soy proteins

A catch-phrase of the meat industry is the need to produce 'value added' items. The producers of soy proteins for meat uses have also concentrated on value added soy protein concentrates and isolates, that are far superior to the soy flours and concentrates used and researched in the 1970s. Soy products developed a poor popular image (McMindes, 1991) that is being overcome by the high quality products being used today. At least one commercially available version of isolated soy protein contains only 0.3% carbohydrate. There is very little or no oligosaccharides, raffinose and stachyose that cause flatulence in these newer soy protein additives (Hoogenkamp, 1992).

Soy proteins will form a gel upon heating in a finely ground product, entrapping liquid and moisture. This gelling action in a low fat/high added water formulation has the potential to return some of the 'bite' and texture often lost when levels of water addition are high. Matulis *et al.* (1995) found that frankfurter juiciness decreased and hardness increased with higher levels of isolated soy protein, which may have been due to the additional water binding and gelation provided by the soy. While soy protein isolates are much blander in flavor than soy flours, Matulis *et al.* (1995) advised against usage levels higher than 3.0% owing to the development of a less desirable flavor. Isolated soy protein (2.0%) increased the bind and cooking yields of low fat, high added water turkey bologna (Dexter, Sofos and Schmidt, 1993).

Decker, Conley and Richert (1986) used reduced levels of textured soy protein with hydrated soy protein isolate to produce an acceptable low fat product. They replaced 22.3% of the meat block with hydrated isolated soy protein. Higher levels of hydration (28:1) were necessary to give juiciness and instrumental hardness comparable to those of a 30% fat control frankfurter. In another experiment, Decker, Conley and Richert (1986) found that 15% fat control frankfurters were most similar to 30% fat controls when >20% of the meat block was replaced with hydrated isolated soy protein. It was also at this level of replacement that soy flavor became more evident. An additional experiment showed that replacement of 31.6% of the meat block with isolated soy protein reduced the hardness of 15% fat frankfurters. The effectiveness of soy protein in a frankfurter-type product seems to vary with level of addition, type and quality of the soy protein and degree of hydration.

A strategy for the successful production of low fat/high added water frankfurters is not to rely solely on the combination of soy products and water to take the place of fat completely. The formulation shown in Table 10.1 takes advantage of the gelling capabilities of soy protein concentrate, combined with the water binding ability of starch and the flavor enhancement of hydrolyzed vegetable protein.

Soy protein isolate is one of the four materials approved as a purge controller. McMindes (1992) pointed out that the high dispersability and functionality of soy protein isolate makes it ideal for cured pork injection systems. Isolated soy protein reduced purge and cooking loss of high

Table 10.1 Lite frankfurter formulation[a]

Ingredient	% of gross weight
Pork blade meat	25.621
Pork 80% trim	21.747
Beef 90% trim	2.641
Water and ice	38.211
Salt	1.911
Food starch-modified	2.000
Dextrose	2.000
Corn syrup solids	1.761
Soy protein concentrate	1.510
Mustard flour	1.000
Hydrolyzed vegetable protein	0.701
Spice	0.501
Sodium tripolyphosphate	0.247
Sodium erythorbate	0.026
Cure (6.25% $NaNO_2$)	0.123
Shrink loss	12.00

[a]Formulation courtesy of Central Soya, Fort Wayne, IN. Final product contains 6 g fat per serving.

water added boneless hams, yet had a slightly undesirable effect on product flavor and overall sensory acceptability, according to Mills (1995). This finding further emphasizes the need to complement the effects of soy proteins with other added ingredients.

10.2.2 Other plant-based proteins

The protein fractions of seeds other than soybeans are sometimes used in processed meat products. Their gelling and water binding capacity make them acceptable as water binding agents in sausage products. However, usage of all of the other plant protein sources combined does not equal the industry applications of soy-based proteins. Some of these protein sources will be discussed, although much of the research in this area has been in higher fat formulations. If yield improvements and textural modifications are noted in higher fat products, we can speculate that the same will hold true in lower fat/higher added water formulations.

(a) Vital wheat gluten

Vital wheat gluten (VWG) is one of the more commonly utilized plant proteins (other than soy) in meat products. Emulsion stability and yield were improved with the addition of 3.5% VWG, according to Keeton, Foegeding and Patana-Anake (1984). However, there was not a VWG effect on total or water losses in the emulsion stability test, no effect on product yield and no effect on instrumental and sensory textural traits. Other researchers have found similar textural and yield results (Patana-Anake and Foegeding, 1985; Ma, Yiu and Khanzada, 1991). In this research, VWG contributed off-flavors to the frankfurters and was rated as less desirable by a sensory panel. These authors noted that the use of VWG would probably have to be accompanied by altering the spice component to produce a desirable product.

(b) Corn germ protein

Corn germ protein (CGP) has also been recently investigated as a functional ingredient in frankfurters. Zayas and Lin (1988) added 3% CGP to 23–25% fat frankfurters. Cooking yields improved with the addition of CGP, illustrating the possible fat and water binding potential of this protein. Frankfurters with CGP exhibited lower shear force values and lower firmness values. There was little effect on sensory quality with CGP, except for a slightly less desirable flavor. In a subsequent study, Zayas and Lin (1989) added CGP as a powder and as a stabilizer in two different emulsions. Yield and water holding capacity were improved in frankfurters with all CGP treatments, with no effect on product firmness. No severe flavor problems were noted in this study. The water binding and texture softening effects of CGP are desirable qualities for a low fat meat ingredient, if the added water levels are low. In higher added water formulations where it is desirable to bind water and form a strong gel, CGP may not be the optimal ingredient, especially if used by itself.

(c) Other plant protein additives

Oat, deamidated oat, soy and pea protein isolates and white navy bean concentrate were compared to an all-meat control in a frankfurter-type product by Ma *et al.* (1989). Each ingredient was added at the 5, 10 and 20% levels. At 5 and 10% addition, all of these ingredients provided cook yields within 94–99% of the all-meat control. There was no difference between these ingredients at any level of addition. However, as the level of addition rose, cook yields, firmness and cohesiveness all declined. Deamidated oat protein isolate did increase cohesiveness slightly over the control. The texture softening characteristics of these ingredients might make them suitable for some low fat products. Again, the use of these ingredients alone is not likely to be the answer if water binding is of great importance.

10.2.3 Starches, flours and fibers

(a) Food starches

Food starches are becoming more accepted and more commonly used in processed meat and poultry items owing to their ability to gel and hydrate at normal processing temperatures, their low sweetness contribution and non-reducing characteristics (avoidance of browning). Additionally, food starches are usually less expensive than soy and dairy based ingredients that also function to bind water. Many processors have found it advantageous to complement a soy protein or concentrate with a modified food starch in order to maximize processing yields, control purge and lend freeze/thaw stability to the products. Food starches can be modified by a variety of approved methods, including hydrolysis, oxidation, cross-linking and substitution. These modifications, used alone or in combination, can improve the functionality of starches (Luallen, 1985).

Modified food starch is also one of the four approved binders for 'water added' and 'X% added ingredient' cured pork items. Mills (1995) found that modified tapioca starch (2.0%) improved the cooking yield of a cook-in-bag ham with 41% added ingredients by 7% over a control ham. Modified waxy maize corn starch was used to decrease cooking loss and purge of low fat, high added water turkey bologna (Dexter, Sofos and Schmidt, 1993). However, this starch did not affect product bind or sensory characteristics. Modified food starch also reduced purge in low fat meat frankfurters (Yang, Trout and Shay, 1995).

Comer *et al.* (1986) used a cross-linked waxy maize corn starch in 22–26% fat wiener formulations. Product yields and textural firmness

Table 10.2 Low fat frankfurter formulation[a]

Ingredient	% of gross weight
Lean pork knuckles	11.81
Pork 80% trim	12.17
Turkey thigh meat	23.98
Water and ice	38.94
Salt	1.93
Food starch-modified	3.00
Dextrose	2.00
Corn syrup solids	2.00
Soy protein concentrate	2.00
Spice	1.00
Flavoring	0.40
Turkey stock	0.25
Sodium phosphate	0.40
Sodium erythorbate	0.20
Cure (6.25% $NaNO_2$)	0.10
Shrink loss	8.00

[a]Formulation courtesy of Grain Processing Company, Muscatine, IA. Final product contains 3 g fat per serving.

increased with the addition of 10.3% cross-linked waxy maize starch in comparison with the all-meat controls. Skrede (1989) found that after a 19 day storage period, sausages with modified potato starch were generally superior in sensory traits and yielded less exudate than sausages with corn or tapioca starch. The quality of sausages with modified potato starch was dependent upon cooking temperature and storage time. Frankfurters (9 and 15% fat) with 2.5% modified waxy maize corn starch were found to have similar sensory qualities to 28% fat control frankfurters, as reported by Odio (1989). The yields of the starch containing frankfurters were slightly lower than that of the control.

As noted previously and demonstrated by the formulations in Table 10.2, many starches are being used most successfully in combination with soy proteins.

(b) Pre-formed gels

Certain flours and starch–protein blends are best utilized by forming a gel prior to incorporation into meat products. This is especially useful in products that receive a relatively short thermal processing treatment. Osburn and Keeton (1994) formed a gel by heating Konjac flour – a product of the Konjac root (Tye, 1991) – with water and sodium citrate. Caramel coloring and seasoning were added to the gel in order to minimize color and flavor dilution problems. This mixture was allowed to cool and form a gel and then ground to appropriate particle size. Higher levels of this pre-formed gel reduced shear force values, while sensory scores became more like controls. The effects of Konjac flour gels can be modified by combining other starches and gums that can assist to form stronger gels (Tye, 1991).

One drawback to the pre-formed gel concept is the additional handling required for such a product. Similarly, pre-cooking ingredients such as rice (Hull, Rogers and Martin, 1992) may also be an effective strategy but one of increased labor inputs. White beans (canned, wet, no pre-cooking required) were added to spice enhanced low fat/high added water frankfurters and resulted in texture and sensory acceptability comparable to those of controls (Hull, Rogers and Martin, 1992). This study emphasizes the vast range of ingredients that can be employed in the production of low fat sausages.

(c) Addition of fiber

The inclusion of fiber can add not only functionality to a meat product, but also the perceived health benefits of a fiber source. Claus and Hunt (1991) used several fibers and starches in an effort to improve the yields and texture of low fat/high added water bologna. While none of the fibers or starches added could mimic a 30% fat/10% added water product for firmness and cohesiveness, all of the materials studied except pea fiber (sugar beet pulp fiber, starch, isolated soy protein, oat fiber and wheat starch) were all more firm than the 10% fat/30% added water control. Cooking losses of bologna with these materials were at least as high or higher than the 10% fat/30% added water control. The proper use and combination of these materials merit further investigation.

An extensive study of several fibers, proteins and gums revealed that a specially processed oat bran was the most functional product for use as a fat replacer in low fat ground meat products, as reported by Pszczola (1991). Additional seasonings were added with this product (LeanMaker) in order to compensate for the loss of flavor in low fat products. Lean meat (90%) and a 3% incorporation of this blend were shown to be optimal in sensory and cooking tests. This product can also be applied to cooked sausages with some modification needed owing to the different cooking procedures involved.

10.2.4 Gums and carrageenan

The use of these hydrocolloids of plant and microbial origin is growing rapidly in low fat meat products. Their ability to thicken and inhibit the release of water has made them very useful in low fat/high added water meat products. They are used extensively as texture-modifying agents in the food industry, most recently becoming commercially acceptable in processed meat applications. Wallingford and Labuza (1983) investigated several gums in low fat meat batters. Xanthan gum was found to be the strongest water binding agent, followed by carrageenan. Foegeding and Ramsey (1986) found that iota- and kappa-carrageenan were the most functional in a low fat, high moisture meat batter. Xanthan gum was similar to the carrageenan for thermal stability, but greatly decreased the textural parameters. Iota-carrageenan increased water holding ability and improved some textural parameters in comparison with kappa-carrageenan, whereas kappa-carrageenan increased hardness to a greater extent, as reported by Foegeding and Ramsey (1987). Xanthan gum also decreased some textural parameters, without any loss in water binding ability. The authors emphasized that the best strategy for gum usage in these products might be a combination of compounds in order to achieve specific goals. Similar to the effects of xanthan gum, Lin *et al.* (1988) found some texture softening in low fat frankfurters upon the addition of carboxymethylcellulose. The success of these agents was found to be product specific. For instance, xanthan gum may not be suited at all for a low fat/high moisture meat batter, yet the matrix-disrupting effect might be ideal in low fat products with less moisture which require a softening effect.

Todd *et al.* (1989) added 3.5 and 7.0% of cellulose, a microcrystalline cellulose–carboxymethylcellulose combination and a mixture of gum arabic and guar gum to restructured pork patties. Products made with cellulose or the cellulose derivative had improved cooking yields in comparison with an all-meat control. The cellulose derivative did not affect textural parameters at the 3.5% level. Addition of 7.0% of any of these materials did not prove to be feasible in this product owing to an unacceptable texture.

Iota-carrageenan was found to be more effective when salt and fat levels of frankfurters were lower, according to Matulis *et al.* (1995). Frankfurter hardness increased with increasing carrageenan levels in low salt frankfurters. Carrageenans have also been approved as purge controllers in cured pork products, as stated by McMindes (1992), along with modified food starches. Carrageenan was the most effective of the approved purge controllers at reducing cook loss and purge of cook-in-bag hams. Kappa-carrageenan is frequently used in turkey breast, a similar processed product.

10.3 Addition of animal products

10.3.1 Deboned poultry, meat and fish

There has been a tremendous amount of change in the regulations concerning mechanically recovered meat and poultry. Regulations concerning advanced lean recovery systems for red meats are in place that allow meat derived from processes that do not crush, grind or pulverize bones or have calcium levels in excess of 0.15% to be incorporated into meat products without additional label declarations (USDA, 1994). These changes were needed to keep up with emerging technologies in this area. Meat from advanced lean recovery (press and/or belt/drum-based separation) is not the same as 'mechanically separated species'. The fine-textured material derived from advanced lean recovery is much lower in calcium and bone substances than 'mechanically separated species'. Additionally, poultry regulations are changing that require poultry meat derived from these systems of crushing and grinding bones to be labeled as 'mechanically separated' (USDA, 1995). It is yet to be determined how these changes will affect formulation of processed meat items. The financial advantage of recovered poultry will probably maintain its favor as a processed meat ingredient. However, the desirable labeling of red meats derived from advanced lean recovery, along with increased pigment concentration and relative leanness (about 80%), will mean that these raw materials will increase in usage.

(a) Mechanically recovered proteins

Owing to the fine texture of mechanically recovered proteins, these materials often have a lower binding ability than regular skeletal muscle. However, this could work to the advantage of a processor in a low fat product, that often experiences textural problems due to excessive bind. Jantawat and Carpenter (1989) reported that smoked sausage with higher levels of mechanically deboned chicken meat was lower in fat, less firm, less chewy and closer to an optimum sensory score. The use of mechanically deboned meat of any species requires special processing attention and steps owing to the increased handling of the material and greater introduction of heme iron and oxygen, as outlined by Judge *et al.* (1989).

Small amounts (10% or less) of meat derived from advanced lean separation do not appear to alter texture and increase color intensity. Beef derived from advanced lean separation technology has been successfully incorporated (10% or less) into ground beef with no effect on product quality. Pork from such systems has been used very successfully in pizza toppings and other sausage items.

In the move to lower fat in processed products, there has been speculation that the demand for lean trimmings will exceed or at least push the extent of supply. It is probably safe to say that if all reduced fat products were reformulated to a higher fat version as of tomorrow, there would not be a shortage of higher fat trimmings. Such raw materials are in abundance. To upgrade these trimmings manually is not feasible owing to man-power inputs, lower yields and final ingredient costs. The process of low temperature rendering offers a means of separating fat from high fat trimmings, yielding a lean trim source at a cost less than skeletal lean. The lower functionality of these materials might make them suitable in some low fat items at low levels of inclusion.

10.3.2 Collagen and gelatin

(a) Collagen

The use of collagen in processed meat items is generally limited because upon heating, collagen molecules shrink and gelatinize, which results in the release of gelatin and fat from the product. However, if processors take advantage of the gelatinization process, it stands to reason that gelatinized collagen might have a place in low fat meat products. The reduced binding properties of collagen in relation to skeletal proteins might decrease the tough, rubbery texture of low fat products.

Jobling (1984) stated that bone collagen has a high absorptive capacity for water and fat, making it a good functional ingredient in meat products. Similarly, enzyme-hydrolyzed beef and pork skin was found to increase emulsion stability in sausage emulsions (Satterlee, Zachariah and Levin, 1973). Research studies have shown that addition of high collagen materials might soften the texture of low fat products (Jones *et al.*, 1982; Chavez, Henrickson and Rao, 1985). Sensory cohesiveness and overall acceptability of meat patties with higher levels of collagen decreased while juiciness increased.

The softening effect with higher levels of collagen noted in these and other research projects gave rise to the hypothesis that increasing levels of collagen, owing to its lower binding characteristics, might be included in lower fat products to soften texture (Jones, 1984; Wirth, 1991). Rao and Henrickson (1983) added 20% food-grade hide collagen to bologna of various fat levels. They found that shear values actually increased with the addition of this collagen source. Their hypothesis was that collagen was becoming integrated into the protein matrix, rather than disrupting it.

(b) Modified beef connective tissue

Quint (1987) added 0, 10 and 20% modified beef chuck connective tissue (MCT) to 15 and 30% fat frankfurters. The connective tissue was frozen and flaked to a small particle size, that allowed for thorough distribution of the connective tissue in the product. The MCT had no effect on instrumental textural measurements or sensory desirability. The addition of MCT increased processing yields of the frankfurters. Eilert *et al.* (1993a) replaced 0–40% of the meat block with beef shank connective tissue, modified in the same manner as Quint (1987). Processing yields increased with the addition of >10% MCT. Cohesiveness of the frankfurters decreased with the addition of MCT, whereas shear values increased. A consumer sensory panel found no differences in product juiciness or flavor due to MCT. Frankfurters with 8% fat and 20% MCT scored near the higher fat frankfurters for texture and overall desirability (Eilert *et al.*, 1993b).

Eilert, Mandigo and Sumner (1996) incorporated MCT into low fat/high added water frankfurter formulations to try and bind the excess water in these products. There was an increase in processing yields of 10% fat frankfurters when beef connective tissue was added at 20% of the formulation. Addition of various phosphates did not seem to alter the connective tissue effects. The direct treatment of connective tissue with acidic and alkaline phosphates was addressed by Eilert, Calhoun and Mandigo (1996). Acidic phosphate treatment dramatically increased the ability of heat to convert collagen to gelatin whereas alkaline phosphate treatment increased the ability of raw collagen to bind water. Calhoun, Eilert and Mandigo (1996a, b) incorporated these preblends into low fat/high added water frankfurters with little improvement in processing yields or final product characteristics. Further research in this area should address the addition of pre-cooked connective tissue collagen, much in the same way that Konjac flour gels were added to pork sausage (Osburn and Keeton, 1994).

(c) Shredded collagen fibers

Graves *et al.* (1993) employed 2% shredded collagen fiber and 8% water in a low fat ground beef patty and compared this with an 18% fat control. The low fat patty with collagen fiber and water exceeded the control for textural desirability, and was equal to the control for flavor, juiciness, overall desirability and instrumental textural measures. These shredded collagen fibers were added to low fat chicken frankfurters by Meullenet *et al.* (1994). A combination of 2% collagen fibers and 20% added water was the optimal formulation, with the collagen fibers offsetting the soft texture contributed by high levels of water addition.

10.3.3 Blood proteins

In the USA, the utilization of blood and blood proteins is essentially non-existent (except for some ethnic sausages and loaves). These highly functional and nutritious proteins are being ignored for no good scientific reasons (Rust, 1988). Knipe (1988) stated that blood proteins can be used to increase the protein content of meat products and water holding capacity. Plasma protein is a very strong binder upon heating. Red cells and whole blood can be used to enhance the color of cured meat products. These products have been safely used in Europe for many years. Their successful usage depends on microbial safety, requiring close inspection and careful collection and handling procedures. Blood proteins are also an alternative in low fat, high moisture meat products. They have a high protein content and excellent water binding ability. They have been proved to be acceptable in sausage products (Caldironi and Ockerman, 1982; Slinde and Martens, 1982). The major limitation of these products is labeling in the USA and the aesthetics of having a blood protein added to product at levels higher than that inherent in meat.

10.3.4 Milk proteins

The inclusion of milk products, including nonfat dry milk (NFDM), whey protein and sodium caseinate into meat products is a common and accepted practice, as these proteins contribute emulsion stability, water immobilization, texture modification, improved color and organoleptic qualities (van den Hoven, 1987). Most of the research with these proteins in meat products has been at more traditional fat levels (30%). However, their usage is being spread commercially to lower fat products. The functionality of these proteins makes them very suitable for low fat, high moisture products. Current cost and usage levels have made some of these milk proteins too costly in certain products.

Sodium caseinate has been approved as a purge controller in cooked pork products (McMindes, 1992). Mills (1995) found that sodium caseinate improved processing yields and reduced purge in both smoked and cook-in-bag hams. However, the processing yields of cured chicken rolls decreased with the addition of sodium caseinate as compared to isolated soy protein (Gillett and Carpenter, 1992).

An improvement in emulsion stability yield was noted with the inclusion of 3.5% calcium reduced NFDM by Keeton, Foegeding and Patana-Anake (1984). The inclusion of NFDM in 28% fat frankfurters resulted in palatability, yield and texture virtually equal to an all meat control. Ensor *et al.* (1987) added whey protein concentrate at different levels, as well as NFDM to knackwurst. Emulsions with these milk proteins were more stable than the all-meat controls. The knackwurst with milk protein was harder and more chewy than the control. It also possessed more desirable flavor, texture and juiciness and less aftertaste as measured by a consumer sensory panel.

Hung and Zayas (1992) compared NFDM, whey protein concentrate and sodium caseinate in frankfurters containing 20% fat. All three of these proteins increased the water holding capacity of the frankfurters and reduced cooking losses. Sodium caseinate and NFDM increased frankfurter shear force and hardness. The juiciness of frankfurters with these proteins was lower, probably because of the water being bound more tightly. Sensory panelists also noted a less meaty flavor with the addition of sodium caseinate and NFDM. The functionality of sodium caseinate, why protein and NFDM milk is well established and these ingredients should be considered for low fat, high moisture meat products.

10.4 Research needs

The research in the area of low fat ingredients for meat products is fairly extensive. US processors are utilizing these ingredients at increasing rates. From a research standpoint, efforts are needed in a 'backward' and 'forward' direction. To go backwards does not mean to redo the excellent work that has been performed to date. Rather, there may need to be more of an effort to delineate the mechanisms by which some of these ingredients alter low fat products. With such an effort, the proper utilization of these ingredients can be assured. Processors can be overwhelmed by the array of ingredients available to them. In some cases, ingredients are not being used to their fullest extent, or are being used for the wrong application. There needs to be a more concerted effort into the optimal combination of various ingredients. This will allow for decreased usage of more expensive ingredients while taking advantage of complementary or even synergistic effects.

As a step forward, meat researchers need to work cooperatively with other areas of food science to find the ingredients that will be successful and functional in meat products. We must continue to find low cost ingredients that can be functional with minimal processing. Such efforts will maximize returns for both the meat processor and the supplier. On a more specific note, almost all of the ingredients reviewed are functional after input of heat. Consideration needs to be given to the growing category of injected, raw meat items in which purge is a major concern. Ingredients that form cold-set gels are sometimes difficult to use in injection processes. This is a category of meat items that will grow rapidly in the near future and requires research attention.

10.5 Summary

Through the proper use of nonmeat additives, many successful low fat and now even 'no fat' products are on the market. This will continue to be a strong market niche. The growth of this market niche will depend heavily on the proper utilization of nonmeat and meat ingredients to produce reduced fat products with quality and processing characteristics similar to those of higher fat controls. The addition of various products of plant and animal origin was discussed in this chapter. This was by no means an inclusive discussion, but an attempt to review the use of some of the more widely applied nonmeat products as well as some of the more novel approaches to low fat meat production.

References

Caldironi, H.A. and Ockerman, H.W. (1982) Incorporation of blood proteins into sausage. *J. Food Sci.*, **47**, 405.

Calhoun, C.M., Eilert, S.J. and Mandigo, R.W. (1996a) Characteristics of low-fat, high added-water frankfurters manufactured with preblends of connective tissue and alkaline phosphate. *J. Muscle Foods,* **7**, 471.

Calhoun, C.M., Eilert, S.J. and Mandigo, R.W. (1996b) Connective tissue/acidic phosphate preblend effects on reduced fat frankfurters. *J. Food Sci.,* **61**, 459.

Camerer, D.L. (1992) Ingredients: enemy or ally? *Proc. Recip. Meats Conf.,* **45**, 75.

Chavez, J., Henrickson, R.L. and Rao, B.R. (1985) Collagen as a hamburger extender. *J. Food Qual.,* **8**, 265.

Claus, J.R. and Hunt, M.C. (1991) Low-fat, high added-water bologna formulated with texture-modifying ingredients. *J. Food Sci.,* **56**, 643.

Comer, F.W., Chew, N., Lovelock, L. and Allen-Wojtas, P. (1986) Comminuted meat products: functional and microstructural effects of fillers and meat ingredients. *Can. Inst. Food Sci. Technol. J.* **19**, 68.

Decker, C.D., Conley, C.C. and Richert, S.H. (1986) Use of isolated soy protein in the development of frankfurters with reduced levels of fat, calories and cholesterol. *Proc. Eur. Mtg. Meat Res. Workers,* **32**, 333.

Dexter, D.R., Sofos, J.N. and Schmidt, G.R. (1993) Quality characteristics of turkey bologna formulated with carrageenan, starch, milk and soy protein *J. Muscle Foods,* **4**, 207.

Eilert, S.J. Blackmer, D.S., Mandigo, R.W. and Calkins, C.R. (1993a) Meat batters manufactured with modified beef connective tissue. *J. Food Sci.,* **58**, 691.

Eilert, S.J., Blackmer, D.S., Mandigo, R.W. and Calkins, C.R. (1993b) Characteristics of low-fat frankfurters manufactured with modified beef connective tissue. *J. Muscle Foods,* **4** 269.

Eilert, S.J., Mandigo, R.W. and Sumner, S.S. (1996) Phosphate and modified beef connective tissue effects on reduced fat, high water-added frankfurters. *J. Food Sci.,* **61**, 1006.

Eilert, S.J., Calhoun, C.M. and Mandigo, R.W. (1996) Phosphate type, concentration and preblend duration to improve water holding capacity of beef connective tissue. *J. Muscle Foods,* **7**, 255.

Ensor, S.A., Mandigo, R.W., Calkins, C.R. and Quint, L.N. (1987) Comparative evaluation of whey protein isolate and calcium-reduced nonfat dry milk as binders in an emulsion-type sausage. *J. Food Sci.,* **52**, 1155.

Foegeding, E.A. and Ramsey, S.R. (1986) Effect of gums on low-fat meat batters. *J. Food Sci.,* **51**, 33.

Foegeding, E.A. and Ramsey, S.R. (1987) Rheological and water-holding properties of gelled meat batters containing iota-carrageenan, kappa-carrageenan or xanthan gum. *J. Food Sci.,* **52**, 549.

Gillett, R.A.N. and Carpenter, J.A. (1992) Effects of binding substrate, type of nonmeat additive and method of tenderizing on cured chicken rolls. *J. Food Qual.,* **15**, 225.

Graves, L.R., Delmore, R.J., Mandigo, R.W. and Calkins, C.R. (1993) Characteristics of low-fat ground beef patties manufactured with collagen fibers. *Proc. Annu. Mtg. Inst. Food Technol.,* **54**, 25.

Hoogenkamp, H.W. (1992) *Vegetable Protein,* Protein Technologies International, St Louis, MO.

Hull, D.H., Rogers, R.W. and Martin, J.M. (1992) Effects of carbohydrate substitution and preblending on the properties of reduced fat-high moisture beef frankfurters. *J. Muscle Foods,* **3**, 323.

Hung, S.C. and Zayas, J.F. (1992) Functionality of milk proteins and corn germ protein flour in comminuted meat products. *J. Food Qual.,* **15**, 139.

Jantawat, P. and Carpenter, J.A. (1989) Salt preblending and incorporation of mechanically deboned chicken meat in smoked sausage. *J. Food Qual.,* **12**, 393.

Jobling, A. (1984) New applications for collagenous proteins in meat products. *J. Sci. Food Agric.,* **35**, 1264.

Jones, K.W. (1984) Collagen properties in processed meats, in *Proceedings of Meat Industry Research Conference, New Orleans, LA,* American Meat Institute, Washington, DC, p. 18.

Jones, K.W., Snyder, W.D., Mandigo, R.W. and Calkins, C.R. (1982) Rheological and textural properties of meat emulsions prepared with collagen. *Proc. Am. Soc. Anim. Sci.,* **74**, 243.

Judge, M.D., Aberle, E.D., Forrest, J.C. *et al.* (1989) *Principles of Meat Science*, 2nd edn, Kendall/Hunt, Dubuque, IA.

Keeton, J.T., Foegeding, E.A. and Patana-Anake, C. (1984) A comparison of nonmeat proteins, sodium tripolyphosphate and processing temperature effects on physical and sensory properties of frankfurters. *J. Food Sci.*, **49**, 1462.

Knipe, C.L. (1988) Production and use of animal blood and blood proteins for human food. *Adv. Meat Res.*, **5**, 147.

Lin, K.C., Keeton, J.T., Gilchrist, C.L. and Cross, H.R. (1988) Comparisons of carboxymethyl cellulose with differing molecular features in low-fat frankfurters. *J. Food Sci.*, **53**, 1592.

Luallen, T.E. (1985) Starch as a functional ingredient. *Food Technol.*, **39**, 59.

Ma, C.-Y., Campbell, C., Khanzada, G. and Modler, H.W. (1989) Functional characteristics of wiener-type products substituted with native and acid-hydrolyzed oat protein isolates. *J. Food Sci.*, **54**, 1450.

Ma, C.-Y., Yiu, S.H., and Khanzada, G. (1991) Rheological and structural properties of wiener-type products substituted with vital wheat gluten. *J. Food Sci.*, **56**, 228.

Mandigo, R.W. (1991) Meat processing: The modification of processed meats, in *Fat and Cholesterol Reduced Foods* (eds C. Haberstroh and C.E. Morris), Portfolio Publishing, The Woodlands, TX, Vol. 12, p. 119.

Mandigo, R.W. and Eilert, S.J. (1994) Strategies for reduced-fat processed meats, in *Low Fat Meats* (eds H. Hafs and R. Zimbelman), Academic Press, Orlando, FL, Chapter 9, p. 145.

Matulis, R.J., McKeith, F.K., Sutherland, J.W. and Brewer, M.S. (1995) Sensory characteristics of frankfurters as affected by salt, fat, soy protein and carrageenan. *J. Food Sci.*, **60**, 48.

McMindes, M.K. (1991) Applications of isolated soy protein in low-fat meat products. *Food Technol.*, **45**(12), 61.

McMindes, M.K. (1992) Building a better bind: USDA (finally) permits purge controllers in ham. *Natl. Provis.*, **206**(11), 55.

Meullenet, J.-F., Chang, H.C., Carpenter, J.A. and Resurreccion, A.V.A. (1994) Textural properties of chicken frankfurters with added collagen fibers. *J. Food Sci.*, **59**, 729.

Mills, E.W. (1995) Nonmeat binders for use in cook-in-bag and smoked ham. *J. Muscle Foods*, **6**, 23.

Odio, E.M. (1989) Chemical, physical, and sensory characteristics of reduced-fat meat batters and products with added carbohydrates. *MS Thesis*, Texas A&M University, College Station, TX.

Osburn, W.N. and Keeton, J.T. (1994) Konjac flour gel as fat substitute in low-fat prerigor fresh pork sausage. *J. Food Sci.*, **59**, 484.

Patana-Anake, C. and Foegeding, E.A. (1985) Rheological and stability transitions in meat batters containing soy protein concentrate and vital wheat gluten. *J. Food Sci.*, **50**, 160.

Pszczola, D.E. (1991) Oat bran based ingredient blend replaces fat in ground beef and pork sausage. *Food Technol.*, **45**(11), 60.

Quint, L.N. (1987) Interaction of fat and connective tissue in meat emulsions. *MS Thesis*, University of Nebraska, Lincoln, NE.

Rao, B.R. and Henrickson, R.L. (1983) Food grade hide collagen in bologna: effect on functional properties, texture and color. *J. Food Qual.*, **6**, 1.

Rust, R.E. (1988) Processing options for improving the nutritional value of animal products, in *Designing Foods*, National Academy Press, Washington, DC., pp. 278–282.

Satterlee, L.D., Zachariah, N.Y. and Levin, E. (1973) Utilization of beef and pork skin hydrolyzates as a binder or extender in sausage emulsions. *J. Food Sci.*, **38**, 268.

Skrede, G. (1989) Comparison of various types of starch when used in meat sausages. *Meat Sci.*, **25**, 21.

Slinde, E. and Martens, M. (1982) Changes in sensory properties of sausages when small amounts of blood replace meat. *J. Sci. Food Agric.*, **33**, 760.

Todd, S.L., Cunningham, F.E., Claus, J.R. and Schwenke, J.R. (1989) Effect of dietary fiber on the texture and cooking characteristics of restructured pork. *J. Food Sci.*, **54**, 1190.

Tye, R.J. (1991) Konjac flour: properties and applications. *Food Technol.*, **45**(3), 82.

USDA (1988) Standards for frankfurters and similar cooked sausages. *Fed. Regist.*, **53**, 8425.

USDA (1992) Use of binders in certain cured pork products. *Fed Regist.*, **57**, 42885.

USDA (1994) Meat produced by advanced meat/bone separation machinery and meat recovery systems. *Fed. Regist.*, **59**, 62551.
USDA (1995) Poultry products produced by mechanical separation and products in which such poultry products are used; final rule. *Fed. Regist.*, **60**, 55962.
van den Hoven, M. (1987) Functionality of dairy ingredients in meat products. *Food Technol.*, **41**(10), 72.
Wallingford, L. and Labuza, T.P. (1983) Evaluation of the water binding properties of food hydrocolloids by physical/chemical methods and in a low fat meat emulsion. *J. Food Sci.*, **48**, 1.
Wirth, F. (1991) Reducing the fat and sodium content of meat products. *Fleischwirtschaft*, **71**, 3.
Yang, A., Trout, G.R. and Shay, B.J. (1995) Evaluation of carrageenan, isolated soy protein and a modified starch in low-fat frankfurters. *Proc. Int. Congr. Meat Sci. Technol.*, **41**, 435.
Zayas, J.F. and Lin, C.S. (1988) Quality characteristics of frankfurters containing corn germ protein. *J. Food Sci.*, **53**, 1587.
Zayas, J.F. and Lin, C.S. (1989) Effect of the pretreatment of corn germ protein on the quality characteristics of frankfurters. *J. Food Sci.*, **54**, 1452.

11 Production of low fat and reduced fat ground beef

D.L. HUFFMAN and R.D. HUFFMAN

11.1 Introduction

The nutritional quality of the food supply has emerged as a major concern of consumers. Dietary fat and calories from fat are among the most important of these consumer concerns. The meat industry and the allied food service and retail industries are addressing these consumer concerns. Food consumption patterns have changed dramatically in the last two decades. Trends show a shift in the consumption of fats, with a decrease in visible, separable fat consumption and an increase in the intake of low fat animal products, such as low fat milk and fish (NRC, 1988). The popular press and consumer studies continue to support the findings of a 1985 consumer study by Yankelovich (1985). In that study, over two-thirds of the consumers surveyed had some concerns about health, with one of their major concerns being the amount of fat in their diets. Media reports also reinforce the 1987 survey by Burke Marketing Research (1987), where most consumers said that they were limiting the amount of fat, calories and cholesterol in their diets. This survey also found that consumers were responding positively to leaner beef cuts with the perception that these cuts were more healthy. Similar findings were reported from a National Consumer Retail Beef Study by Cross, Berry and Wells (1980). In that study, consumers perceived that closely trimmed (0.76 cm external fat) or completely trimmed beef cuts were lower in fat and cholesterol than cuts with 1.27 cm external fat. In a National Restaurant Association consumer attitude and behavior study, approximately half of those interviewed were making a conscious effort to restrict their consumption of fat and cholesterol (NRA, 1987). The American Cancer Society (ACS, 1984) and American Heart Association (AHA, 1986) have recommended restriction of calories from fat to less than 30% of total caloric intake.

Ground beef constitutes nearly half of the total beef consumed in the USA. Therefore, reduction in the fat content of ground beef would be a mechanism for repositioning beef as a healthful food. The fact that fat plays a major role in the important palatability attributes of flavor and juiciness coupled with economic considerations provides a major challenge for production of low fat ground beef.

11.2 Demand for low fat and reduced fat ground beef

11.2.1 Consumer demand

Consumers are increasingly aware of the need for consumption of a healthier diet. Based on a study by the Food Marketing Institute (FMI, 1990), consumer concerns for the fat content in food products increased from only 9% in 1983 to 46% in 1990. A similar trend was shown for concern about cholesterol content of foods; however, concern for calorie content showed only a modest increase from 1983 to 1990 (FMI, 1990). Introduction of low fat ground beef into the fast food industry by McDonald's (Egbert *et al.*, 1991) and the National School Lunch Program (Berry, 1991a, b) provide evidence that there is a significant demand for low fat ground beef in the marketplace.

11.2.2 Trends in production – historical perspective

Over seven billion pounds (3.1752 billion kg) of ground beef products are consumed annually in the USA. Ground beef represented approximately 43% of the total fresh beef consumption in 1990 (AMI, 1990). By 1996, the National Cattleman's Beef Association (NCBA, 1996) estimated that ground beef had increased to more than 47% of total beef consumption. Hamburgers and cheeseburgers accounted for more than 75% of beef servings in the food service sector (NCBA, 1996). Labeling regulations are constantly changing and these changes in per cent fat labeling will probably impact the production of leaner ground beef products.

The simple reduction of fat would be the most efficient method of producing low fat ground beef products. However, the palatability of ground beef is directly related to fat content. Early research (Law *et al.*, 1965; Glover, 1968; Huffman and Powell, 1970; Mize, 1972) showed that sensory properties, particularly juiciness, of higher fat ground beef formulations were more desirable than those of reduced fat products. Those studies had as their major objective the interaction of soya products and sensory properties of ground beef. Fat content in the early studies ranged from 15 to 35%. The lower level (15%) would not be considered 'low' by current standards, and the higher level (35%) is now above the legal limit. Current labeling regulations governing reduced fat ground beef may be paraphrased as follows: (1) 'fat free' as 'less than 0.5 g of fat per serving'; (2) 'low fat' as '3 g of fat or less per serving'; (3) 'reduced fat' as 'at least 25% less fat per serving than the traditional item'; (4) 'light' as 'at least 33% fewer calories or 50% less fat per serving than the traditional item'; (5) 'lean' as 'less than 10 g of total fat, 4.5 g of saturated fat and less than 95 mg of cholesterol per serving'; and (6) 'extra lean' as 'less than 5 g of total fat, 2 g of saturated fat and less than 95 mg of cholesterol per serving'. For more details on labeling requirements, see Chapter 3.

11.2.3 Consumer studies

A consumer study reported by Glover (1968) included 300 households and compared 15, 20, 25 and 30% fat with no nonmeat additives. Generally consumer panelists preferred the product containing 20% fat. A second study reported by Glover (1968) involved 250 households and compared 15 and 20% fat levels and 0, 1 and 3% added soya. Consumers did not discriminate between the levels of fat, but preferred products containing a small amount of added soya.

A study of Huffman and Powell (1970) compared three fat levels (15, 25 and 35%) with and without 2% added soya. Trained sensory panelists preferred the lower levels of fat (15 or 25%) with 2% added soya. A consumer research report edited by Mize (1972) summarized research (649 households) from five southeastern states. All products (15, 25 and 35% fat, with and without 2% soya) were prepared in one processing plant. The presence of soya in all three levels of fat increased the consumer acceptance scores. Flavor scores were highest for ground beef containing 15% fat and 2% added soya. These results confirmed the earlier report of Huffman and Powell (1970) with trained sensory panelists using identical formulations.

Drake *et al.* (1975) conducted a ground beef consumer study, comparing fat content (15, 20, 25 and 30% fat) and level of TSP (textured soy protein at 0, 15, 20 and 25%) and concluded that there was no interaction between TSP and fat levels. Furthermore, any addition of TSP resulted in lower quality and acceptance ratings regardless of fat content.

Nutritional considerations were considered in Chapters 1 and 2. However, a report by Ono, Berry and Paroczay (1985) on nutrient retention of ground beef patties containing approximately 18, 22 and 27% fat is relevant here. These researchers concluded that for most nutrients, differences due to fat level were too small to be of practical nutritional significance.

11.3 Sensory properties of low fat and reduced fat ground beef

The principal chemical components of ground beef are protein, moisture and fat. Since protein is relatively constant, the finding that an inverse relationship existed between fat and moisture was expected (Kregal, Prusa and Hughes, 1986; Huffman and Egbert, 1990; Troutt *et al.*, 1992a). Thus, if either too much fat or moisture is lost during the cooking process, the ground beef will be dry. Since fat contributes to juiciness, flavor and textural properties of ground beef, if fat content is significantly reduced then sensory properties are adversely affected.

11.3.1 Juiciness and textural properties

The National Live Stock and Meat Board Beef Industry Council funded projects in 1988 at three universities and two private laboratories. The broad objective of these projects was to develop a ground beef product with less than 10% fat that had physical and sensory properties comparable to those of 'typical' ground beef. The first step in this collaborative research was to decide the level of fat that could be considered 'typical', which would provide a baseline for future research. Earlier reports had shown that as fat content of ground beef decreases, there is a significant decrease in product juiciness and tenderness (Kendall, Harrison and Dayton, 1974; Cross, Berry and Wells, 1980; Berry and Leddy, 1984; Kregel, Prusa and Hughes, 1986). Work at Auburn University (Huffman and Egbert, 1990) and Kansas State University (Troutt *et al.*, 1992b) established the fact that ground beef patties containing approximately 20% fat have the highest sensory ratings among patties containing 5, 10, 15, 20, 25 and 30% fat.

Berry (1992) reported an in-depth study of ground beef patties containing 0, 4, 8, 12, 16 and 20% fat cooked under varying conditions. He concluded that alterations in processing and cooking would probably be necessary to achieve acceptance of extremely low fat patties. This study confirms other reports (Huffman and Egbert, 1990; Troutt *et al.*, 1992b) which indicated that production of low fat ground beef products through simple fat reduction would result in a substantial decrease in product palatability, flavor intensity, juiciness and tenderness. In a later study, Berry (1993) found that juiciness scores for ground beef patties containing 6 and 20% fat were not different. Other studies (Huffman and Egbert, 1990; Liu, Huffman and Egbert, 1991; Liu *et al.*, 1991; Troutt *et al.*, 1992b) with patties ranging in fat content from 5 to 30% found that patties containing 10% fat or less had a denser structure and were generally less acceptable to panelists than patties containing up to 30% fat. From these reports it may be concluded that textural properties must be improved to have palatability attributes similar to those of ground beef patties containing 20% fat, the level that provided a standard for future studies.

11.3.2 Flavor enhancement

Huffman and Egbert (1990) reported that beef flavor intensity was highly correlated with overall acceptability ($r = 0.69$; $P < 0.05$). Berry and Leddy (1984) had reported earlier that trained panelists could detect greater beef flavor intensity in ground beef containing 19% fat than in that containing 14% fat. Interestingly, this report showed a decline in beef flavor intensity in ground beef patties containing 24% fat. These studies found that consumer acceptance of beef flavor peaked at about 20% fat. In the study reported by Huffman and Egbert (1990), sensory panelists rated patties containing iota-carrageenan, water, hydrolyzed vegetable protein (HVP) and encapsulated salt higher ($P < 0.05$) in beef flavor intensity than all-beef patties containing 20% or 8% fat. The two flavor enhancers, encapsulated salt and HVP, were probably responsible for the increased beef flavor intensity in this ground beef product. The cholesterol content of cooked patties containing 8% fat, added water, encapsulated salt and HVP was lower ($P < 0.05$) than that of 8 or 20% fat all-beef patties. This was probably due to the addition of water to the product resulting in a dilution of cholesterol.

Huffman and Egbert (1990) and Kasaback (1991) evaluated the effects of various flavor-related additives (e.g. black, white and red pepper, monosodium glutamate, onion and garlic powder, flaked salt, encapsulated salt and HVP) on the enhancement of beef flavor intensity in low fat ground beef. The results suggested that the use of 2:1 ratio of salt to HVP at concentrations of 0.25 and 0.125%, respectively, produced a beef flavor intensity similar to the 20% fat control.

There has been some interest in substituting vegetable fat for beef fat in ground beef patties. Therefore, a study designed to evaluate compositional properties and overall consumer acceptance of ground beef patties containing 10% fat (Liu, Huffman and Egbert, 1991), from beef or partially hydrogenated plant oils (corn, cottonseed, palm, peanut and soybean) is of interest. Addition of hydrogenated plant oils did not affect the proximate composition of raw or cooked patties. Patties containing hydrogenated corn or palm oil were not different ($P > 0.05$) from all-beef patties in flavor, cooking loss or overall acceptability. These researchers concluded that substitution of hydrogenated oils for beef fat in production of lean ground beef patties may be feasible if cost and marketing considerations are favorable.

11.4 Production of low fat and reduced fat ground beef

11.4.1 Beef raw materials

Raw material for ground beef is generally manufactured by blending lean cow beef and fat trimmings from grain fed cattle. Bullock *et al.* (1994) compared the sensory properties of low fat patties manufactured from six combinations of cow beef and Choice beef trimmings. They concluded that low fat ground beef patties manufactured using the carrageenan-based technology had equal or improved eating quality compared with commercially prepared 20% fat all-beef patties. While the raw material source can affect storage stability (color and lipid oxidation) of low fat ground beef patties, the source of raw materials had little effect on the sensory properties.

(a) Effects of pre- and post-rigor beef
Processing beef carcasses pre-rigor could produce hot-boned raw materials for ground beef production. Williams, Johnson and Reagan (1994) investigated the properties of ground beef patties containing 10 and 20% fat from pre- and post-rigor beef. They concluded that ground beef could be prepared from hot fat raw materials with almost no reduction in quality. Use of pre-rigor lean and fat raw materials decreased cooking losses and improved textural properties of both the 10 and 20% fat ground beef patties. Based on this study, reduced fat ground beef patties manufactured from hot processed carcasses have palatability traits equivalent to those of conventional ground beef patties. More recently, research by Berry, Joseph and Stanfield (1996) compared the use of electrical stimulation (ES) vs none (NS) and cold processing (CP) vs hot processing (HP) on the properties of low fat (10%) ground beef patties. They found that NSHP raw materials provided ground beef that had higher pH and cooking yields, less shrink in patty dimensions during cooking and lower shear force values than patties formulated from NSCP raw materials. The use of ES improved the tenderness and juiciness of patties from cold processed raw materials. Raw materials from beef that received electrical stimulation and hot processed (ESHP) had higher cooking yields than that from control (NSCP) raw materials. Hot processing of beef raw materials did result in a reduction of beef flavor intensity in the present study. However, hot processing did not affect beef flavor intensity in an earlier study in the same laboratory (Berry and Leddy, 1984). Based on these studies, Berry, Joseph and Stanfield (1996) recommended the use of electrical stimulation and hot processing for low fat ground beef production.

(b) Influence of pressure applied during processing
Several studies (Huffman and Egbert, 1990; Roth *et al.*, 1990; Kasaback, 1991) have reported that overall palatability can be improved by final grinding through a fine (2–3 mm) plate. Research conducted by Roth *et al.* (1990) suggested that pressure applied during processing greatly affected the textural properties of ground beef patties. The implications of these findings are that minimal pressure should be used during patty formation to avoid compaction.

11.4.2 Non-meat ingredients

The literature contains reports of the successful use of many nonmeat proteins (e.g. soy and milk proteins), carbohydrate-based materials (e.g. carrageenans, maltodextrins, starches and oat fiber) and functional blends of these products in the manufacture of low fat ground beef. The functional properties of these nonmeat ingredients will vary considerably, and therefore studies should be undertaken to evaluate these nonmeat ingredients in low fat ground beef systems. These nonmeat additives are discussed further in Chapter 10.

(a) Addition of soy proteins and oil

There have been significant refinements in soy protein technology over the last three to four decades. In earlier studies soy was generally incorporated into ground beef as soya bits of varying mesh sizes or as soy flour. Today products such as soy protein isolate, textured vegetable protein (TVP) and hydrolyzed vegetable protein (HVP) are in common use. A study reported by Liu *et al.* (1991) had the objective of elucidating the effect of addition of soy protein isolate, frozen textured soy protein isolate and hydrogenated soy oil on compositional, physical, microbial and sensory stability of low fat ground beef. Hydrogenated soy oil was used to replace beef fat at varying levels. They concluded that the use of a soy protein and oil emulsion was an acceptable method for incorporation of oil into ground beef products. Sensory panelists found the beef flavor of low fat ground beef products with 25% frozen textured soy protein isolate (10% fat from beef) as intense ($P > 0.05$) as the flavor of an all-beef control product. Addition of soybean oil and/or isolated soy protein decreased ($P < 0.05$) lipid oxidation and meat pigment discoloration. Soy protein and/or oil addition did not affect the microbial properties of the low fat ground beef formulations.

(b) Effects of carrageenans

Moisture retention during manufacture and frozen storage prior to cooking is imperative for desired juiciness and tenderness of the product. The food gum carrageenan forms a complex with water and protein and, therefore, should retain moisture in ground beef patties. There are three basic types of carrageenan: kappa, iota and lambda. Each of these carrageenans imparts different properties to the products to which they are added. The most important function of iota-carrageenan in a meat system, unlike other binders that function to bind meat pieces together, is that of moisture retention. Iota-carrageenan also has various other intrinsic properties that are beneficial in the production of low fat ground beef. Some, but not all, iota-carrageenans exhibit cold solubility, whereas the kappa form does not. Some functionality is achieved with higher concentrations of iota-carrageenan at lower temperatures. The cold solubility of some iota-carrageenans enhances the machinability of low fat ground beef. Iota-carrageenans have very good freeze/thaw stability compared with the kappa or lambda forms, which is essential in the commercial manufacturing of beel patties. Not only are all carrageenans not alike, but not all iota-carrageenans are alike. Therefore, the proper blend(s) of iota-carrageenan must be used to produce successfully high quality low fat ground beef. Various blends and concentrations of carrageenan were evaluated in a low fat ground beef product by Egbert *et al.* (1991). Juiciness and tenderness scores similar to those of the 20% fat control were achieved through the addition of 0.5% of an iota-carrageenan and 10% added water to the low fat product. With greater moisture retention, it was necessary to adjust the concentrations of salt and HVP to 0.4 and 0.2%, respectively.

(c) Improving sensory acceptance of low fat ground beef
As stated earlier, the major objective in the development of low fat (<10% fat) ground beef was to produce a low fat product with sensory properties equivalent to those of ground beef containing 20% fat. Trained sensory panel evaluation of beef patties containing 20% fat, 8% fat and 8% fat with added water, carrageenan, HVP and encapsulated salt suggested that the last product was more ($P < 0.05$) juicy and tender than either the 20 or 8% fat all-beef patties, according to Egbert *et al.* (1991). Lower juiciness scores were expected for low fat patties; however, this sensory problem was corrected by the addition of iota-carrageenan and water. Mealiness, a common problem in low fat ground meat products, was prominent for the 8% fat all-beef patties. Sensory panelists found these patties to be more ($P < 0.05$) mealy than the 20% fat patties and the reduced fat patties containing iota-carrageenan, water, salt and HVP. Improvements in textural properties were probably related to the addition of iota-carrageenan and water. Shear force values for patties containing 8% fat and added nonmeat ingredients were lower ($P < 0.05$) than those for 8% fat all-beef patties. The tenderness difference can be attributed to the addition of carrageenan and water.

11.4.3 Morphology

(a) Influence of iota-carrageenan
The morphology of the ground beef products discussed above was examined by light and transmission electron microscopy before and after cooking by Egbert *et al.* (1991). Fiber width decreased 10–15% upon cooking of the patties. In samples containing 8% fat with no additives, the muscle fiber bundles became closely pressed together after cooking. Particles of iota-carrageenan were homogeneously distributed in the patties containing 8% fat, water, iota-carrageenan, encapsulated salt and HVP. The size and shape of the particles remained unchanged after cooking. These particles were similar in size and shape to the lipid droplets observed in cooked 8 and 20% fat patties. At the ultrastructural level, the particles had the appearance of a gel or network. Based on the obser-

vations of Egbert *et al.* (1991), it was concluded that (1) given the similarity in size and shape of carrageenan particles in the low fat product to lipid droplets in the 8 and 20% products, it could be argued that the hydrated carrageenan might be similar in organoleptic perception to lipid droplets, and in that sense functioned as a replacement, and (2) the carrageenan particles had the ultrastructural morphology of fully hydrated gels after cooking and these gels probably helped retain moisture within the system.

(b) Effects of modified potato starch

Berry and Wergin (1993), using low fat ground beef patties manufactured from modified pregelatinized potato starch (MPPS), studied their ultrastructural morphology. These workers made the observation that MPPS became swollen due to hydration, and if this observation is accurate, the improved tenderness of the patties containing MPPS may be due to the swollen and hydrated MPPS particles. They further observed that hydration of the MPPS particles helped to open the structure and vary the fibrous texture of the patties. These conclusions are consistent with the earlier report by Dylewski (Egbert *et al.*, 1991) working with patties containing hydrated iota-carrageenan. Scanning electron photomicrographs from studies by Egbert *et al.* (1991) and Berry and Wergin (1993) provide evidence that nonmeat additives, such as carrageenan and potato starch, produce patties that are similar in morphological properties to patties containing 20% fat.

11.4.4 Addition of water binders

(a) Carrageenans

Dunkelberger *et al.* (1991a, b) tested consumer responses to the low fat ground beef patties containing iota-carrageenan, water, encapsulated salt and HVP that were developed at Auburn University. They found a positive response from consumers for the patties. This suggested a strong potential for consumer acceptance of the developed low fat ground beef product in the retail market.

Recent studies by Berry *et al.* (1996) comparing the use of iota- and kappa-carrageenan in a low fat (10% fat) ground beef system suggested that use of iota-carrageenan provided improvements in tenderness and juiciness over patties containing no carrageenan or kappa-carrageenan. This research agreed with an earlier report by Huffman and Egbert (1990) and concluded that iota-carrageenan should be recommended for commercial production of low fat or reduced fat ground beef products.

(b) Starch and fibers

Troutt *et al.* (1992b) found that added ingredients had the potential for improving the palatability of 5 and 10% fat ground beef, especially in comparison with the firmer, dryer, more crumbly low fat patties. A three-way combination of polydextrose, potato starch and either sugar beet, oat or pea fiber reduced firmness, cohesiveness, cohesiveness of mass and sustained cohesiveness in 5 and 10% fat patties so that they were similar to the 20% fat patties. These ingredients reduced beef flavor intensity, but the flavor scores were not significantly different from 20 to 30% fat controls. The juiciness traits of low fat products were reduced by the added nonmeat ingredients.

(c) Combination systems

Bullock *et al.* (1995) investigated nine different water binding systems, including five different types of carrageenan supplied by FMC (Viscarin SD 389, Viscarin ME 389, Gelcarin GP 379, 50:50 ME 389 and GP 379, and Gelcarin ME 621) and a combination of xanthan/locust bean gum, pea flour, algin and modified food starch. The only differences ($P < 0.05$) in water holding capacities were between patties containing Viscarin ME 389 and patties containing algin. Sensory panelists found only Viscarin ME 389 patties to be less juicy and tender than patties containing Viscarin SD 389, while patties containing algin were more ($P < 0.05$) tender than patties from other treatments. There were no differences ($P > 0.05$) between treatments for off-flavor or perceptible connective tissue. Patties containing algin received the lowest flavor intensity scores ($P < 0.05$). This study confirmed the fact that the formulation containing 10% water, Viscarin SD 389, HVP and encapsulated salt provided the most acceptable low fat ground beef patties.

Desmond, Troy and Buckley (1995) compared 16 blends of nonmeat ingredients (soy and whey proteins, tapioca starch, oat fiber, carrageenan, locust bean gum and pectin). These workers concluded that low fat burgers containing blends of tapioca starch, oat fiber and whey protein were superior in overall quality. They reported that blends containing pectin and a mixture of carrageenan and locust bean gum were lowest in overall quality. There are many types of hydrocolloids available and each has specific activity in a food system. Caution must be exercised when evaluating a generic nonmeat ingredient. Pectin may be derived from many sources and each may have different functional properties.

(d) Whey protein concentrate

Whey protein concentrate (WPC) has proved useful as a fat replacement in some food systems because of its gelation characteristics and high water binding ability. A report by El-Magoli, Laroia and Hansen (1995) showed that the addition of 1–4% of WPC improved cook yield, decreased shrinkage and increased the moisture content of the cooked patties. Higher levels (4%) of whey protein concentrate improved the flavor of low fat ground beef to the same level as the control patties containing

27% fat. The inclusion of calcium chloride with WPC did not enhance moisture retention or the sensory attributes of low fat ground beef patties. Based on the results of this study, whey protein concentrate is a nonmeat ingredient that should receive further study.

(e) Konjac flour

Konjac flour has the same cold gel setting properties displayed by iota-carrageenan. The possibility of taking advantage of the functional properties of iota-carrageenan and konjac flour in low fat ground beef was the focus of a report by Dickson *et al.* (1995). They used ground beef patties containing 20% fat and no konjac/carrageenan gel (k/c) as the control and patties containing 15% fat and 5% k/c, 10% fat and 10%k/c and 5% fat and 15% k/c as fat mimetics. They concluded that there were no significant differences in cooking yields between the control and the test samples, thus the k/c blend held moisture throughout the cooking process. There were no significant differences in cooking times among the treatments. Results showed that additions of the k/c gels could be used to reduce fat levels in a ground meat system to 10 and 15% without any significant effect on consumer acceptability. However, when the fat level was reduced to 5% by the addition of k/c gel, there was a significant reduction in consumer acceptability. This study is innovative because it approached fat replacement by using a cold setting gel rather than the simple addition of moisture to the low fat system.

11.4.5 Method of cookery

The method of cookery for low fat ground beef patties has been standardized in many research reports; however, studies by Troutt *et al.* (1992a, b) suggested that longer cooking times may be required with lower fat patties to reach internal temperatures equivalent to those of higher fat patties. Thus, some reduction in the sensory traits of low fat patties may be partially due to the longer cooking times required. Because of the food safety considerations prompted by the *Escherichia coli* 0157:H7 outbreak associated with inadequate cooking of beef patties, various regulatory agencies and trade associations have issued regulations or made suggestions regarding the cooking of beef patties. The traditional approach in ground beef cookery 'cook until the juices run clear' may not be relevant for low fat ground patties. Pink color may still be present at any of the minimum internal temperatures and holding times regulated by the USDA (Mendenhall, 1989). Further, food service establishments are likely to use longer cooking schedules to lessen the presence of pink color and the concern for food safety. As indicated earlier, the longer cooking schedule may adversely impact the sensory properties of low fat or reduced fat ground beef patties. Studies on method of cookery of low fat ground beef patties by Berry (1994) evaluated the effect of high temperature cookery to two visual degrees of doneness. This study included two methods of cookery, one designed to provide a temperature of 160 °C and the second designed to provide a temperature of 260 °C. Cooking of low fat (4% fat) patties to a well-done degree by high temperature cookery lessened tenderness, juiciness and flavor quality. However, cooking the low fat patties to a medium degree of doneness resulted in similar sensory properties to cooking 20% fat patties to the well-done degree. The concerns raised by this researcher are that while cooking of low fat patties to the medium degree of doneness assures palatability, this medium endpoint cookery may not assure food safety.

(a) Premature browning

Premature browning is a phenomenon where beef patties have a brown interior color typical of cooked ground beef at endpoint temperatures that should still be red or pink, as explained by Warren, Hunt and Kropf (1996). These researchers postulated that premature browning may be due to the oxidation state of the internal myoglobin prior to cooking. Patties with interior myoglobin in an oxidized state immediately prior to cooking are more likely to develop premature brown, whereas patties with interior myoglobin in a reduced state prior to cooking are more likely to be normal. Lavelle, Hunt and Kropf (1995) studied the internal cooked color of 10% fat patties made from vitamin E supplemented steer raw materials. Vitamin E did not affect internal cooked color or expressible juice color of the patties, and it did not cause premature browning. Further studies are under way which may provide new evidence regarding the relationship between preprocessing treatment and premature browning of ground beef patties.

11.4.6 Storage stability

Low fat ground beef technologies have been developed that have the desired sensory and physical properties. If low fat ground beef systems are to be successful in the retail trade, there must be an assurance that the addition of nonmeat ingredients and water does not decrease storage stability during normal retail and food service distribution. Egbert *et al.* (1992a) designed a study to learn the effect of low fat ground beef production using water, carrageenan, encapsulated salt and hydrolyzed vegetable protein on the microbial growth and oxidative (color and lipid) stability of low fat ground beef during simulated retail distribution. No differences were found in the aerobic plate counts, populations of psychrotropic bacteria or yeast and mold populations for low fat carrageenan-based ground beef and a low fat all-beef control. Retail display of the low fat products in oxygen-permeable (aerobic) packaging resulted in greater product discoloration than that of the same products in oxygen-impermeable packaging. The color stability of low fat ground beef products decreased with increased storage time. Few differences were found in oxidative stability and microbial populations of the two low fat products.

Other studies on refrigerated storage stability of low fat, carrageenan-based ground beef patties by Egbert *et al.* (1992b) evaluated the effect of potassium lactate on the aerobic storage stability and sensory properties of the patties. Bacterial growth in low fat carrageenan-based patties was not different from low fat, all-beef patties and was retarded by 2–3% potassium lactate with no deleterious effect on the sensory properties. Low fat carrageenan-based patties underwent greater discoloration and lipid oxidation during refrigerated storage, probably as the result of sodium chloride promoting oxidation.

An interesting technology involving the addition of tart cherries for low fat ground beef production was evaluated by Liu, Booren and Gray (1995). They compared the cherry formulation (Prevalean) with low-fat products containing iota-carrageenan and oat bran. These authors concluded that the addition of tart cherries reduced caloric content, had an antioxidant effect during 6 months of storage and decreased warmed-over flavor. They suggested that anthocyanins and polyphenols may be responsible for the antioxidant effects observed on incorporating the red tart cherries into ground beef.

11.5 Research needs

Acceptance of low fat and reduced fat ground beef products has been slow. The greatest challenge facing the research community is the development of a fat mimetic that has the satiety value of beef fat. There may be a psychological dimension to the consumer's desire for fat in a ground beef burger! Meat scientists should team with psychologists to explore this area. Further research is needed on nonmeat ingredients to achieve moisture retention and provide a juicy, tender cooked product. Flavor enhancement of low fat ground beef products is a dynamic area. New science and technology must be constantly evaluated. Research that has been initiated related to cookery and premature browning must be expanded to assure the safety of ground beef. Low fat ground beef will achieve wider acceptance only when a product that has equivalent safety and sensory properties to the high fat product can be produced for a competitive price.

11.6 Summary

A chronology of the development of low fat ground beef has been presented. Early research was largely with only modest reductions in fat content and generally involved addition of a soya product. Consumer awareness related to fat in the diet during the 1980s resulted in increased activity in the scientific community to develop technologies for reduction of fat in ground beef products.

Technology has been developed which includes the addition of nonmeat additives to low fat formulations. These technologies have received in-depth scientific scrutiny for biochemical, physical and sensory properties. Studies have been completed on oxidative and microbiological stability during fresh and frozen storage. Furthermore, consumer studies have documented the viability of some proposed concepts. Some developed technologies have been successfully transferred to the private sector. There is a need for further research on fat mimetics that will provide the satiety value and sensory properties of fat at a competitive price.

References

ACS (1984) Nutrition and cancer, cause and prevention. An American Cancer Society special report. *Ca-A Cancer J. Clinicians*, **34**(2), 121.

AHA (1986) Dietary guidelines for healthy adult Americans. *Am. Heart Assoc. Circ.*, **74**, 1465A.

AMI (1990) *Meat Facts '90*. American Meat Institute, Washington, DC.

Berry, B.W. (1991a) *Report to AMS, USDA on Evaluations of Lean Beef Patties (10% Fat) for Future Procurement and Use in the National School Lunch Program*, Meat Science Research Laboratory, ARS, USDA, Beltsville, MD.

Berry, B.W. (1991b) *Report to AMS, USDA on Evaluations of a Second Production of Four Formulations of Lean Beef Patties (10% fat) Selected for Testing in the National School Lunch Progran*, Meat Science Research Laboratory, ARS, USDA, Beltsville, MD.

Berry, B.W. (1992) Low fat level effects on sensory, shear, cooking and chemical properties of ground beef patties. *J. Food Sci.*, **57**, 537.

Berry, B.W. (1993) Fat level and freezing temperature affect sensory, shear, cooking and compositional properties of ground beef patties. *J. Food Sci.*, **58**, 34.

Berry, B.W. (1994) Fat level, high temperature cooking and degree of doneness affect sensory, chemical and physical properties of beef patties. *J. Food Sci.*, **59**, 10.

Berry, B.W. and Leddy, K.F. (1984) Effects of fat level and cooking method on sensory and textural properties of ground beef patties. *J. Food Sci.*, **49**, 870.

Berry, B.W. and Wergin, W.P. (1993) Modified pregelatinized potato starch in low-fat ground beef patties. *J. Muscle Foods*, **4**, 305.

Berry, B.W., Joseph, R.L. and Stanfield, M.S. (1996) Use of electrical stimulation, hot processing and carrageenan for processing of low-fat ground beef patties. *Meat Sci.*, **42**, 111.

Bullock, K.B., Huffman, D.L., Egbert, W.R. *et al.* (1994) Storage stability of low-fat ground beef made with lower value cuts of beef. *J. Food Sci.*, **59**, 6.

Bullock, K.B., Huffman, D.L., Egbert, W.R. *et al.* (1995) Nonmeat ingredients for low-fat ground beef patties. *J. Muscle Foods*, **6**, 7.

Burke Marketing Research (1987) *The Consumer Climate for Meat Study*, Prepared for the National Live Stock and Meat Board, Chicago, IL, and the American Meat Institute, Washington, DC.

Cross, H.R., Berry, B.W. and Wells, L.H. (1980) Effects of fat level and source on the chemical, sensory and cooking properties of ground beef patties. *J. Food Sci.*, **45**, 91.

Cross, H.R., Stansfield, M.S., Green, E.C. *et al.* (1975) Effect of fat and textured soy protein content on consumer acceptance of ground beef. *J. Food Sci.*, **40**, 1331.

Desmond, E.M., Troy, D.J. and Buckley, D.J. (1995) Eating quality of low fat beefburgers containing fat replacing functional blends. *Proc. Int. Congr. Meat Sci. Technol.*, **41**, 64.

Dickson, R.L., Pearson, A.M., Anglemeir, A.F. and Holmes, Z. (1995) Konjac flour carrageenan gel as a suitable fat replacer in a ground meat system. *Proc. Int. Congr. Meat Sci. Technol.*, **41**, 433.

Drake, S.R., Hinnergardt, L.C., Kluter, R.A. and Prell, P.A. (1975) Beef patties: the effect of soy protein and fat levels on quality and acceptability. *J. Food Sci.*, **40**, 1065.

Dunkelberger, J.E., Summerford, D.H., Adrian, J.L. *et al.* (1991a) Consumer home-use evaluation of a developed lean ground beef product. *J. Food Distrib. Res.*, **22**, 1.

Dunkelberger, J.E., Summerford, D.H., Huffman, D.L. and Egbert, W.R. (1991b) Consumer response to a new lean ground beef product: results from a home-use, comparative test of AU lean. *Ala. Agric. Exp. Stn. Bull.*, 612.

Egbert, W.R., Huffman, D.L., Chen, C.M. and Dylewski, D.P. (1991) Development of low fat ground beef. *Food Technol.*, **45**(6), 64.

Egbert, W.R., Huffman, D.L., Chen, C.M. and Jones, W.R. (1992a) Microbial and oxidative changes in low-fat ground beef during simulated retail distribution *J. Food Sci.*, **57**, 1269.

Egbert, W.R., Huffman, D.L., Bradford, D.D. and Jones, W.R. (1992b) Properties of low-fat ground beef containing potassium lactate during aerobic refrigerated storage. *J. Food Sci.*, **57**, 1033.

El-Magoli, S.B., Laroia, S. and Hansen, P.M.T. (1995) Quality characteristics of low-fat ground beef patties formulated with whey protein concentrate. *Proc. Int. Congr. Meat Sci. Technol.*, **41**, 449.

FMI (1990) *Food Marketing Institute Survey –1990*, Food Marketing Institute, Washington, DC.

Glover, R.S. (1968) Consumer acceptance of ground beef. *Proc. Recip. Meat Conf.*, **21**, 353.

Huffman, D.L. and Egbert, W.R. (1990) Advances in lean ground beef production. *Ala. Agric. Exp. Stn. Bull.*, 606.

Huffman, D.L. and Powell, W.E. (1970) Fat content and soya level effect on tenderness of ground beef patties. *Food Technol.*, **24**, 100.

Kasaback, C.M. (1991) Developments in lean ground beef technology. *MS Thesis*, Auburn University, Auburn, AL.

Kendall, P.A., Harrison, D.L. and Dayton, A.D. (1974) Quality attributes of ground beef on the retail market. *J. Food Sci.*, **39**, 610.

Kregal, K.K., Prusa, K.J. and Hughes, K.V. (1986) Cholesterol content and sensory analysis of ground beef as influenced by fat level, heating and storage. *J. Food Sci.*, **51**, 1162.

Lavelle, C.L., Hunt, M.C. and Kropf, D.H. (1995) Display life and internal cooked color of ground beef from vitamin E supplemented steers. *J. Food Sci.*, **60**, 1175.

Law, H.M., Beeson, M.S., Clark, A.B. *et al.* (1965) Consumer acceptance studies. II. Ground beef of varying composition. *La. Agric. Exp. Stn. Bull.*, 597.

Liu, M.N., Huffman, D.L. and Egbert, W.R. (1991) A research note: replacement of beef fat with hydrogenated plant oil in lean ground beef patties. *J. Food Sci.*, **56**, 861.

Liu, M.N., Huffman, D.L., Egbert, W.R., McCaskey, T.A. and Liu, C.W. (1991) Soy protein and oil effects on chemical, physical and microbial stability of lean ground beef patties. *J. Food Sci.*, **56**, 906.

Liu, H.F., Booren, A.M. and Gray, J.I. (1995) Quality attributes of low-fat cherry ground beef patties. *Proc. Int. Congr. Meat Sci. Technol.*, **41**, 427.

Mendenhall, V.T. (1989) Effect of pH and total pigment concentration on the internal color of cooked ground beef patties. *J. Food Sci.*, **54**, 1.

Mize, J.J. (1972) Factors affecting meat purchases and consumer acceptance of ground beef at three levels with and without soyabits. *Southern Coop. Ser. Bull.*, 173.

NCBA (1996) *The Beef Brief*, National Cattleman's Beef Association, Englewood, CO.

NRA (1987) *National Restaurant Association Food Service Industry Forecast*, National Restaurant Association, Washington, DC.

NRC (1988) Current trends in the consumption of animal products, in *Designing Foods: Animal Product Options in the Marketplace*, National Research Council, National Academy Press, Washington, DC, pp. 18–44.

Ono, K., Berry, B.W. and Paroczay, E. (1985) Contents and retention of nutrients in extra lean, lean and regular ground beef. *J. Food Sci.*, **50**, 701.

Roth, D.M., McKeith, F.K., McLaren, D.G. *et al.* (1990) Effect of temperature, pressure, particle size, mixing time and fat content on the textural properties of ground beef. *Proc. Recip. Meat Conf.*, **43**, 171 (Abstr.).

Troutt, E.S., Hunt, M.C., Johnson, D.E. *et al.* (1992a) Chemical, physical and sensory characterization of ground beef containing 5 to 30 percent fat. *J. Food Sci.*, **57**, 25.

Troutt, E.S., Hunt, M.C., Johnson, D.E. *et al.* (1992b) Characteristics of low-fat ground beef containing texture modifying ingredients. *J. Food Sci.*, **57**, 19.

Warren, K.E., Hunt, M.C. and Kropf, D.H. (1996) Myoglobin oxidative state affects internal color development in ground beef patties. *J. Food Sci.*, **61**, 513.

Williams, S.E., Johnson, S.P. and Reagan, J.O. (1994) Hot processed raw materials and fat level affect physical and sensory properties of ground beef. *J. Food Sci.*, **59**, 707.

Yankelovich, (1985) *The Consumer Climate for Meat Products*, Yankelovich, Skelly and White, Inc., prepared for the American Meat Institute, Washington, DC and the National Live Stock and Meat Board, Chicago, IL.

12 Low fat/salt cured meat products

J.F. PRICE

12.1 Introduction

Spurred by US Federal (USDA and HHS) 'dietary guidelines' and generalized concerns about fat intake from meats and reduction of dietary sodium (Chapters 1, 4 and 5), there has been a marked increase in the appearance of 'reduced fat', 'low fat' (<6% fat) and 'no fat or fat free' (<0.5% fat) intact muscle or restructured–brine added meat products in the marketplace and more choices relative to reduced salt items. It is notable how quickly the meat processing industries presented 95–98% 'fat free' products as the demand was acknowledged. The trend for 'no fat' or 'fat free' has been more recent as processors developed or adapted systems for selection of low fat meats or fat removal and for applying restructuring technology to produce cured meats, which may or may not be intact muscles or pieces, but that have the appearance and texture of whole muscle rather than that of ground, reformed or jellied meat.

Manufacture of low fat cured or brine added products is relatively easy from a production standpoint, but one must reckon with the roles that fat plays in appearance, texture and palatability, and thus the acceptance of low fat meats by the consumer (Pearson *et al.*, 1987). The palatability importance of fat rests mainly with its positive association with juiciness or succulence, the flavor contribution or flavor carrying effect and texture smoothness or mouth-feel, as well as satiety value. The visual contrast between lean and fat in a cut surface of a meat product makes it more attractive and also may affect marketability. Owing to the low demand for fat meat, fat impacts the cost/value or economic factors.

Eliminating the salt (minimizing sodium) in the manufacture of brine/additive extended cured meats, however, presents a more complex and difficult challenge. Salt is relied upon as the primary ingredient to control or adjust ionic strength of the fluid phase to the optimum level for protein extraction, bind or adhesion during gelation and water-holding capacity, and to provide flavor enhancement. Therefore, the more productive approach for this class of product has been to lower the target salt content to the minimum necessary for function by either leaving out a portion or replacing it with other ingredients (with a lower sodium level) that contribute to maintaining ionic strength or providing salty flavor, or do both simultaneously. In high salt-preserved or dried cured meats, the role of salt in lowering water activity and providing shelf stability and safety also needs to be considered.

Cured or brined, cooked and uncooked meat products may be classed traditionally as 'whole cut' (meaning entire primal or subprimal), 'whole muscle' (all or part of individual and adjacent muscles) or 'restructured'. Restructured products may be envisioned as 'sectioned and formed' (S&F), 'chunked and formed' (C&F), 'flaked (or sliced) and formed' (F&F) and/or 'ground (chopped or minced) and formed'. The last type often implies a finely chopped, emulsion-type product. The distinction between any two adjacent categories listed above is often arbitrary or obscure, blurring with low fat/extended products the difference between sausages and minced and formed restructured meats and between S&F and C&F or 'whole muscle' and S&F groups. Even with S&F and C&F products a minor portion may be a minced or emulsion phase, which may contain moderately high to very low levels of fat.

The principles of salt reduction in each of the classes (except for dried) is much the same and, to avoid redundancy with later chapters, this chapter is restricted to consideration of whole cut and whole muscle cured meats and S&F and C&F brined–extended items.

The author recognizes that the groupings by level of nonmeat ingredient addition and retention are fairly well defined and specifically limited for products falling within the purview of the USDA 'Protein Fat Free' or PFF regulation for cured pork items and turkey ham, and that the same definitions do not apply for dry cured items and bacon, raw or cooked beef, turkey breast or surimi products. Therefore, there is some disparity in what processes and manipulation of fat and salt can be prescribed for products made with meat from different sources. Often the distinctions made in one country do not apply internationally.

12.2 Low fat cured meats

12.2.1 Selection of low fat muscles and cuts

Generally, lower fat levels in final products are achieved by either starting out with less (Chapters 4, 7 and 8) or by diluting the fat present in meat raw material by extending the product with added ingredients (Chapter 10). By selecting low fat meat and extending it with ingredients and water, very low fat levels are possible. In most meat producing animals, with or without consideration of genetic selection for fatness/leanness and/or plane of nutrition or maturity factors, certain muscles or cuts have lower levels of adipose deposits as cover, seam or intramuscular fat than others and may be selected and trimmed to suitable fat contents. For processing whole cuts, such as bone-in hams, loins and picnics, those from animals bred or reared to have small amounts of cover and seam adipose must be selected for the reduced fat designation.

With chicken and turkey, the large breast muscles have little intramuscular fat, while the skin and seam fat are easily removed. Therefore, those muscles, which contain as little as 0.3% fat, are a logical choice to make lightly or heavily extended items that fit within the 'no fat' or 'fat free' labeling restrictions. Also, such meat is fit for manufacture of 95–98% fat free products, even with a small portion of emulsified skin in an S&F or C&F product style and when non-fat added ingredients with water are added up to 38% of product weight.

Meat from game and nontraditional animal species has been claimed to be low in fat. Fat levels of 1% or less have been reported in bison muscles by Berge *et al.* (1995), with little difference between individual muscles. In selected ostrich muscles, the fat content of broiled meat ranged from a low of 2.1% in gastrocnemeus to 3.8% in obturatorius medialis according to Morris *et al.* (1995). Assuming most of the cooking losses to be moisture, the fat content of raw muscle would probably be at or less than 3% in all muscles. This makes the meat suitable as a low fat raw material.

The larger limb and back muscles from red meat species tend to have small amounts of intramuscular fat, and the seam and cover fats can be removed by hand or machines. Such muscles from grass fed steers were shown to contain 2% or less intramuscular fat according to Garcia *et al.* (1995). The cutaneous (a superficial chest muscle) had the lowest (1.2%) intramuscular fat content, but the larger leg (semitendinosus and semimembranosus) and loin (longissimus and psoas) muscles were also fairly lean (1.5–1.8% fat). The larger foreleg muscles, including the biceps and triceps brachii, pectoral, infraspinatus, supraspinatus and serratus ventralis, can also be separated from seam fat and may contain 3–4% or less of intramuscular fat from selected carcasses. Almost any lean piece, which is trimmed of larger fat deposits, will contain less than 6–8% fat and can be used in reduced fat whole muscle or restructured cured items.

12.2.2 Removal of excess fat

A straightforward way to remove or avoid fat in processed meat raw materials is to produce, through management schemes or selection, breeding animals that have a low fat content throughout the edible tissues (Chapters 7 and 8). However, the cover (subcutaneous, skin or surface adipose) and seam (heavy adipose deposits within and between tissue membranes) fats are more accessible for physical separation by mechanical means (Chapter 4). Hand-knife trimming is suitable for adipose–lean separation, but may be expensive to accomplish, making it more useful for high value raw materials. Hand-operated electric or pneumatic 'trimmer' knives add speed, efficiency and completeness of surface fat removal.

Skinners or membrane removal 'skinning' machines assist in adipose tissue separation in certain circumstances. Using these techniques, meat cuts may be 'denuded', such that only the marbling (intramuscular) fat remains. The concepts and machines used for meat–bone and meat–membrane separation (e.g. some fish deboner styles) have been adapted as desinewing devices, allowing the separation of heavy connective tissues.

The fat in ground or minced lean–fat mixtures may be separated by virtue of differences in density if the mixture can be 'fluidized'. Centrifugal separators are used for 'partial defatting', and the floating fatty layer can be skimmed from minced tissue wash solution or cold meat–brine slurries to reduce the fat content in order to yield a leaner raw material for dewatering and further processing.

12.2.3 Low fat cured meats

(a) Pork ham, other pork cuts and turkey ham

These products in the USA may contain 0–50% added ingredients or nonmeat and must be processed and labeled to fit specific categories defined by the PFF of the product (per cent meat protein of the fat-free soft tissue). Similar products are being manufactured in other countries with varying ingredient restrictions. The 20.5 PFF assumed for raw ham and loin muscle (20.0 for other cuts) is based on the supposition that the moisture to protein ratio in these pork cuts is 3.8 : 1. Generally, the protein concentration in raw pork or turkey thigh muscle is slightly higher than this (PFF range <20.5–23.5) and about $6 \pm 3\%$ of product weight can be added ingredients (salt, sugar, cure, etc.) without triggering a label designation implying that the product had been extended with water, e.g. 'Product' with Natural Juices; 'Product' Water Added; 'Product' and water item – with declaration of the percentage (*XX%*) of the item that is added ingredients. The US regulation requires processors to target to (design a process for) a specific label category.

One hundred units (weight) of lean meat can be diluted with brine up to about 119 units of cooked product with minimum PFF of 18.5, 129 units with minimum PFF of 17.0 or 146 units with minimum PFF of 15.0 (natural juices, added water and product and water categories, respectively). Table 12.1 gives data ranges. Greater than 42–45% brine addition with a salt concentration within an acceptable range of palatability generally requires the addition of some other ingredient (Chapter 10) that will bind water and gel to produce an acceptable texture and to avoid excessive brine exudation during distribution.

The brine–extender matrix is primarily water and low fat ingredients. The dilution effect is to reduce the fat percentage in the extended products relative to that in unprocessed meat, yielding products 0.5–6% lower in percentage fat than the initial meat.

Table 12.1 Range of potential yields within PFF[a] groups for injection curing lean muscle per 100 units

			Raw meat with PFF[a] level		
PFF label group	Product PFF target		High 23	Moderate 21.5	Low 20
No qualifier	Wet	20.5	111.5	104.5	97.5
	Medium	21.5	106.5	100.0	93.5
	Dry	22.5	102.0	95.5	89.5
With natural juices	Wet	18.5	123.0	115.5	107.5
	Medium	19.3	118.0	111.0	103.5
	Dry	20.2	113.0	106.0	99.0
Water added	Wet	17.0	133.5	125.4	117.0
	Medium	17.7	128.5	120.5	112.5
	Dry	18.2	125.0	117.4	109.5
Product and water item	Wet	15.0	151.0	141.5	132.5
	Medium	16.0	142.0	133.0	124.0
	Dry	16.7	136.0	127.5	119.0

[a]PFF = protein fat free.

The effect is opposite (fat concentration) in dried or dry-cured products and those experiencing weight loss as moisture evaporation or drip. For such products, it is essential to select raw materials with a much lower fat level (%) than is targeted in the final (low fat) product depending upon the proportion of moisture lost during processing. To extend or dilute the fat in drier products, often low fat or nonfat proteinaceous extenders (collagen, soy, blood components) or grain flours and fiber ingredients (Chapter 10) may be used to satisfy reduced fat labeling definitions.

Muller (1989) has provided a diagram or flow outline of several schemes applied in preparing cured meat products, as shown in Figure 12.1. For most low fat products (<5% fat in product), a wet curing system with mechanical working (Figure 12.1) is applied. Any of the schemes may be applied to result in 'reduced fat' hams, provided that the raw material is sufficiently lean. Generally, for whole cuts and larger intact muscles, the brine is applied by pumping or injection, and these may or may not be subjected to mechanical forces. A combination of injection (pumping) and direct addition of brine into a mixer or massager is applied to turkey thigh muscles or smaller (often macerated) pieces and chunked meat. For chunked and formed types of product, brine addition to a massager or mixer is an obvious choice. Pieces could be brine-soak cured prior to the mechanical action.

The amount of brine to add per unit weight of raw meat and the concentration of the ingredients in the brine are governed by (1) the concentration of each ingredient desired in the final product per unit (nitrite and ascorbate salts are generally targeted to raw or ingoing meat weight

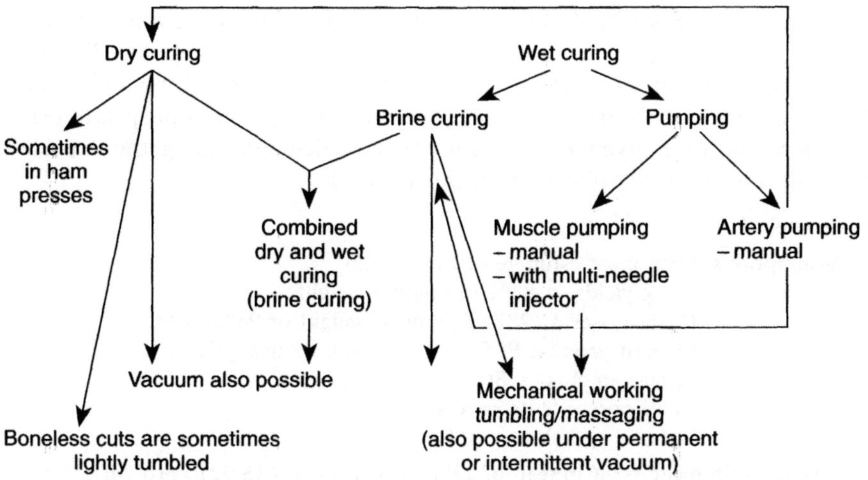

Figure 12.1 Schematic diagram showing various curing methods and treatments. (Reproduced with permission from Muller, 1989.)

as opposed to % in product), (2) the amount of brine retained in the meat (vs drainage losses), (3) smoking/cooking yield and (4) the targeted PFF range for the product (Brauer, 1987; Muller, 1989).

Brine uptake and retention are improved by appropriate mechanical action, i.e. gentler tumbling or massaging of larger pieces or more vigorous mixing–blending of smaller pieces. Mechanical working minimizes drain-out losses in most cases. Generally, more dilute brines are applied at the higher pumping/addition rates for products targeted for lower PFF ranges (i.e. greater extender %), and the mechanical working needed for brine retention is of greater vigor or greater duration or both. Therefore, no generally applicable formula for applying mechanical action can be given for all products and/or circumstances. The vigor of mechanical action (speed and machine action) and the duration (time) of application should be balanced to yield the desired results not only with brine uptake, retention and yield, but also with texture (rubberiness) and appearance of cut surfaces. Meat structure is lost with overworking and often longer times at slower speeds (or with intermittent action) do less damage to structure or textural appearance (Muller, 1989). There is some evidence that holding for 10–48 h after working prior to heating improves cook yields. Few studies have shown any advantage of raising the temperature during or due to mechanical agitation. Temperatures of 0–4°C are generally recommended for microbial control and optimum protein swelling. Heating to 10°C or above may produce detrimental effects (foaming, brine release, etc.). Vacuum or pressure during massaging helps avoid foaming and air bubble entrapment and may produce a firmer gel.

Estimation of pump% for PFF target. If, from experience and historical analytical records, one can estimate the protein percentage (as PFF) in raw lean tissue, the proportion of nonlean but brine absorbing soft tissues (adipose) and nonabsorbing tissues (bone and skin), the appropriate brine addition rate or a given PFF target can be calculated using the expected brine drain-out and cook-shrink losses (if any).

Example 1

Assumptions: Lean meat estimated PFF = 22.0
Cook yields = 88% of ingoing weight
Drain losses = 2% of pumped weight or 98% yield
Desired product PFF = 18.7 (ham, natural juices)
% lean in cuts = 86%
% adipose in cuts = 4%
% bone and skin = 10%

Given: 100 lb meat = 86 lb lean of 22% protein or has 18.92 lb protein

18.92 lb protein = 18.7% of what weight?
Answer = 18.92/0.187 = 101.1 + lb

Thus, one can raise 86 lb lean to 101.1 lb,
add fat and bone weight 101.1 + 10 + 4 = 115.1 lb target product weight

115.1 = 88% of what weight?
Answer = 115.1/0.88 = 130.8 lb estimated retained brine weight

130.8 = 98% of what weight?
Answer = 130.8/0.98 = 133.5 = target pump to weight
Thus, pieces should receive 33 ± 1% pump of raw cut weight.

Even if this were a boneless product, the estimated pump percentage is about the same (or slightly greater) when one corrects for bone/skin content and a lower drain loss attributable to massaging of boneless pieces.

The logic of this process becomes less practical in application if drain losses are large and/or if cook losses are extreme, and process procedure modifications are needed to control the losses. For high pump percentage levels associated with low PFF products, the brine ingredient concentrations are usually lower and oven cooking losses may be higher, necessitating the use of common sense in applying the arithmetic formula for establishing pump percentage. 'Cook-in' ham products (cooked in an adhering impervious container) or canned hams (Schied, 1984) experience no cooking loss and the procedure is reasonable.

Estimation of brine ingredient levels or brine formula to fit pump percentage computed for PFF target. To formulate a brine for the process, ingredient levels in the finished product should be targeted to fit palatability and regulatory requirements, and the brine should be of sufficient ionic strength or composition to allow brine uptake and retention within the muscle. For brines containing no extender/water binding ingredients, pump percentages ranging from 10 to 40% (6.5–15% salt in brine) are reasonable (Muller, 1989).

Example 2: Continue previous or bone-in product

Assumptions: 86% lean, 4% fat, 10% bone
Desired salt in cooked product = 2.0% (soft tissue only)
Desired sucrose in cooked product = 0.8%
Desired phosphate blend, product basis = 0.6%
Sodium nitrite, 156 ppm of meat weight = 0.0136% of product weight
Cook shrink (12%) is primarily moisture evaporation

Given: 115 lb product from 100 lb raw meat = 10 lb bone + 105 lb soft tissue (lean, fat, brine)

1. Target salt = 2% of 105 = 2.1 lb should be delivered in the 30.8 lb of retained brine or 2.1 (100)/30.8 = 6.8% by weight salt in brine
2. Sucrose = (0.008 × 105) (100)/30.8 = 2.7+%
3. Phosphate = (0.006 × 105) (100)/30.8 = 2.0%
4. Nitrite = (0.000136 × 105) (100)/30.8 = 0.46 or 0.05%

For boneless product, the desired brine concentration may be recomputed or just increased slightly, because the estimated finished product from 100 units of boneless meat will be 116.8 vs 115.1 owing to deletion of non-absorbing bone. Also, drain losses should be less.

When computations result in salt concentrations approaching 5.0% or less in the pumping brine without water binding additive(s), the ionic strength may be insufficient for adequate brine retention and protein swelling.

(b) Corned beef and pastrami

Traditionally these items were made from briskets and navels, which are the fatter cuts from fed cattle. Reduced fat products can be prepared with these traditional cuts by selecting lean pieces, separating larger fat deposits and extending the products with brine and binders. Raw corned beef brisket usually contains up to 20% added brine, while corned beef round is regulated for a lower brine extrusion (ca 10%). Cooked corned beef and pastrami may be pumped (massaged) and cooked to retain the upper limit of brine, ingredients (see ham section) or allowed to dehydrate from the pumped state during cooking producing a drier product.

Therefore, to produce a low fat item (<5% fat), only the lean cuts and pieces can be used. Dilution of the fat by high brine levels does not result in a great lowering of product fat percentage owing to limits on the weight gain imposed, particularly in whole muscle products. Restructured (and extended) beef lean matrices can be made with low fat levels, seasoned as corned beef or pastrami and cooked to produce products with fairly low fat levels. The brine addition level estimation and formulation principles given for ham in the previous section may also be applied in processing cured beef products.

Dried beef or dried cured beef may be made with whole muscles or from a restructuring process prior to drying. To target for low fat in these products, one must select very lean raw material.

(c) Turkey breast products

Turkey breast muscle usually contains little fat (Calcin, 1996) and is generally more fragile or easily disrupted at the muscle bundle level by mechanical working than pork or beef muscle and is reputed to be able to take up and hold a greater proportion of brine than red muscle provided appropriate levels of salt, phosphates and/or co-gelling extenders are included. Many, if not most, of the brine added turkey breast products are not 'cured' or do not contain nitrite for both oven roasted (in permeable film or open) and water cooked products (in impervious container in oven, steam or hot water).

Most nitrite-cured breast meat items are oven roasted or smoked in a humidity-controlled convection chamber (smokehouse). Surface browning after cooking may be by dipping in hot oil, exposure to a broiler or a gas flame or by hot air 'impingement'. Colored (caramel) coatings are often used to similate or enhance the browning effect.

The principles of formulating and adding brine in the manufacture of turkey breast products differ only slightly from those for ham. The final product composition in terms of moisture to protein ratio or PFF is not regulated, however, in the same manner as with ham. Attributable to tradition and regulatory prescription (salmonella control), turkey breast (cured–smoked) is cooked to a higher endpoint than ham [71°C (160°F) for turkey vs 62°C (143°F) to 70°C (158°F) for ham]. Thus, there is a tendency for a higher cook loss (evaporative and exudate) and increased muscle structure fragility. The latter is observed as loss of slice integrity and crumbliness. To offset these effects, a co-gelling extender, such as modified starch, may be used effectively (Chapters 10 and 15). The amount is generally about 2% of the product weight.

Intact (half) major breast muscles are usually injected with brine and tumbled with a measured amount of brine under vacuum with appropriate time and vigor for full brine absorbance without muscle disintegration.

On the other hand, restructured, C&F turkey breast can be made to appear like a whole muscle product and may contain a small portion of skin mixed with breast and wing trimmings made into an 'emulsion'. The batter may be shaped or molded into logs, loaves or simulated whole breasts and smoked. A sample recipe might be as follows:

100 lb turkey meat (<3% fat preferable) – chunked breast muscle (including up to 10 lb emulsion)
31 lb water or ice to remain as part of product weight (cook-in impervious material)
2.8 lb modified starch
2.2 lb salt
0.7 lb sugar
0.55 lb phosphate blend
0.25 lb curing salt with 6.25% nitrite
0.03 lb sodium erythorbate

137.5 lb = estimated yield per 100 lb meat
Add 15–30 lb ice or water to offset expected oven–smokehouse cooking shrink if cooked in pervious container.

The mixture should be blended/massaged under vacuum until the brine is fully incorporated and the muscle structure of chunks remains visible. It is then formed into the desired shape and smoke–oven cooked at 69–70°C (158–160°F).

(d) Surimi products
The washing process applied in the manufacture of surimi or a surimi-like base from muscle generally results in a lowering of the fat content even when the meat is from a fatter fish material (Smyth, O'Neill and Buckley, 1995; Kim *et al.*, 1996). Similar protein gels extracted from beef or pork have potential for use as ingredients in the manufacture of low fat products. The gel hardness and water-holding characteristics of beef/pork surimis are similar to those of fish surimis (Park *et al.*, 1996). Repeated washings of beef/pork minces resulted in low fat surimi but did not remove all of the color, according to Park *et al.* (1996). Surimi made from chicken meat mechanically recovered from bone by Smyth, O'Neill and Buckley (1995) was shown to be white in color with low fat levels and excellent gelling properties. It is difficult, however to visualize high volume production of low fat consumer products with either red meat or fish surimi alone, aside from those products already successfully marketed as specialty products (e.g. simulated crab meat and traditional dishes). However, surimi-like, high moisture, high protein gels have good potential for use as ingredients or binders in restructured low-fat meats and poultry products (Chapter 10).

12.3 Low salt cured meats

As noted in the Introduction, the aim of lowering the salt (sodium chloride) level in meats is to reduce the sodium intake for those individuals sensitive to dietary sodium (Kolari, 1980; Wirth, 1989, 1991). The inclusion of nutritional information on labels of processed meats has alerted the public to the high sodium levels attributable to the salt in brine cured meats and resulted in greater sales of lowered salt products. The limit to which salt or a replacer–salt combination can be reduced in the meat products considered in this chapter lies in the region of 1% ($\pm$0.2%) of product weight (Wirth, 1989, 1991) owing to the lessening of water binding capacity and the resulting exudate or jelly release during the cooking phase of manufacture. The substitution for the flavor effects of salt is another challenge and may or may not contribute to the ionic strength or sodium reduction.

12.3.1 Salt replacers – substitutes

(a) Other chloride salts

Water binding, protein swelling and some other functions of salt are attributable to their contribution to the ionic strength (μ) of the brine phase (water + salt) of the meat–brine matrix. In a lean meat mass each 1% of sodium chloride adds about 0.27 units to μ of the blend and NaCl generally ionizes completely. For these functions, depending on product style, the optimum μ falls in the range 0.6–1.4 (2.2–5.0% NaCl equivalent) and is achieved by using a combination of chloride salts and polyphosphate blends. Other chloride salts (Ca, K, Li, Mg) have been suggested and tried as replacers for NaCl and their contribution to the ionic strength (particularly KCl) closely matches that of NaCl. Several studies done or reported in the 1980s assessed the quality effects of the substitution for NaCl by other chlorides (Hand, Terrill and Smith, 1982a–c; Marsh, 1983; Terrill, 1983; Terrill *et al.*, 1983). It was during that period that NaCl and other chlorides were re-evaluated for their influences on flavor, water binding factors and microbial inhibition (Marsden, 1980; Maurer, 1983; Terrill *et al.*, 1983; Smith, 1988). Partial replacement of NaCl with KCl, up to about 50% replacement, was generally successful without the major negative effects of KCl on flavor. Increasing the KCl proportion was accompanied by increased bitterness and 'off-flavor' (Olson and Terrill, 1981; Marsh, 1983). Hand, Terrill and Smith (1982b) and Terrill (1983) found that LiCl replacement of NaCl came closest to producing the same flavor as NaCl with little off-taste. However, the saltiness scores for KCl- and LiCl-treated products were lower than that for NaCl use. LiCl is not approved as a food ingredient and is considered to be toxic, so has not been applied in commercial practice. Thus, 50:50 blends of NaCl and KCl (as 'lite-salt' or 'half-salt') are often sold and used as a salt replacer/substitute in several applications, including meat curing. At levels used in cured meats, 50% replacement by KCl often results in detectable flavor defects. Research indicates that 25–40% replacement appears to be the range at which the flavor impact is not as noticable (Terrill, 1983; Askar *et al.*, 1994). As the intensity of some added flavors increases, more salty, more acid or spicier, a higher proportion of KCl substitution may be tolerated.

(b) Salty flavor substitutes and enhancers
It is widely accepted that the presence of greater than taste threshold levels of food acids, particularly lactic, acetic or citric acid, enhances or intensifies the saltiness impact of the salt or contributes to saltiness itself. Also, the sodium or potassium salts of these acids may add to saltiness at the pH levels in cured meats (Chapters 13 and 14). Therefore, the use of sodium or potassium lactate with a corresponding reduction in the NaCl level tends to maintain a certain saltiness while reducing the sodium content in products to some degree (Anders, Cerveny and Milkowski, 1989; Askar *et al.*, 1994). Further reduction in sodium can be accomplished by up to 30% substitution of the NaCl in a given recipe by potassium lactate and using a KCl–NaCl blend for the remaining 70%.

Hofmann and Marggrander (1989) demonstrated that a particular collagen hydrolysate enhanced saltiness in certain products and might have potential for salt reduction. Turk and others associated with the Michigan Biotechnology Institute have experimented with salty flavor replacement by mixtures of lysine and succinic acid, eXalt, which may be engineered with varying sodium or potassium contents (Hause, 1962; Turk, 1993; Turk and Price, 1994; Turk *et al.*, 1996). The replacement ratio for similar saltiness is from 0.75:1 to 1:1 eXalt : NaCl. Although the flavor is slightly different, the replacer imparts saltiness to meat and may be used to replace up to 75% of the NaCl from a flavor perspective. Phosphates, gums or starches can be used to maintain the water binding function lost due to salt deletion.

12.3.2 Other approaches to salt replacement

The potassium salts of polyphosphates and phosphate blends that provide for optimum water binding, gelation or ionic strength effects are often equally as effective as the sodium salts. Several commercially available series of polyphosphate blends are available as the potassium salt (often in liquid) solubilized form. A 1% level in a lean meat product of a tetra- or penta-base polyphosphate (e.g. $K_4P_2O_7$, tetrapotassium pyrophosphate; $K_5P_3O_{10}$, potassium tripolyphosphate) has the potential to contribute up to 0.48 toward the μ of the brine phase. Even though they do not completely ionize in solution, in the process of mechanical working of brined meats the polyphosphates seem to have the indicated boosting effect on ionic strength. Phosphates are already common ingredients in cured meats up to 0.5% of product weight. Thus, the potential is with the use of potassium forms to replace the sodium contribution rather than replacing the ionic strength influence of NaCl (1 unit of phosphate potentially replaces 1.5–1.8 units of salt).

12.3.3 Examples of calculations

Brine recipe manipulation options can modify a standard brine targeted for 2.0% NaCl and 0.3% phosphate in the finished product as follows:

Brine recipe for 'cook-in' ham to yield 115 lb per 100 lb meat:
NaCl = 15.3% of brine weight
Phosphate = 2.3% of brine weight
(sweetener, nitrite and ascorbate levels common to all).

(a) Reduced chloride–potassium lactate substitution
Targeted for 1% NaCl, 0.5% KCl, 2.0% potassium lactate, 0.3% potassium polyphosphate in product.

Brine recipe:
NaCl = 7.67% of brine weight (50% less)
KCl = 3.83% of brine weight
K lactate = 15.3% of brine weight
K polyphosphate = 2.3% of brine weight.

(b) Reduced salt with sodium lactate and increased phosphate
Targeted for: 1.0% NaCl, 2.0% sodium lactate, 0.45% potassium polyphosphate.

Brine recipe:
NaCl = 7.67% of brine weight (50% less)
Na lactate = 3.83% of brine weight (adds some sodium)
K polyphosphate = 3.47% of brine weight.

(c) Reduced salt with increased lactate and phosphate
Targeted for: 0.5% NaCl, 3% sodium or potassium lactate, 0.5% potassium polyphosphate in product.

Brine recipe:
NaCl = 3.83% of brine weight (75% less)
Na or K lactate = 23% of brine weight
K polyphosphate = 3.83% of brine weight.

12.4 Research needs

The trend for market success of reduced fat and no-fat products will probably continue. Either ongoing or further research will be needed to assess the economic viability of traditional and newer methods and machinery for extracting fat or lean from meats and tissues. As a consequence of manipulating NaCl levels, safety questions relating to foodborne pathogens, chemical contaminants and unnatural proportions of unwanted components with fat separation methods will merit investigation as changes are adopted.

Further, the real or actual, as opposed to claimed, benefits to the general public from fat or salt reduction should remain under scrutiny by the scientific community. Fat substitutes or replacers (Chapter 9), their application in meats and safety and health aspects also merit ongoing investigation. Although not cited here, there exists a body of research results addressing questions relating comparative health impacts of any significant increase in potassium intake resulting from replacement of sodium compounds with their potassium counterparts.

12.5 Summary

There are few recent research reports in the traditional scientific journals on the subject of processing low fat and reduced sodium cured meats. The bulk of the work in developing low fat products, separating fat and lean and in reducing salt has been proprietary and applied in nature, and carried out by meat or equipment supplier plants and development laboratories. The principles of fat and salt reduction have been reiterated and most recent reports have been noted. Meat tissue and other food ingredients included in cured meats often contribute a significant portion of the sodium content, so that removal of salt *per se* may not be the sole means of accomplishing the goal of sodium reduction.

References

Anders, R.J., Cerveny, J.G. and Milkowski, A.L. (1989) Method for delaying *Clostridium botulinum* growth in fish and poultry. *US Pat.* 4 888 191.

Askar, A., El-Samahy, S.K., Shehata, H.A. and Tawfik, M. (1994) Pasterma and beef bouillon. The effect of substituting KCl and K-lactate for sodium chloride. *Fleischwirtschaft,* No. 1, 50.

Berge, P., Picard, B., Touraille, C. and Micol, D. (1995) Composition and sensory traits of bison meat. *Proc. Int. Congr. Meat Sci. Technol.,* **41**, 44.

Brauer, H. (1987) Cooked ham from the practical angle. *Fleischwirtschaft,* No. 3, 53.

Calcin, J. (1996) Only "lite" turkey in the U.S.: Jaindl Grand Champion all natural turkey. *Turkey World,* May–June, 10.

Garcia, P.T., Casal, J.J., Pensel, N.A. and Margaria, C.A. (1995) Lipids in lean beef. *Proc. Int. Congr. Meat Sci. Technol.,* **41**, 58.

Hand, L.W., Terrill, R.N. and Smith, G.C. (1982a) Effects of chloride salt, method of manufacturing and frozen storage on sensory properties of restructured pork roasts. *J. Food Sci.,* **47**, 1771.

Hand, L.W., Terrill, R.N. and Smith, G.C. (1982b) Effects of complete or partial replacement of sodium chloride on processing and sensory properties of hams. *J. Food Sci.*, **47**, 1776.

Hand, L.W., Terrill, R.N. and Smith, G.C. (1982c) Effects of chloride salts on physical, chemical and sensory properties of frankfurters. *J. Food Sci.*, **47**, 1800.

Hause, N.L. (1962) Process for enriching the L-lysine content of food and the resulting product. *US Pat.*, 3 015 567.

Hofmann, K. and Marggrander, K. (1989) Reducing common salt content of meat products by using collagen hydrolysates. *Fleischwirtsch. Int.*, No. 3, 34.

Kim, J.M., Liu, C.H., Eun, J.B. *et al.* (1996) Surimi from fillet frames of channel catfish. *J. Food Sci.*, **61**, 428.

Kolari, O.E. (1980) Salt dietary concerns, in *Proceedings of Meat Industry Research Conference*, American Meat Institution Foundation, Arlington, VA, pp. 89–100.

Marsden, J.L. (1980) The importance of sodium in processed meats, in *Proceedings of Meat Industry Research Conference*, American Meat Institution Foundation, Arlington, VA, pp. 77–87.

Marsh, A.C. (1983) Processes and formulations that affect the sodium content of foods. *Food Technol.*, **37**(7), 45.

Maurer, A.J. (1983) Reduced sodium usage in poultry muscle foods. *Food Technol.*, **37**(7), 60.

Morris, C.A., Harris, H.D., May, S.G. *et al.* (1995) Nutrient content of cooked ostrich meat. *Proc. Int. Congr. Meat Sci. Technol.*, **41**, 38.

Muller, W.-D. (1989) Technologie der Kochpokelwaren. *Fleischwirtsch. Int.*, **69**, 164.

Olson, D.G. and Terrill, R.N. (1981) Sensory properties of processed meats using various sodium salt substitutes, in *Proceedings of Meat Industry Research Conference*, American Meat Institution Foundation, Arlington, VA, pp. 59–66.

Park, S., Brewer, M.S., Novakofski, J. *et al.* (1996) Process characteristics for a surimi-like material made from beef or pork. *J. Food Sci.*, **61**, 422.

Pearson, A.M., Asghar, A., Gray, J.I. and Booren, A.M. (1987) Impact of fat reduction on consumer acceptance of processed meat. *Proc. Recip. Meat Conf.*, **40**, 105.

Schied, D. (1984) Manufacture of cook-in ham. *Fleischwirtschaft*, **64**, 1077.

Smith, D.M. (1988) Meat proteins: functional properties in comminuted meat products. *Food Technol.*, **42**(4), 116.

Smyth, A., O'Neill, E. and Buckley, D.J. (1995). Heat-induced gelation of chicken surimi. *Proc. Int. Congr. Meat Sci. Technol.*, **41**, 489.

Terrill, R.N. (1983). Reducing the sodium content of processed meats. *Food Technol.*, **37**(7), 66.

Terrill, R.N., Quintanilla, M., VanderZant, C. and Gardner, F.A. (1983). Effects of reduction or replacement of sodium chloride on growth of *Micrococcus, Moraxella* and *Lactobacillus* inoculated ground pork. *J. Food Sci.*, **48**, 122.

Turk, R. (1993). Metal free and low metal salt substitutes containing lysine. *US Pat.*, 5 229 161.

Turk, R.S. and Price, J.F. (1994). Use of "salty-flavor" salt replacer in Polish sausage and marinated chicken breasts, unpublished work.

Turk, R.S., Stitley, J.I., Dulebohn, L.Z. and Carlotti, R.J. (1996). The development of low sodium foods. *Abstr. 26th Annu. Spring Am. Assoc. Cereal Chem., Mgt., May 8–10, 1996, Wichita, KS.*

Wirth, F. (1988) Technologies for making fat-reduced products. *Fleishchwirtschaft Int.*, **68**, 1153.

Wirth, F. (1989). Reducing common salt content of meat products. *Fleischwirtschaft Int.*, No. 3, 25.

Wirth, F. (1991). Reducing the fat and sodium content of meat products. What possibilities are there? *Fleischwirtschaft.*, No. 2, 3.

13 Overcoming sensory problems in low fat and low salt products

F.J. MONAHAN and D.J. TROY

13.1 Introduction

Dietary recommendations for decreasing intake of fat and salt have led to the development of processed meat, poultry and fish products which meet these specifications. Meat products can contribute significantly to dietary fat and salt and, like other foods, low fat and low salt meat products are now available to consumers. However, both the removal of fat, with its associated decrease in calories and cholesterol, and the removal of salt have implications for the quality of meat products. Fat imparts succulence and the characteristic texture to meat products, while excessive reduction of fat results in products that are bland and have a rubbery or mealy texture (Keeton, 1994). Products with reduced fat may also appear darker than normal.

Removal of salt also has implications for the sensory quality of meat products. The flavor, juiciness, functionality and cook yield of meat products are reduced by removal of salt (McAuley and Mawson, 1994).

13.2 Low fat meat products

Fat replacement in low-fat products can be achieved in a number of ways: (1) use of leaner meats (fat reduced beef/pork, mechanically separated meats); (2) addition of water; (3) inclusion of protein-based substitutes (blood proteins, egg proteins, caseinates, soy proteins, wheat proteins, whey proteins); (4) inclusion of carbohydrate-based fat substitutes (fiber, cellulose, starches, maltodextrins, hydrocolloids, gums, konjac flour); and (5) inclusion of synthetic fat substitutes (Keeton, 1994; McAuley and Mawson, 1994).

13.2.1 Flavor problems in low fat products

Flavor problems associated with low fat meat products can be classified as those that are derived from both the absence of the fat component and those from the replacement of the fat by non-meat substitutes.

(a) Effects of fat reduction on flavor

Research into the relationship between fat level and quality in meat products has shown that the overall acceptability is affected by the fat level. However, the effects of fat level on flavor *per se* are not well established. If there are flavor effects, these may be confounded by the often more obvious effects of fat on the textural properties of meat products. In beef patties, studies conducted by Cross, Berry and Wells (1980), Berry and Leddy (1984), Kregel, Prusa and Hughes (1986) and Troutt *et al.* (1992) showed no significant effect of fat level on flavor intensity at fat levels in the range 10–30%. However, Troutt *et al.* (1992) found some evidence of a reduction in beef flavor when fat in patties was reduced to 5%. Huffman and Egbert (1990) also found no significant difference in beef flavor intensity of patties in the range 5–20% fat, even though overall acceptability was significantly higher in patties containing 15–20% fat. Nevertheless, it is clear that the lipids make a significant contribution to the flavor of meats, although their contribution relative to other meat components remains to be defined. Lipid reactions are believed to contribute ~90% of the flavor volatiles in meat (Reineccius, 1994).

Earlier studies (Wood and Bender, 1957; Hornstein and Crowe, 1960, 1963; Kramlich and Pearson, 1960; Wasserman and Talley, 1968) demonstrated that fat was needed to give the unique species-specific flavor notes to meat while the characteristic 'meaty' flavor of different species was associated with the water-soluble fraction. There is little doubt that fat contributes to the species-specific flavor components. For example, the presence of branched-chain fatty acids (4-methyloctanoic, 4-methylnonanoic) in the adipose tissue of mutton and goat contributes to the characteristic flavor of these meats (Wong, Johnson and Nixon, 1975).

(b) Effect of level and type of fat on flavor

Potential effects of fat on the flavor of meat products are dependent on both the level of fat and the type of fat (inter- or intramuscular, storage or structural) in the product. In the manufacture of low fat meat products, it is largely the subcutaneous or intermuscular fat that is reduced while the intramuscular fat remains. Fostier *et al.* (1993) showed that flavor intensity increased with increasing intramuscular fat and not with intermuscular fat level. Thus, the reduction in the lipid content (especially triglyceride lipid) in low fat meat products may not affect the potential 'meaty' flavor development in the cooked product. In addition, it appears that it is the phospholipid fraction that contributes substantially to basic 'meaty' flavor (Mottram and Edwards, 1983; Whitfield *et al.*, 1987; Farmer and Mottram, 1990). The phospholipid component is not significantly reduced in low fat meat products. In this context, the flavor problems in low fat meat products may not be due so much to the absence of specific flavor precursors as to the loss of the physical characteristics of the fat.

Fat contributes to taste sensation by controlling the release of desirable flavor in the mouth, affecting retention of flavor components in the mouth or affecting the interaction between flavor compounds and flavor receptors on the tongue (Kinsella, 1990). Mela (1992) stressed the importance of both texture and flavor in the acceptance of foods. Thus, the absence of the texture of fat may contribute to the perception that low fat products are less flavorful. Furthermore, Imafidon and Spanier (1994) pointed to nonvolatile water soluble products of proteolysis and Reineccius (1994) to Maillard reaction and thiamine degradation products as important determinants of flavor. These components should be present in both normal and low fat meat products, again suggesting that it is the loss of fat structure and texture in low fat products that makes them less flavorful. This may be a physicochemical phenomenon arising from the slow release of flavor from fat during mastication and the lubricant characteristic of the fat.

Successfully overcoming flavor problems associated with low fat products requires an understanding of exactly what is missing in such products. Flavors that are smooth and lingering on the palate in full fat products come across as harsh and biting in reduced fat products and even more so in nonfat products (Shamil, Wyeth and Kilcast, 1991–92). Various fat substitutes in combination with flavor and flavor-enhancing ingredients have found application in low fat products and these will be dealt with in the following section. There is no ingredient formulation that successfully meets all the needs of low fat products, and in some cases the fat substitutes themselves can cause flavor problems.

(c) Effects of fat replacers on flavor

The range of fat substitutes used in low fat meat products has been discussed by Keeton (1994) and McAuley and Mawson (1994). The fact that there is a range of products in use demonstrates that there is no single ingredient or blend of ingredients available to meet all the needs of manufacturers of low fat meats. Among the problems that are associated with low fat meat products containing fat substitutes are (1) a decrease in meat flavor due to dilution by nonmeat ingredients, (2) a decrease in the formation of flavor compounds between indigenous meat components, (3) development of flavors specific to the fat substitutes themselves and (4) altered flavor release and mouth retention characteristics.

Flavor dilution

Many nonmeat ingredients provide a matrix for incorporation of additional water into products. In ground beef products, coarse sausage and emulsified products, levels of addition of nonmeat fat substitutes and water are in the ranges 0.25–5% on a dry weight basis (dwb) and 7–25%, respectively (Keeton, 1994). It is not surprising, therefore, that 'meaty' flavor can be diluted by the use of nonmeat ingredients. In low fat ground beef patties containing algin, Bullock *et al.* (1995) attributed the low flavor intensity to the increased moisture content in the product after cooking. Ground beef patties containing a starch gel were also found to be lower in flavor than controls (Berry and Wergin, 1990). In addition to the dilution effect, water-soluble meat flavor compounds in products containing added water may be lost as a result of water loss during cooking. This may be a problem with carrageenan (melting point 49 °C) used in ground beef products (Keeton, 1994).

Decreased meat flavor formation

The expected meaty flavor of low fat products may be reduced because less flavor compounds are formed. For example, fewer Maillard browning reactions may occur because the meat constituents that contribute to these reactions are diluted more or react differently in the presence of fat substitutes. In addition, the flavor compounds derived from the fat itself, especially species-specific compounds, are likely to be lower in low fat products.

Flavors arising from fat substitutes

Low fat meat products may deviate in flavor from the expected flavor of their full fat counterparts as a result of particular flavors brought to products by the fat substitutes. Soy proteins can impart a 'beany' or 'cereal-like' flavor to products which contain them (Williams and Zabik, 1975). In low fat ground beef containing soy flours and concentrates researchers have reported the presence of increased soy flavor (Drake *et al.*, 1975; Kotula, Twigg and Young, 1976; Gardze, Bowers and Caul, 1979; Berry, Leddy and Bodwell, 1985; Liu *et al.*, 1991). When a combination of textured soy protein and soy protein isolate was used in wiener-type products, flavor scores decreased with increasing soy content (Sofos, Noda and Allen, 1977). Some of the newer soy protein preparations have good flavor profiles with little or no 'beany' flavor. They may require slightly increased seasoning for optimal flavor. Defatted corn germ protein flour used as an extender in beef patties gave rise to a significant increase in off-aroma in these products (Brown and Zayas, 1990).

Sub-optimal flavor release and retention

Perhaps the greatest challenge facing manufacturers of low fat products is the development of low fat substitutes that mimic the texture and mouthfeel of fat and have similar flavor release properties to fat. Several fat substitutes are claimed to have textural attributes that approach those of fat. For example, Simplesse, a low calorie protein-based fat substitute, is produced by microparticulation of milk proteins and/or egg whites. The resultant spheroidal particles, ranging in size from 0.1 to 2.0 mm, have a smooth and creamy texture attributed to their ability to roll easily over one another. However, effectively mimicking the textural attributes of fat in fat substitutes does not imply that the flavor release characteristics will mimic those of fat. The binding affinities of lipids for flavor compounds differ from those of other food components (Kinsella, 1990). For example, starch-based food ingredients bind flavor compounds by forming inclusion complexes, while some proteins with hydrophobic pockets have flavor binding properties (Plug and Haring, 1993).

(d) Overcoming flavor problems in low fat products

In most low fat meat products, seasonings or flavorings are included to improve palatability. In products, such as hamburgers, where the expected flavor is the natural flavor of the meat, adjustment of product formulations for low fat products involves the use of seasoning blends and ingredients such as hydrolyzed vegetable protein, hydrolyzed yeast extract, hydrolyzed milk proteins, beef extract and meat-based flavor concentrates to provide acceptable meat-like flavor (Anon., 1991, 1995; Giese, 1994a, b; Keeton, 1994). In the case of products which have a distinctive and expected flavor derived from the use of seasonings, adjustment of the product formulation for low fat products may be easier because the more subtle meat flavors are masked by added flavorings. In both cases, however, the loss of the fat matrix may have a detrimental effect on consumer perception of 'meaty' flavor and on flavor from seasonings.

Many researchers have investigated approaches to improving the flavor intensity of low fat products. In most reports of the successful application of fat substitutes in meat products, the substitutes are used in combination with added flavorings. In ground beef patties, Egbert *et al.* (1991) used hydrolyzed vegetable protein (0.2%) and encapsulated salt (0.4%) to compensate for reduced flavor when 0.5% iota-carrageenan and 10% water were used to replace fat in the formulation. Pszczola (1991) reported the use of an oat bran-based ingredient with added flavorings and seasonings to replace fat in burgers and sausages and overcome taste and texture problems in the low fat product. Taki (1991) reported the successful use of a blend consisting of modified starch, rice flour, salt, emulsifier and flavor in low fat beef patties and pork sausage. The functional blend was claimed to be as flavorful as or more flavorful than fat products containing additive blends of soy isolate, hydrolyzed vegetable protein, carrageenan and salt. Some nonmeat ingredients used to decrease the fat content of meat products may themselves bring desirable flavor characteristics to the product, which can mask any dilution of meat flavor. For example, ground beef containing wild rice, which has a 'smoky, tea-like, toasted' flavor (Withycombe, Lindsay and Stuiber, 1978), was rated more desirable than beef without rice (Minerich *et al.*, 1991).

As flavor chemists learn more about particular compounds that impart flavors to meat products, the possibility of synthesizing these 'natural' flavor components becomes more likely (Kinzel, 1992).

There are numerous reports in the literature of ingredient blends found to be suitable for particular low fat meat applications. However, a more thorough understanding of how these ingredients interact with each other and with the meat matrix is urgently needed. Low fat ingredient formulations that are successful in the laboratory following chemical and sensory evaluation of their flavor effects may not stand up to the rigors of the marketplace. It is clear that the physical phenomena relating to flavor release in low fat meats are as important as the presence of the flavor compounds themselves (Plug and Haring, 1993). Thus, the retention and release of flavor are undoubtedly influenced by the presence of fat in meat products. An understanding of the how specific fat replacers in specific low fat products influence flavor release and retention in the mouth is needed.

The level of addition of fat replacers is also important in assessing their effect on flavor. For example, in the case of hydrated soy protein concentrates, incorporation at 20–30% in beef patties had a detrimental effect on flavor (Kotula *et al.*, 1976), while no significant effect on flavor was reported at 10% incorporation (Ziprin *et al.*, 1981).

13.2.2 Texture problems in low fat products

The texture of food systems is multi-faceted in nature and its underlying basis and quantitative measurement are of immense interest to researchers (Moskowitz, 1987). The contribution of fat to the texture of a food system is dependent not only on the level and source of fat but also on the interactions of the fat with other food components. Meat product texture is greatly affected by fat reduction and much research effort is directed towards lessening this effect. Descriptors of food texture include hardness, softness, juiciness, chewiness, greasiness, viscosity, slipperiness, creaminess, crispiness, crunchiness and brittleness, all of which are influenced by the fat content of foods (Drewnowski, 1992). Fat also imparts essential mouth-feel and plays a major role in the mechanism of fracture in semisolid and solid foods (Marshall, 1990). In ground meat products, such as beefburgers, fat imparts a soft texture which is in contrast to the harder texture of the fibrous muscle proteins. Low fat beefburgers are harder, denser and less cohesive during initial biting (Berry and Leddy, 1984). In finely comminuted products, such as frankfurters, fat is present in the form of small oil droplets or particles surrounded by a protein film. When the fat level is reduced in these emulsified products, the structure is disrupted and the meat juices or added water are less effectively immobilized.

(a) Overcoming texture problems in low fat products

The approaches to fat replacement in meat products have already been mentioned (Keeton, 1994; McAuley and Mawson, 1994). These approaches can be classified broadly as involving either the addition of extra lean meat to product formulations or the addition of extra water. Cost has a major influence on the approach taken, since replacing fat with lean instead of water alters profit margins considerably. It goes without saying that if added water is increased in a formulation it should not be lost excessively during subsequent processing.

Replacing fat with lean meat can result in a product with a fibrous, hard, rubbery texture while addition of water can result in a product which is soft and pasty. The water may be added as is or it may be added with extra physical processing of the product or it may be added with other nonmeat ingredients. The latter is the more common approach and requires the judicious selection of ingredients to ensure sufficient water binding during processing, storage and cooking.

Replacing fat with lean meat

While it is possible to produce low fat meat products simply by replacing the fat with lean, it is not normally commercially viable, nor does it result in products with desirable attributes. For example, tenderness and juiciness in beef patties were shown to decrease as fat was reduced from 28 to 16% (Cross, Berry and Wells, 1980). Nevertheless, lean for fat substitution can yield products with acceptable palatability if fat reductions are small and if a moderate cooking regime is used. Troutt *et al.* (1992) demonstrated that the palatability of low fat beef patties cooked to 71 °C was greater than that of those cooked to 77 °C. However, if the cost of a low fat product is substantially higher than its full fat counterpart and if the consumer must follow very precise cooking instructions to make it palatable, with due concern for food safety, such a product will not be commercially viable.

In finely comminuted products, such as frankfurters and bolognas, fat reduction achieved by increasing lean meat content has similar effects as in the coarsely ground products, i.e. they are more rubbery, harder, firmer and more chewy (St John *et al.*, 1986; Hand *et al.*, 1987; Marquez *et al.*, 1989). These effects on texture have been attributed to increased protein–protein interactions and excessive binding in the higher protein products. One approach to preventing this problem is to reduce protein functionality in these high lean, low fat meat products. Freezing causes a reduction in protein functionality (Miller, Acherman and Palumbo, 1980). Brauer (1993) described a process whereby altering the ratio of frozen to nonfrozen lean meat in low fat frankfurter formulations produced an acceptable product. Mechanically deboned meat also has a lower binding ability and could be used similarly. Particle size effects on low fat precooked pattie texture were investigated by Lin and Keeton (1994). They concluded that particle size, shape and density affected yield and textural traits and recommended that these products should be flaked during manufacture.

Addition of water

Adding water to substitute for fat increases profit margins but national regulations regarding water addition must be followed. Generally, reducing fat and adding water results in a less firm, softer product. However, Claus (1991), working with a bologna product, showed that as fat was decreased and added water increased at a constant protein level only a slight decrease in product hardness was detected. It was concluded that in order to achieve firmness similar to that of the full fat product, the percentage increase in added water should be less than the percentage decrease in fat. Barbut and Mittal (1995) recently demonstrated in breakfast sausage that a 13% fat product was as acceptable as 23% fat sausage in terms of tenderness when the protein content was kept constant at 16.4%. Thus, the texture of meat products with water added to replace fat depends greatly on the protein content of the meat system and its ability to bind the added water. It appears that, in contrast to the substitution of fat by lean, when added water replaces fat, the functionality of the proteins must be sufficient to maintain a higher water binding capacity, lower cook yield and texture. A limited amount of extra water can be added to low fat formulations with no major detrimental effects on texture once the protein is available and functional.

Methods to increase the functionality of the protein have been examined. Claus (1991) suggests that, since pre-rigor meat has higher protein functionality owing to the large spaces available for water retention between the myofilaments, its use in low fat/high added water meat product manufacture could be advantageous. However, its potential has not been fully examined to date.

Addition of extra water with physical processing

Tumbling and massaging are already used widely in the meat industry to extract salt soluble proteins in muscle. In low fat bologna manufacture, massaging the meat ingredients for an extended period led to an improvement in emulsion thermal stability and sensory and instrumental cohesiveness, but had no effect on cooking yields (Claus *et al*., 1990; Gregg *et al*., 1993).

High pressure and ultrasound treatments have also been proposed to increase protein–protein interactions and increase protein functionality

(Mandava, Dilber and Fernandez, 1995; Reynolds *et al.*, 1978). However, neither of these techniques has been examined in the context of low fat meat products.

Addition of extra water with functional ingredients
If adding extra lean meat, adding extra water or the use of physical processes are not feasible, either commercially or technologically, functional nonmeat ingredients may be included in low fat meat formulations. These ingredients should permit the meat system to retain added water without adversely affecting cooking loss, texture or other sensory attributes. There are over 30 such ingredients and many reviews dealing with them (Giese, 1992; Mandigo and Eilert, 1994).

Fat replacer systems utilize properties of extenders to increase and maintain water binding capacity during processing, storage and cooking of the products. These nonmeat ingredients may be protein or carbohydrate based or functional blends.

Protein-based fat replacers are derived from both animal and plant materials and include soy protein flours, concentrates and isolates, nonfat dried milk, caseinates and whey proteins and collagen. Less widely used proteins include wheat flour, vital wheat gluten, cotton and oat flours, surimi, blood plasma and egg protein. These proteins have all proved useful in overcoming the effects of fat reduction in meat products. Soy protein flours and concentrates decrease total cooking loss, moisture loss and shear force but also decrease meat aroma and flavor intensity. A wealth of information exists on the use of soy proteins in different forms and in different products (Keeton, 1994). It appears that soy proteins can offset many of the textural effects of fat reduction, but may require additional seasonings to overcome palatability problems. This is not an uncommon problem with fat replacers. Milk proteins can act as emulsifiers and stabilizers and as water and fat binders. It is not surprising, then, that they can help alleviate textural defects in low fat meat products. One added advantage is that they have no strong associated off-flavors. Inclusion of dairy proteins in low fat (10%) beefburgers gave products with similar textural characteristics to full fat (23%) controls without extenders (Desmond, Troy and Buckley, 1994).

Collagen is becoming more widely used as a meat protein extender and fat replacer. Mandigo and Eilert (1994) describe its origin for this purpose. It has been shown to improve the textural attributes of low fat patties but its performance is affected by its source, form, level of inclusion and mode of application and further work is required to optimize its use (Eilert *et al.*, 1993).

Other protein sources with the potential to enhance the textural attributes of low fat meat products include blood plasma, surimi and egg (Colmenero, 1996). Corn germ protein has water-binding and texture-softening properties in frankfurters, according to Zayas and Lin (1989).

Carbohydrate-based fat replacers are probably the most widely used class of replacers in low fat meat product manufacture (Shand *et al.*, 1990). They consist of water-soluble carbohydrate polymers of gums, starches, cellulosics, dextrins and maltodextrins and are derived from cereals, grains and plant exudates. They possess good water-binding abilities and are widely used as texture-modifying agents throughout the food industry.

Gums have received considerable attention in relation to low fat meat product manufacture. Carrageenans, especially the iota form, retain moisture and give good textural characteristics in low fat ground beef (Egbert *et al.*, 1991). When iota-carrageenan was included in a low fat formulation (Mittal and Barbut, 1994), low fat frankfurters (14%) were similar in hardness, springiness and chewiness to full fat controls (28%). Hoogenkamp (1991) described some problems encountered on using carrageenan in low fat patties, including the formation of a skin on the patty surface. Nevertheless, carrageenans have excellent properties in meat product fat replacement systems.

Xanthan gum has a stronger water-binding ability than carrageenan (Wallingford and Labuza, 1983). Its inclusion led to an improvement in texture parameters in meat batters (Foegeding and Ramsey, 1987) and in low fat frankfurters (Mittal and Barbut, 1994) without any adverse effects on water-binding capacity.

Other common gums also have potential as functional components in fat replacement systems based on their properties as texture-modifying agents. These include locust bean gum, guar gum and alginates. However, it is more likely that they will be used in conjunction with other ingredients in order to achieve product-specific goals (Goutenfongea and Dumont, 1990; Mandigo and Eilert, 1994).

Starches are the most commonly used food hydrocolloid and Shand *et al.* (1990) provided a good summary of their properties. Most starches used in low fat meat technology are pre-gelatinized to permit cold water swelling. They can stabilize meat batters by absorbing excess water and maintaining texture. Starches (or their maltodextrins) come from a variety of sources including potato, oat tapioca, wheat, corn and rice. Texture improvement in low fat beef patties was obtained with potato starch (Berry and Wergin, 1990) and rice starch (Minerich *et al.*, 1991). However, starches from different plants do not all perform equally well (Goutenfongea and Dumont, 1990). Oat-based maltodextrins are used on a commercial basis in some low fat meat products, such as ground beef (Giese, 1992). Hull, Rogers and Martin (1992) demonstrated that low fat frankfurters with acceptable texture and taste could be manufactured with added carbohydrate in the form of white beans. Similarly Liu, Booren and Gray (1995) showed that red tart cherries and oat bran could be used in the manufacture of low fat beef patties.

Functional blends containing a number of ingredients have been formulated for different products. These have emerged because there is no single fat replacer which imparts quality attributes to low fat meat products equivalent to those of their full fat counterparts (Desmond, Troy and Buckley, 1994). Advanced Oat Fibre (Williamson Fibre Products) is a functional blend containing carbohydrates, egg albumin and flavorings (Desmond, Troy and Buckley, 1994) while Primo-O-Lean (Devro) contains tapioca starch, partially hydrogenated canola oil, hydrolyzed beef plasma, sodium alginate and salt (Giese, 1992). Taki (1991) described an ingredient blend composed of modified food starch, rice flour, salt, emulsifier and flavor. The nature of the interactions between these ingredients and meat matrices is not understood.

13.2.3 Color problems in low fat products

Reducing the fat/lean ratio in meat products can increase product darkness and redness, particularly if red meats are used and if compensatory measures are not undertaken to modify color (Kregal, Prusa and Hughes, 1986; Hand *et al.*, 1987; Reitmeier and Prusa, 1987; Claus and Hunt, 1991; Troutt *et al.*, 1992). Mechanically separated meats, if included in a meat product formulation to increase lean content, can give rise to color problems because of their susceptibility to oxidation reactions (Ockerman and Hansen, 1988). On the other hand, replacing fat by adding water, directly or incorporated into hydrated nonmeat substitutes, can lead to a decrease in the color of uncooked meat products, since myoglobin is solubilized in the water (Ahmed *et al.*, 1990; Brown and Zayas, 1990; Reitmeier and Prusa, 1991). Some nonmeat ingredients may impart an uncharacteristic color to the product. Wet-milled corn germ used in pork patties led to an increase in yellowness of these products (Reitmeier and Prusa, 1991).

(a) Overcoming color problems in low fat products

Selection of meat ingredients

In meat products, the color quality of the starting ingredients influences the final product color (Boles, 1995). Thus, if a fresh red meat color is required in the final product, the meat used in manufacture should be high in oxymyoglobin. This may be particularly important in low fat meat products where the diluting effect of fat on color is absent and consumers expect a more intense red color.

Packaging of products

Storage of fresh meat products in modified atmosphere packs can reduce the rate of discoloration and extend the shelf-life of low fat products which may otherwise be prone to oxymyoglobin oxidation. The suitability of different gas atmospheres for color extension is product dependent. Legarreta, Usborne and Ashton (1988) showed that in minced pork sausage flushing with high oxygen concentrations gave a more acceptable initial color, but the color rapidly deteriorated during storage. Minced beef color can benefit from storage in oxygen-rich atmospheres (80% O_2–20% CO_2) (O'Grady, 1995). Product color may also benefit from vacuum packaging in mother packs during transit followed by exposure to oxygen-rich atmospheres for display on the intended day of sale (Gill and Jones, 1994).

Colorants

Dilution of color in low fat products may be overcome by the use of permitted colorants. Red 2G and cochineal carmine are colorants which may be used in meat products (Craven, 1995). Where color dilution is a problem arising from added water and nonfat ingredients, these colorants may be used to supplement the natural meat color. Cochineal carmine is a natural alternative to Red 2G, and therefore has a more attractive image with consumers. These dyes may be added directly to the meat formulation via the seasoning mix or they may be incorporated into the casings (collagen) during manufacture.

13.3 Low salt meat products

13.3.1 Flavor problems in low salt products

Sodium chloride has a flavor enhancing effect in meat products in addition to its functions in solubilization of meat proteins and inhibition of pathogens. The level of addition of sodium chloride to meat products presents the meat processor with a dilemma. On the one hand, salt has desirable preservative and flavor effects in meat products while, on the other, dietary sodium contributes to hypertension and sodium chloride can act as a potent pro-oxidant, leading to the formation of rancid flavors in meat products.

The level of salt required for an acceptable product depends on the saltiness expected in the product. For example, in frankfurter-type sausages, 1.6% salt is regarded as 'mildly salty' and 2.4% as 'very salty' (Wirth, 1991). Thus, simply reducing salt levels in order to reduce sodium may adversely affect product flavor.

Many reports illustrate the problem facing the meat processor wishing to reduce sodium chloride yet maintain the desirable characteristics which salt imparts. Huffman, Ly and Cordray (1981) found that restructured pork chops manufactured without salt were rated lower by sensory panelists and that the inclusion of salt (0.5–1.5% NaCl) in their manufacture significantly increased flavor and textural properties. However, sodium chloride, even at low levels of addition, accelerates lipid oxidation and off-flavor development in products lacking curing salts. Huffman, Ly and Cordray (1981) and Matlock *et al.* (1984) showed a significant acceleration of lipid oxidation at 0.5% NaCl addition to restructured pork chops and pork sausage patties, respectively, following freezer storage.

Other authors have also found problems with low salt products. Gardze, Bowers and Caul (1979) found that when up to 10% hydrated, texturized soy protein was included in beef patties, meaty flavor and aroma decreased and that the undesirable 'cereal-like' flavor associated with the soy protein was most pronounced when NaCl was omitted from the patties. Smith and Brekke (1985) found that low salt frankfurters (0.5%) were rated significantly lower for flavor than control frankfurters (2.3%). Barbut, Maurer and Lindsay (1988) also found that frankfurters made from 20 and 49% reduced salt formulations had less desirable sensory properties. Huffman, Ly and Cordray (1981) showed that both lipid oxidation and flavor in restructured pork chops increased with increasing salt concentration from 0 to 1.5% NaCl. At higher salt levels (1.0 and 1.5%) and after freezer storage of the products, the positive effect of salt on flavor appeared to be masked by oxidative rancidity.

(a) Overcoming flavor problems in low salt products

Alternatives to NaCl

Overcoming the loss of desirable flavor associated with reduction in NaCl may be achieved by the use of KCl and $MgCl_2$. Complete replacement of NaCl with $MgCl_2$ or KCl causes flavor problems, but partial replacement can effectively enhance the flavor of products with low levels of NaCl while still fulfilling the low sodium requirements of these products. Wheeler *et al.* (1990) reported that replacing 35% of the NaCl with KCl at 0.375 or 0.750% salt levels did not have any detrimental effects on the sensory properties of restructured beef steaks.

Seman, Olson and Mandigo (1980) investigated the effect of reduction of NaCl and partial replacement with KCl and $MgCl_2$ on bologna characteristics and acceptability. They found that a 50% reduction in sodium resulted in a bologna product that was as acceptable as the control product, but reduction of sodium to one third of the control level adversely affected the flavor of the product.

In fermented sausages and dry-cured pork loin, Gou *et al.* (1996) reported taste defects in products when more than 30% of the NaCl was replaced by KCl or potassium lactate. Gou *et al.* (1996) also investigated the use of glycine as a potential NaCl substitute. Glycine reduces a_w and could thereby improve the microbial stability of low salt products. Noticeable flavor effects were apparent when more than 30% NaCl was replaced with glycine.

Other alternatives with a salt-like taste include formulations with calcium ascorbate as the major constituent and NaCl, sodium ascorbate, ascorbic acid and KCl as minor constituents (Gregory and Singh, 1994) or RiboKCl a KCl-based salt substitute in which KCl is coated with a 50:50 mixture of disodium 5′-inosinate and disodium 5′-guanylate (Anon., 1991).

Antioxidants

Flavor problems arising from NaCl-induced oxidation of lipids may be prevented by the use of appropriate antioxidants. The level of antioxidants needed to inhibit oxidation effectively depends on the degree of oxidative stress in the product. The degree of unsaturated lipids in the product, the presence of heme iron and the extent of comminution all affect susceptibility to lipid oxidation. Both synthetic and natural antioxidants can be used to inhibit oxidation but natural antioxidants, which are more acceptable to consumers, are being increasingly used. Synthetic antioxidants such as BHA, BHT, TBHQ and propyl gallate can be used to offset the problems associated with salt-catalyzed oxidation. Of the natural antioxidants, tocopherols and herbs, such as rosemary, have been shown to be particularly effective in meat products (Gray and Weiss, 1988). Indeed, the natural antioxidant properties of the phenolic components of plant materials such as spices, tea extracts, peppers and grapes are now receiving considerable attention for application in the manufacture of healthier food products (Gordon, 1987; Gray and Weiss, 1988; Kinsella *et al.*, 1993).

Most studies of the effectiveness of antioxidants in preventing lipid oxidation use meat systems without salt and, thus, the added stress imposed by the salt is absent. In studies of meat products containing salt, it appears that the antioxidants are less effective than in controls without salt (Buckley *et al.*, 1989).

Phosphates

Reduced flavor in low salt meat products can be offset by the inclusion of phosphates. Like NaCl, phosphates have a number of functions in meat products, including water binding (Shults, Russell and Wierbicki, 1973), emulsification (Knipe, Olson and Rust, 1985) and retardation of lipid oxidation (Tims and Watts, 1958). Specifically regarding flavor, reduced salt products may be improved by the inclusion of phosphates in meat formulations (Barbut, Maurer and Lindsay, 1988).

Encapsulation
Another approach to overcoming the problem of salt-catalyzed oxidation reactions in meat products is the use of encapsulated salts, in which the salt is slowly released into the product or released on cooking (Dziezak, 1988; Anon., 1991). This assumes that the salt is not required for preservative purposes or for protein extraction in the product prior to cooking.

13.3.2 *Texture problems in low salt products*

In addition to its role as an antimicrobial agent and flavoring agent in meat products, salt facilitates the binding of meat components together. Salt is critical for texture in many meat products because it solubilizes the structural proteins (Gillett *et al.*, 1977) and permits them to function as binding agents. The chloride ion is believed to be the major contributor to the enhanced binding ability of muscle proteins, while the sodium ion imparts flavor (Whiting and Richards, 1978). In restructured meat products the salt-solubilized myofibrillar proteins form a tacky layer on the surface of meat pieces. During cooking this layer forms a matrix of heat-coagulated protein which entraps free water and binds meat pieces together. In finely chopped or emulsified products the solubilized protein in the continuous phase forms a protein film around fat globules. On cooking, the fat is retained within the coagulated meat protein matrix. Thus, through its effects on adhesion of meat pieces, on fat retention and on water-binding capacity salt is critical to the texture of meat products.

Microstructural effects of NaCl on meat structure have been described by Voyle *et al.* (1984), who showed that marked swelling of peripheral myofibrils and disintegration of Z-lines occurred in muscle treated with 3.5% NaCl. Increased binding of meat pieces occurred when the amount of extracted myosin increased (Siegal and Schmidt, 1979). The increased water-binding capacity (WBC) of meat in the presence of NaCl is believed to be due to binding of the chloride ions to the positively charged amino groups of muscle proteins. The resultant increase in electrostatic repulsion between muscle myofilaments permits increased water binding to the charged myofilaments (Wismer-Pedersen, 1987). Sensory evaluation of sausages manufactured without NaCl showed them to be significantly more crumbly than those with 1.5% NaCl (Matlock *et al.*, 1984), an effect that could be attributed to decreased meat binding and decreased WBC in the absence of salt.

(a) Overcoming texture problems in low salt products
To retain the texture of low salt meat products relative to their full fat counterparts, it is critical that meat binding and WBC are maintained.

Raw materials

It is well established that pre-rigor meat has a higher water-holding capacity (WHC) and its salt soluble proteins are more extractable than meat that has undergone rigor mortis (Bard, 1965). Use of pre-rigor meat in low-salt products may reduce the normal salt requirement (Poulanne and Terrell, 1983). Similar effects would also be expected with high pH muscle. As pH is highly correlated to WHC (Wismer-Pedersen, 1987) and WHC increases with even small increases in pH. However, from a safety viewpoint a low salt, high pH product is not desirable.

Physical processes

Many salt-containing meat products are tumbled or massaged in order to extract the salt soluble proteins. It is established that both increased time and severity of treatment increase protein extraction (Schmidt, 1987), although excessive physical treatment causes protein denaturation. Binding strength between meat pieces also increases over time with increased physical processing, but only up to a point. Increased tumbling time or severity may permit a reduction in the level of NaCl required for sufficient protein solubilization. The temperature at which protein extraction takes place is also important, and Schmidt (1987) suggested 0 °C as the optimum extraction temperature while Gillett *et al.* (1977) suggested it to be 7.2 °C. These experiments were carried out under different conditions and optimum extraction temperatures are probably system specific. However, determination of an optimum extraction temperature might permit some reduction in salt requirements for adequate protein extraction and binding.

Pre-blending is a process which allows more time for salt extraction of meat proteins in the manufacture of meat products. Pre-blending led to an improvement in emulsion stability in 0.5% NaCl sausages (Ching-Cheng *et al.*, 1993) but did not affect the textural properties in low salt (1.5%) frankfurters (Hand *et al.*, 1987). It appears that pre-blending can enhance the extraction properties when the level of salt is low (less than 1%).

As mentioned earlier, high pressure and ultrasound applications could have potential in low salt meat product manufacture. High pressure treatment (1500 bar) increased binding strength in beef patties (MacFarlane, McKenzie and Turner, 1984). It was postulated that the increase was due to pressure-induced disaggregation and conformational changes of proteins. Mandava, Dilber and Fernandez (1995) reported enhanced protein extraction in pressure-treated hams and concluded that high pressure treatment might be a good alternative to additives for low salt products in order to obtain acceptable texture and yield. Reynolds *et al.* (1978) concluded that ultrasound treatment of cured ham rolls increased protein extractability and binding strength, while Vimini, Kemp and Fox (1983) reported that ultrasound treatment of restructured beef rolls had similar results. The latter authors suggested that ultrasound treatment may offer a full or partial alternative to NaCl.

Non-meat ingredients

Alkaline phosphates act synergistically with NaCl to enhance protein extractability, increase WBC and increase meat binding (Sofos, 1986). Thus, reduced NaCl levels can be achieved by the addition of phosphates without adverse effects on cooking yields and sensory quality. However, it is important to remember that most phosphates approved for food use are sodium salts. Therefore, the sodium content may not be substantially reduced when phosphates are used in conjunction with NaCl. Potassium phosphates are also available and, although they are more expensive, they are more soluble (Schnee, 1996).

As stated earlier, it is the chloride ion which is primarily responsible for the functional efficiency of NaCl in meat products (Hamm, 1970). Consequently, the use of alternative chloride salts has been investigated (Seman, Olson and Mandigo, 1980; Hand, Terrell and Smith, 1982). Full replacement of NaCl with KCl or $MgCl_2$ decreased firmness of frankfurters manufactured from mechanically deboned turkey, but 35% replacement with these salts did not affect firmness (Hand, Terrell and Smith, 1982). It was concluded that a 35% replacement of NaCl by KCl was the most viable alternative. The texture of meat batters was significantly affected by the type of chloride salt (Gordon and Barbut, 1989). In general, batters containing divalent chloride salts had lower brittleness and formed less stable batters than those containing monovalent salts. Texture profiles of batters containing KCl (3.2%) were similar to those of NaCl (1.5%) but a higher level of NaCl (2.5%) gave a different texture profile to all other salts. It was concluded that partial replacement of NaCl by KCl may be feasible under certain conditions. Optimization of partial substitution was examined by Lin, Mittal and Barbut (1991). It appears that the problems associated with partial replacement of NaCl relate less to textural aspects than other sensory attributes.

Glycine and potassium lactate have also been investigated as potential NaCl substitutes (Gou *et al.*, 1996). In fermented sausages, low levels (10%) of NaCl substitution with glycine affected product texture but in dry-cured pork loin no significant effect was detected by instrumental analysis with up to 40% substitution. In both products potassium lactate had detrimental effects on texture at 10–20% substitution.

Other nonmeat ingredients have been studied for their potential to replace salt soluble proteins as binders in restructured meat products (Means and Schmidt, 1991). These include an algin/calcium system and various gums similar to those discussed earlier. Blood clotting proteins, fibrinogen and thrombin have also been used successfully to bind meat pieces in restructured meat products (Buitink, Den Reijer and Wijngaards, 1994).

It appears that to reduce the requirement for NaCl in meat products approaches involving judicious selection of raw material (e.g. pre-rigor meat), additional processing (extra tumbling and massaging) and use of other nonmeat ingredients (KCl, gums) will need to be investigated and optimized.

13.3.3 Color problems in low salt products

NaCl, even at low levels, in meat products accelerates the formation of metmyoglobin, the pigment responsible for the brown discoloration of meat. Beef is higher in myoglobin than other meats and is particularly prone to discoloration during storage if certain conditions prevail. In beef steaks during frozen storage Trout (1990) found that the rate of metmyoglobin formation doubled when the salt concentration was increased from 0 to 1% and increasing the salt level from 1 to 3% resulted in a further 20% increase in the rate of metmyoglobin formation. Chu *et al.* (1987) showed that oxymyoglobin oxidized to metmyoglobin in the presence of 1.5% salt. The effect was particularly pronounced after restructuring or after a freeze–thaw–freeze cycle.

(a) Overcoming color problems in low salt products

Phosphates

Schwartz and Mandigo (1976) demonstrated that inclusion of sodium tripolyphosphate in restructured pork containing 0.75–2.25% NaCl off-set somewhat the color deterioration caused by salt in these products. Chu *et al* (1987) also found that sodium tripolyphosphate partly overcame the color deterioration caused by salt.

Reducing agents

A reducing environment favors the retention of oxymyoglobin in fresh meats. For example, Harbers, Harrison and Kropf (1981) showed that beef steak color was stabilized by dipping the beef steaks in ascorbic acid. Manu-Tawiah *et al.* (1991) found a combination of 0.05% ascorbic acid, 0.025% citric acid and 0.3% tetrasodium pyrophosphate to be effective in maintaining the color of ground beef.

Sulfites are also effective in maintaining myoglobin in the reduced form. Although not permitted in the USA, sulfite is used elsewhere in processed meat products such as fresh sausage. In addition to their color-stabilizing effect, sulfites have an antimicrobial effect which may also contribute to the color-stabilizing effect (Cornforth, 1994).

Lactates

Sodium and potassium lactates are used in processed meats for their flavor enhancing and antimicrobial effects, but they can also have a positive effect on color. Bradford *et al.* (1993) found minced pork sausage containing 3% potassium lactate to be darker and more color stable than controls without lactate.

Packaging

Packaging regimes discussed for low fat products may also benefit low salt products. In fresh meat products the objective in both cases is maintenance of myoglobin in its oxygenated form.

13.4 Research needs

13.4.1 Assessing texture and flavor characteristics

The need for more informative methods of assessing texture and flavor characteristics is a major limitation to the development of healthier meat products. Correlations between instrumental measurements and sensory evaluations are often weak, especially with regard to comminuted products. Modern techniques applied to meats, such as infrared analysis, elastography and ultrasound, may provide a better understanding of meat structure and eating quality.

Research into the mechanism of flavor binding and release in full fat meat products is necessary. Similarly, the flavor binding and release properties of low fat substitutes need to be investigated. It remains uncertain whether the flavor binding capacity of fat replacers can mimic the binding and, particularly, the flavor release behavior of fat.

13.4.2 Flavor perception

It is also apparent that flavor perception cannot be gauged in terms of flavor intensity alone but that the timed release of flavors from foods in the mouth is an important determinant of product acceptability. This is an area that requires urgent investigation and the development of techniques to mimic and monitor the release of flavor compounds in the mouth. Simulating the flavor release and time–intensity flavor characteristics of full fat products in their low fat counterparts presents a challenge to meat processors but, if successful, will significantly improve the sensory quality of low fat meat products.

13.4.3 Factors affecting sensory characteristics

A thorough understanding of the factors affecting individual sensory aspects (flavor, texture, color) of full fat meat products is necessary to explain what is missing in low fat and low salt products. Only then can strategies be put in place to compensate for the missing attributes. However, in adopting approaches to improving the sensory quality of low fat and low salt products, the meat processor must be aware of interactions between the different sensory traits. While flavor, texture and color problems associated with low fat and low salt meat products are distinct problems in themselves, they cannot be tackled in isolation. This means, for example, that a solution to a texture problem must take into account its effect on flavor. More research should be directed towards the meat system itself, with less emphasis on the multitude of potential nonmeat ingredients. Methods of enhancing the functionality of the meat system to deal with low fat and low salt formulations merit investigation.

13.5 Summary

Adjusting the formulation of any product to meet new needs requires the assessment of the impact of the modifications on the product quality and a comparison of the attributes of the modified product with those of the unmodified product. In this chapter, some of the problems associated with fat and salt reduction in meat products, and ways of overcoming these problems, have been discussed. Fat reduction affects product texture and flavor perception primarily. Salt (sodium) reduction affects flavor and, depending on the product, can also affect texture.

While low fat and low salt meat products have been dealt with in different sections in this chapter, some of the strategies for improving flavor, color or texture can be applied to both low fat and low salt products. The flavor and texture problems associated with low fat and low salt meat present meat processors with the greatest challenges. To overcome these problems, approaches that take into account molecular interactions between different components and particularly the flavor binding and release properties of different components are necessary.

References

Ahmed, P.O., Miller, M.F., Lyon, C.E. *et al.* (1990) Physical and sensory characteristics of low-fat fresh pork sausage processed with various levels of added water. *J. Food Sci.*, **55**, 625.

Anon. (1991) Ingredients for low-sodium/low-fat foods. *Food Technol.*, **45**(10), 124.

Anon. (1995) Meat flavors from heat processing. *Perfum. Flavor.*, **20**(1), 16.

Barbut, S. and Mittal, G.S. (1995) Physical and sensory properties of reduced fat breakfast sausages. *J. Muscle Foods*, **6**, 47.

Barbut, S., Maurer, A.J. and Lindsay, R.C. (1988) Effects of reduced sodium chloride and added phosphates on physical and sensory properties of turkey frankfurters. *J. Food Sci.*, **53**, 62.

Bard, J.C. (1965) Some factors influencing extractability of salt soluble proteins, in *Proceedings of the Meat Industry Research Conference*, American Meat Institute Foundation, Arlington, VA, p. 96.

Berry, B.W. and Leddy, K.F. (1984) Effects of fat level and cooking method on sensory and textural properties of ground beef patties. *J. Food Sci.*, **49**, 870.

Berry, B.W. and Wergin, W.P. (1990) Effects of fat level, starch gel usage and freezing rate on various properties of ground beef patties. *J. Anim. Sci.*, **68** (Suppl. 1), 204.

Berry, B.W., Leddy, K.F. and Bodwell, C.E. (1985) Sensory characteristics, shear values and cooking properties of ground beef patties extended with iron- and zinc-fortified soy isolate, concentrate or flour. *J. Food Sci.*, **50**, 1556.

Boles, J.A. (1995) Colour of processed meat products: a literature review. *Technical Report No. 951*, Meat Industry Research Institute of New Zealand, Hamilton.

Bradford, D.D., Huffman, D.L., Egbert, W.R. and Mikel, W.B. (1993) Potassium lactate effects on low-fat fresh pork sausage chubs during simulated retail display. *J. Food Sci.*, **58**, 1245.

Brauer, H. (1993) Fat reduced frankfurter-type sausage. A technology for preventing too firm and rubbery bite. *Fleischwirtschaft*, **73**, 64.

Brown, L.M. and Zayas, J.F. (1990) Corn germ protein flour as an extender in broiled beef patties. *J. Food Sci.*, **55**, 888.

Buckley, D.J., Gray, J.I., Asghar, A. *et al.* (1989) Effects of dietary antioxidants and oxidized oil on membranal lipid stability and pork quality. *J. Food Sci.*, **54**, 1193.

Buitink, S., Den Reijer, M.W.H. and Wijngaards, G. (1994) Effect of meat binding by Fibrimex on the weight loss and sensoric evaluation after heating, in *Proceedings of the 40th International Congress of Meat Science and Technology*, The Hague, File No. S-VIB.30.

Bullock, K.B., Huffman, D.L., Egbert, W.R. *et al.* (1995) Nonmeat ingredients for low-fat ground beef patties. *J. Muscle Foods*, **6**, 37.

Ching-Cheng, J., Yuu-Chu, W. and Yie-Shuing, C. (1993) Effects of various salt, phosphate levels and pre-blending on properties and keeping quality of hotdogs. *J. Chinese Soc. Anim. Sci.*, **22**, 97.

Chu, Y.H., Huffman, D.L., Trout, G.R. and Egbert, W.R. (1987) Color and color stability of frozen restructured beef steaks: effect of sodium chloride, tripolyphosphate, nitrogen atmosphere, and processing procedures. *J. Food Sci.*, **52**, 869.

Claus, J.R. (1991) Fat reduction in comminuted meat systems. *Proc. Recip. Meat Conf.*, **44**, 93.

Claus, J.R. and Hunt, M.C. (1991) Low-fat, high added-water bologna formulated with texture-modifying ingredients. *J. Food Sci.*, **56**, 643.

Claus, J.R., Hunt, M.C., Kastner, C.L. and Kropf, D.H. (1990) Low-fat, high added water bologna. Effects of massaging, pre-blending and time of addition of water and fat on physical and sensory characteristics. *J. Food Sci.*, **55**, 338.

Colmenero, F.J. (1996) Technologies for developing low-fat meat products. *Trends Food Sci. Technol.*, **7**, 41.

Cornforth, D. (1994) Color – its basis and importance. *Adv. Meat Res.*, **9**, 34.

Craven, G. (1995) Color applications in meat and fish products. *Food Market. Technol.*, April 1995, p. 18.

Cross, H.R., Berry, B.W. and Wells, L.H. (1980) Effects of fat level and source on the chemical, sensory and cooking properties of ground beef patties. *J. Food Sci.*, **45**, 791.

Desmond, E.M., Troy, D.J. and Buckley, D.J. (1994) Fat replacers and the eating quality of low-fat beefburgers, in *Proceedings of the 40th International Congress of Meat Science and Technology*, The Hague, File No. W-3.22.

Drake, S.R., Hinnergardt, L.C., Kluter, R.A. and Prell, P.A. (1975) Beef patties: the effect of textured soy protein and fat levels on quality and acceptability. *J. Food Sci.*, **40**, 1065.

Drewnowski, A. (1992) Sensory properties of fats and fat replacements. *Nutr. Rev.*, **50**(4), 17.

Dziezak, J.D. (1988) Microencapsulation and encapsulated ingredients. *Food Technol.*, **42**(4), 136.

Egbert, W.R., Huffman, D.L., Chen, C. and Dylewski, D.P. (1991) Development of low-fat ground beef. *Food Technol.*, **45**(6), 64.

Eilert, S.J., Blackmer, D.S., Mandigo, R.W. and Calkins, C.R. (1993) Meat batters manufactured with modified beef connective tissue. *J. Food Sci.*, **58**, 691.

Farmer, L.J. and Mottram, D.S. (1990) Interaction of lipid in the Maillard reaction between cysteine and ribose: the effect of triglyceride and three phospholipids on the volatile products. *J. Sci. Food Agric.*, **53**, 505.
Foegeding, E.A. and Ramsey, S.R. (1987) Rheological and water-holding properties of gelled meat batters containing iota carrageenan, kappa carrageenan or xanthan gum. *J. Food Sci.*, **52**, 549.
Fostier, B., Turin, F., Chatelin, Y. and Cartier, P. (1993) Relationship between fat level and flavour in minced beef, in *Proceedings of the 39th International Congress of Meat Science and Technology*, Calgary, File No. S7PO8.WP.
Gardze, C., Bowers, J.A. and Caul, J.F. (1979) Effect of salt and textured soy level on sensory characteristics of beef patties. *J. Food Sci.*, **44**, 460.
Giese, J. (1992) Developing low-fat meat products. *Food Technol.*, **46**(4), 100.
Giese, J. (1994a) Spices and seasonings blends: a taste for all seasons. *Food Technol.*, **48**(4), 88.
Giese, J. (1994b) Modern alchemy: use of flavors in food. *Food Technol.*, **48**(4), 105.
Gill, C.O. and Jones, T. (1994) The display life of retail-packaged beef steaks after their storage in master packs under various atmospheres. *Meat Sci.*, **38**, 385.
Gillett, T.A., Meilring, D.E., Brown, C.L. and Simon, S. (1977) Parameters affecting meat protein extraction and interpretation of model system data for meat emulsion formation. *J. Food Sci.*, **42**, 1606.
Gordon, A. and Barbut, S. (1989) The effect of chloride salts on the texture, microstructure and stability of meat batters. *Food Microstruct.*, **8**, 271.
Gordon, M.H. (1987) Novel antioxidants, *Food Sci. Technol. Today*, **1**, 172.
Gou, P., Guerrero, L., Gelabert, J. and Arnau, J. (1996) Potassium chloride, potassium lactate and glycine as sodium chloride substitutes in fermented sausages and in dry-cured pork loin. *Meat Sci.*, **42**, 37.
Goutenfongea, R. and Dumont, J.P. (1990) Developments in low-fat meat and meat products, in *Reducing Fat in Meat Animals* (eds J.D. Wood, and A.V. Fisher), Elsevier Applied Science, London, p. 398.
Gray, J.I. and Weiss, G.M. (1988) Warmed-over flavor in meat, in *Research and Nutrition Information*, National Live Stock and Meat Board, Chicago.
Gregg, L.L., Claus, J.R., Hackney, C.R. and Marriott, N.G. (1993) Low-fat, high added water bologna from massaged, minced batter. *J. Food Sci.*, **58**, 259.
Gregory, G.S. and Singh, H. (1994) Palatable low salt substitutes. *US Pat.*, 5 288 510.
Hamm, R. (1970) Interactions between phosphates and meat proteins, in *Symposium: Phosphates in Food Processing* (eds J.M. Desman and P. Melnychyn), AVI, Wesport, CT, p. 65.
Hand, L.W., Terrell, R.N. and Smith, G.C. (1982) Effect of chloride salts on physical, chemical and sensory properties of frankfurters. *J. Food Sci.*, **47**, 1800.
Hand, L.W., Hollingsworth, C.A., Calkins, C.R. and Mandigo, R.W. (1987) Effects of preblending, reduced fat and salt levels on frankfurter characteristics. *J. Food Sci.*, **52**, 1149.
Harbers, C.A.Z., Harrison, D.L. and Kropf, D.H. (1981) Ascorbic acid effects on bovine muscle pigments in the presence of radiant energy. *J. Food Sci.*, **46**, 7.
Hoogenkamp, H.W. (1991) Low fat beef patties, in *Proceedings of the Food Ingredients Europe Conference*, Paris, p. 311.
Hornstein, I. and Crowe, P.F. (1960) Flavor studies on beef and pork. *J. Agric. Food Chem.*, **8**, 494.
Hornstein, I. and Crowe, P.F. (1963) Meat flavor: lamb. *J. Agric. Food Chem.*, **11**, 147.
Huffman, D.L. and Egbert, W.R. (1990) Advances in lean ground beef production. *Ala. Agric. Exp. Stn. Bull.*, No. 606.
Huffman, D.L., Ly, A.M. and Cordray, J.C. (1981) Effect of salt on quality of restructured pork chops. *J. Food Sci.*, **46**, 1563.
Hull, D.H., Rogers, R.W. and Martin, J.M. (1992) Effects of carbohydrate substitution and pre-blending on the properties of reduced fat, high moisture beef frankfurters. *J. Muscle Foods*, **3**, 323.
Imafidon, G.I. and Spanier, A.M. (1994) Unraveling the secret of meat flavor. *Trends Food Sci. Technol.*, **5**, 315.
Keeton, J.T. (1994) Low-fat meat products – technological problems with processing. *Meat Sci.*, **36**, 261.

Kinsella, J.E. (1990) Flavor perception and binding. *INFORM*, **1**, 215.

Kinsella, J.E., Frankel, E., German, B. and Kanner, J. (1993) Possible mechanisms for the protective role of antioxidants in wine and plant foods. *Food Technol.*, **47**(4), 85.

Kinzel, B. (1992) Flavor gains over flavor loss. *Agric. Res.*, **40**(1), 19.

Knipe, C.L., Olson, D.G. and Rust, R.E. (1985) Effects of selected inorganic phosphates, phosphate levels and reduced sodium chloride levels on protein solubility, stability and pH of meat emulsions. *J. Food Sci.*, **50**, 1010.

Kotula, A.W., Twigg, G.G. and Young, E.P. (1976) Evaluation of beef patties containing soy protein, during 12-month frozen storage. *J. Food Sci.*, **41**, 1142.

Kramlich, W.E. and Pearson, A.M. (1960) Separation and identification of cooked beef flavor components. *Food Res.*, **25**, 712.

Kregal, K.K., Prusa, K.J. and Hughes, K.V. (1986) Cholesterol content and sensory analysis of ground beef as influenced by fat level, heating, and storage. *J. Food Sci.*, **51**, 1162.

Legarreta, I.G., Usborne, W.R. and Ashton, G.C. (1988) Extending the retail storage time of pork sausage using modified atmospheres and freezing. *Meat Sci.*, **23**, 21.

Lin, G.C., Mittal, G.S. and Barbut, S. (1991) Optimization of tumbling and KCl substitution in low sodium restructured hams. *J. Muscle Foods*, **2**, 71.

Lin, K.W. and Keeton, J.T. (1994) Determination of optimum particle size for low-fat, precooked ground beef patties. *J. Muscle Foods*, **5**, 63.

Liu, M.N., Huffman, D.L., Egbert, W.R. *et al.* (1991) Soy protein and oil effects on chemical, physical and microbial stability of lean ground beef patties. *J. Food Sci.*, **56**, 906.

Liu, H.F., Booren, A.M. and Gray, J.I. (1995) Quality attributes of low-fat cherry ground beef patties, in *Proceedings of the 41st International Congress of Meat Science and Technology*, San Antonio, TX, p. 427.

MacFarlane, J.J., McKenzie, I.J. and Turner, R.H. (1984) Binding of comminuted meat: Effect of pressure. *Meat Sci.*, **10**, 307.

Mandava, R., Dilber, E. and Fernandez, I. (1995) Meat and meat products: use of high pressure treatment. *Meat Focus Int.*, **4**(4), 147.

Mandigo, R.W. and Eilert, S.J. (1994) Strategies for reduced fat processed meats, in *Low-fat Meats: Design Strategies and Human Implications* (eds H.D. Hafs and R.G. Zimbelman), Academic Press, San Diego, p. 145.

Manu-Tawiah, W., Ammann, L.L., Sebranek, J.G. and Molins, R.A. (1991) Extending the color stability and shelf life of fresh meat. *Food Technol.*, **45**(3), 94.

Marquez, E.J., Ahmed, E.M., West, R.L. and Johnson, D.D. (1989) Emulsion stability and sensory quality of beef frankfurters produced at different fat or peanut oil levels. *J. Food Sci.*, **54**, 866.

Marshall, R.J. (1990) Composition, structure, rheological properties and sensory texture of processed cheese analogues. *J. Sci. Food Agric.*, **50**, 237.

Matlock, R.G., Terrell, R.N., Savell, J.W. *et al.* (1984) Factors affecting properties of precooked-frozen pork sausage patties made with various NaCl/phosphate combinations. *J. Food Sci.*, **49**, 1372.

McAuley, C. and Mawson, R. (1994) Low-fat and low-salt meat product ingredients. *Food Aust.*, **46**, 283.

Means, W.J. and Schmidt, G.R. (1991) Restructuring without salt or phosphate. *Adv. Meat Res.*, **3**, 469.

Mela, D. (1992) The perception and acceptance of dietary fat: what, who, why? *Br. Nutr. Found. Bull.*, **17**(1), 74.

Miller, A.J., Acherman, S.A. and Palumbo, S.A. (1980) Effects of frozen storage on functionality of meat for processing. *J. Food Sci.*, **45**, 1466.

Minerich, P.L., Addis, P.B., Epley, R.J. and Bingham, C. (1991) Properties of wild rice/ground beef mixtures. *J. Food Sci.*, **56**, 1154.

Mittal, G.S. and Barbut, S. (1994) Effects of carrageenans and xanthan gum on the texture and acceptability of low-fat frankfurters. *J. Food Process. Preserv.*, **18**, 201.

Moskowitz, H.R. (1987) *Food Texture, Instrumental and Sensory Measurement*, Marcel Dekker, New York.

Mottram, D.S. and Edwards, D.A. (1983) The role of triglycerides and phospholipids in the aroma of cooked beef. *J. Sci. Food Agric.*, **34**, 517.

Ockerman, H.W. and Hansen, C.L. (1988) Edible tissue from bone, in *Animal By-Product Processing*, Ellis Horwood, Chichester, p. 158.
O'Grady, M.N. (1995) Factors affecting oxymyoglobin oxidation in beef. *PhD Thesis*, Faculty of Agriculture, University College Dublin.
Plug, H. and Haring, P. (1993) The role of ingredient-flavour interactions in the development of fat-free foods. *Trends Food Sci. Technol.*, **4**, 150.
Poulanne, E.J. and Terrell, R.N. (1983) Effects of salt levels in pre-rigor blends and cooked sausages on water binding, released fat and pH. *J. Food Sci.*, **48**, 1022.
Pszczola, D.E. (1991) Oat-bran-based ingredient blend replaces fat in ground beef and pork sausage. *Food Technol.*, **45**(11), 60.
Reineccius, G. (1994) Flavor and aroma chemistry. *Adv. Meat Res.*, **9**, 184.
Reitmeier, C.A. and Prusa, K.J. (1987) Cholesterol content and sensory analysis of ground pork as influenced by fat level and heating. *J. Food Sci.*, **52**, 916.
Reitmeier, C.A. and Prusa, K.J. (1991) Composition, cooking loss, color and compression of ground pork with dry- and wet-milled corn germ meals. *J. Food Sci.*, **56**, 216.
Reynolds, J.B., Anderson, D.B., Schmidt, G.R. *et al.* (1978) Effects of ultrasonic treatment on binding strength in cured ham rolls. *J. Food Sci.*, **43**, 866.
Schmidt, G.R. (1987) Fractional behaviour of meat components in processing, in *The Science of Meat and Meat Products*, 3rd edn (eds J.F. Price and B.S. Schweigert), Food and Nutrition Press, Westport, CT, p. 413.
Schnee, R. (1996) Phosphates: a key ingredient. *Food Technol. N.Z.*, **31**(5), 13.
Schwartz, W.C. and Mandigo, R.W. (1976) Effect of salt, sodium tripolyphosphate and storage on restructured pork. *J. Food Sci.*, **41**, 1266.
Seman, D.L., Olson, D.G. and Mandigo, R.W. (1980) Effect of reduction and partial replacement of sodium on bologna characteristics and acceptability. *J. Food Sci.*, **45**, 1116.
Shamil, S., Wyeth, L.J. and Kilcast, D. (1991–92) Flavor release and perception in reduced-fat foods. *Food Qual. Prefer.*, **3**, 51.
Shand, P.J., Schmidt, G.R., Mandigo, R.W. and Claus, J.R. (1990) New technology for low-fat meat products. *Proc. Recip. Meat Conf.*, **43**, 37.
Shults, G.W., Russell, D.R. and Wierbicki, E. (1973) Effect of condensed phosphates on pH, swelling and water-holding capacity of beef. *J. Food Sci.*, **37**, 860.
Siegel, D.G. and Schmidt, G.R. (1979) Crude myosin fractions as meat binders. *J. Food Sci.*, **44**, 1129.
Smith, D.M. and Brekke, C.J. (1985) Characteristics of low-salt frankfurters produced with enzyme-modified mechanically deboned fowl. *J. Food Sci.*, **50**, 308.
Sofos, J.N. (1986) Use of phosphates in low sodium meat products. *Food Technol.*, **40**(9), 52.
Sofos, J.N., Noda, I. and Allen, C.E. (1977) Effects of soy proteins and their levels of incorporation on the properties of weiner-type products. *J. Food. Sci.*, **42**, 879.
St John, L.C., Buyak, C., Keeton, J.T. *et al.* (1986) Sensory and physical attributes of frankfurters with reduced fat and elevated mono-unsaturated fats. *J. Food Sci.*, **51**, 1144.
Taki, G.H. (1991) Functional ingredient blend produces low-fat meat products to meet consumer expectations. *Food Technol.*, **45**(11), 70.
Tims, M.J. and Watts, B.M. (1958) Protection of cooked meats with phosphates. *Food Technol.*, **12**, 240.
Trout, G.R. (1990) The rate of metmyoglobin formation in beef, pork, and turkey meat as influenced by pH, sodium chloride, and sodium tripolyphosphate. *Meat Sci.*, **28**, 203.
Troutt, E.S., Hunt, M.C., Johnson, D.E. *et al.* (1992) Chemical, physical, and sensory characterization of ground beef containing 5 to 30 percent fat. *J. Food Sci.*, **57**, 25.
Vimimi, R.J., Kemp, J.D. and Fox, J.D. (1983) Effects of low frequency ultrasound on properties of restructured beef rolls. *J. Food Sci.*, **48**, 1572.
Voyle, C.A., Jolley, P.D. and Offer, G.W. (1984) The effect of salt and pyrophosphate on the structure of meat. *Food Microstruct.*, **3**, 113.
Wallingford, L. and Labuza, T. (1983) Evaluation of the water binding properties of food hydrocolloids by physical/chemical methods and in a low-fat meat emulsion. *J. Food Sci.*, **48**, 1.
Wasserman, A.E. and Talley, F. (1968) Organoleptic identification of roasted beef, veal, lamb and pork as affected by fat. *J. Food Sci.*, **33**, 219.

Wheeler, T.L., Seideman, S.C., Rolan, T.L. and Davis, G.W. (1990) Effects of mechanically separated beef with various chloride salts in restructured beef steaks. *J. Food Sci.*, **55**, 342.

Whitfield, F.B., Mottram, D.S., Brock, S. *et al.* (1987) Effect of phospholipid on the formation of volatile heterocyclic compounds in heated aqueous solutions of ribose and amino acids. *J. Sci. Food Agric.*, **42**, 261.

Whiting, R.C. and Richards, J.F. (1978) Influence of divalent cations on poultry meat emulsions and sausages. *J. Food Sci.*, **43**, 312.

Williams, C.W. and Zabik, M.E. (1975) Quality characteristics of soy substituted ground beef, pork and turkey meat loaves. *J. Food Sci.*, **40**, 502.

Wirth, F. (1991) Reducing fat and sodium content of meat products. What possibilities are there? *Fleischwirtschaft*, **71**, 294.

Wismer-Pedersen, J. (1987) Water, in *The Science of Meat and Meat Products*, 3rd edn (eds J.F. Price, and B.S. Schweigert). Food and Nutrition Press, Westport, CT, p. 141.

Withycombe, D.A., Lindsay, R.D. and Stuiber, D.A. (1978) Isolation and identification of volatile components from wild rice grain (*Zizania aquatica*). *J. Agric. Food Chem.*, **26**, 816.

Wong, E., Johnson, C.B. and Nixon, L.N. (1975) The contribution of 4-methyloctanoic (hircinoic) acid to mutton and goat meat flavour. *N.Z. J. Agric. Res.*, **18**, 261.

Wood, T. and Bender, A.E. (1957) Analysis of tissue constituents. Commercial ox-muscle extract. *Biochem. J.*, **67**, 366.

Zayas, J.F. and Lin, C.S. (1989) Effect of the pre-treatment of corn germ protein on the quality characteristics of frankfurters. *J. Food Sci.*, **54**, 1452.

Ziprin, Y.A., Rhee, K.S., Carpenter, Z.L. *et al.* (1981) Glandless cottonseed, peanut and soy protein ingredients in ground beef patties: effect on rancidity and other quality factors. *J. Food Sci.*, **46**, 58.

14 Reducing salt (sodium) levels in processed meat, poultry and fish products

JANET E. COLLINS

14.1 Introduction

Nutritional concerns represent only one consideration in making decisions about foods generally, and meats specifically, that consumers choose to purchase and enjoy. The nutritional benefits and consequences of food choice have been discussed in previous chapters. Within the past decade, from the Surgeon General's Report (USDHHS, 1988) through the latest Dietary Guidelines for Americans (USDA/USDHHS, 1995), regulatory and health-concerned agencies have made recommendations regarding sodium intake for optimal health. Intake of dietary sodium has been linked to increased incidence of hypertension, and dietary sodium restriction has been widely recommended. However, most recently Midgley *et al.* (1996) suggested that the scientific evidence in the normotensive population does not support current recommendations for universal dietary sodium restriction.

Sodium chloride, common table salt, is the most abundant source of sodium in the American diet. The physiological requirement for dietary sodium is 0.2–0.5 g per day (Svetkey, Shusterman and Lin, 1996), yet average daily sodium intake of US adults is approximately 3.9 g per day (Dietary Guidelines Advisory Committee, 1995), with the Food and Drug Administration (FDA) establishing the daily reference value for sodium as 2.4 g per day (Federal Register, 1991).

Meat products (including red meat, poultry and fish) contribute about 21% of the sodium in the average US diet (NRC/NAS, 1989). The high proportion of dietary sodium is accounted for in processing and processed meat products. The sodium content of fresh muscle foods is relatively low at approximately 70 mg per 100 g meat (USDA, 1990, 1992), but approximately 1000 mg per 100 g in processed meat (Breidenstein and Williams, 1985). Therefore, concern about sodium in meat is related to added sodium during the processing and curing of meat products. The meat industry has made a concerted effort, with some success, over the past 10–15 years to reduce the sodium content in processed meats. However, with the concomitant reduction in fat in processed meat products, sodium reduction has become sodium chloride substitution – and the products rarely typify 'standard' processed meat products. Murphy (1995) reported that one can achieve some sodium reduction in processed meats, but one cannot cut back dramatically – owing to flavor and textural concerns. It is critical to recognize the myriad of functions of sodium in processed meat products, from product quality and functionality to safety.

14.2 Stability and spoilage

Sodium chloride (NaCl) and other sodium salts were first added to meats for their preservative effects. These salts are food additives that are preservatives not defined by law or generally recognized as safe (GRAS) under the Federal Food, Drug and Cosmetic Act as amended by the 1958 Food Additives Amendment (Ziporyn, 1985). Sodium chloride is the oldest known and most widely utilized curing agent for meats. Sodium chloride is used in brines and curing solutions or is applied directly to meat surfaces. A sufficient quantity is applied or added to decrease growth of spoilage and pathogenic microbes and/or to encourage growth of fermentative microbes. According to Judge *et al.* (1989), a salt content of roughly 9–11% would be required in the finished product for effective preservation. From 1 to 2.5% of salt is most commonly found in commercially produced processed meats. Meat products in the marketplace today use a variety of preservation methods so that sodium chloride is used principally for functional qualities other than preservation.

14.2.1 Microbial considerations – spoilage microbes

According to Sofos (1986a), sodium chloride is inhibitory against many spoilage and pathogenic microbes in meat because of its ability to reduce water activity. Relatively higher concentrations of salt could provide shelf stability, but quality factors are reduced when salt concentrations in cured meat products are above the 2–3% range. Brine (NaCl solution) for curing meats must be sufficient to inhibit the growth of molds, yeasts and other spoilage and pathogenic microbes (Rust, 1988). It is expected that reducing NaCl levels below those typically used, without other preservative measures, would result in shortened product shelf-life and reduced safety (Sofos, 1986b). Microbiological data and quality assessments from research studies with reduced NaCl meat products may conflict, owing to variations in test product composition, pH, handling, processing variables, inoculum size or microbial load and type, as well as storage conditions used for these studies.

(a) Effects of salt on various microorganisms

It is known that different organisms tolerate different levels of NaCl and combinations of ingredients, including other chloride salts and nitrites, further affect NaCl tolerance of microbes. *Staphylococcus aureus* is resistant to NaCl, whereas *Moraxella* spp. in chilled poultry have been found to be more NaCl tolerant than *Pseudomonas* spp. (Juven and Getshovki, 1976). A 50% sodium reduction (to 1.25% NaCl) slightly increases the growth of *Lactobacillus* spp. in ground pork, but neither type nor level of chloride salt (NaCl, KCl, $MgCl_2$) has an effect on *Moraxella* or *Micrococcus* (Terrell *et al.*, 1984). When 60% of the NaCl in a formulation (2.5% NaCl) is eliminated, growth of the natural microflora reportedly is more rapid in meat batters (Whiting *et al.*, 1984). Similarly, reducing the NaCl level to 1.5% also results in slightly more rapid growth of natural flora in frankfurters stored at 11°C for 11 days.

Sofos (1985) and Madril and Sofos (1985) found more rapid microbial growth and spoilage in comminuted meat products formulated with reduced NaCl levels when inoculated with *Clostridium sporogenes* spores before or after thermal processing and stored at 20 or 27°C. In a study by Terrell *et al.* (1984), replacement of NaCl (2.5%) at 25% increments with a mixture of other chloride salts did not influence microbial growth. Equivalent amounts (2.5%) of various chloride salts (NaCl, KCl and $MgCl_2$) did not affect the microbial growth of refrigerated pork sausage in a study by Terrell *et al.* (1982), although more trichina larvae were killed with the addition of KCl and NaCl when compared with equivalent amounts of $MgCl_2$ or $CaCl_2$.

14.2.2 Microbial considerations – pathogenic microbes

Whiting *et al.* (1984) reported a one log higher total plate count in inoculated frankfurter mixtures containing 1.5% versus 2.0 or 2.5% NaCl. In the study, growth of staphylococci was unrelated to salt content and *Clostridium sporogenes* did not grow. In a subsequent study, Whiting *et al.* (1985) demonstrated that NaCl (2.5 and 1.5%) had no influence on *Clostridium sporogenes* or *Staphylococcus aureus* at 5–30°C. However, Nielsen and Zeuthen (1985) reported that whereas growth of *Staphylococcus aureus* was unaffected by NaCl level, brine levels of 2.8–6.0% were highly influential on growth of various pathogenic bacteria in bologna type products.

According to Maurer (1983), higher levels of sodium chloride provided better control of *Clostridium botulinum* in processed meats, i.e. less sodium nitrite was required to supplement the NaCl. Robinson, Gibson and Roberts (1982) demonstrated that brine (NaCl) levels were involved in the probability of toxin production by *Clostridium botulinum*. Reduction of NaCl in wieners from 3.5 to 2.0% accounted for a reduction in the time to develop botulinal toxin from 2.5 weeks to 5 days. There are significant antimicrobial interactions between NaCl and nitrite in cured meats. Raw hams (without addition of sodium nitrite) are unsuitable for any range of salt-reduced products (Wirth, 1991), i.e. at least 4.5% salt (1800 mg sodium per 100 g) is needed to ensure the microbial stability of these products. Sofos (1986b) reported that use of potassium sorbate (0.26%) in low sodium (0.65 and 1.30% NaCl) thermally processed turkey products delayed the initiation and lowered the rate of microbial growth and gas production in inoculated processed products that were temperature abused. Sodium lactate has been studied recently as a potential antilisterial agent in various processed meats and meat products (Miller and Acuff, 1994; Buncic *et al.*, 1995).

(a) Microbial shelf-life and safety
It is important to examine microbial shelf-life and safety of individual processed meat products before NaCl levels are reduced or replaced by other ingredients. It is also important to be reminded, however, that NaCl generally is not used at concentrations high enough to effect preservation (Judge *et al.*, 1989), although some preservative effect may exist at NaCl levels currently used. The main functions of sodium chloride in processed meats are to solubilize the proteins, develop flavor and control fermentation rates in some sausages. Meat curing with salts and other additives was originally a means of preserving meats. Today, most cured meats are mild-cured and must be stored under refrigeration (Pearson and Tauber, 1984).

In general, processed meats are not a major growth category. However, jerky, slim jims and summer sausages, which are shelf stable at ambient temperatures, are an exception and have grown in volume rapidly in recent years. These intermediate moisture (IM) products usually contain high levels of salt, although some are dried to a low water activity (a_w = 0.70–0.80) by application of heat (Chang, Huang and Pearson, 1996).

14.3 Reducing sodium levels

14.3.1 Functions of sodium chloride in processed meats

Sodium chloride is essential in processed meats to solubilize meat proteins that affect the water holding capacity, fat binding and texture of the products, typical salty flavor development and microbial stability. Decreasing salt content, therefore, has many implications for muscle food products apart from shelf-life and safety issues. The product implications include textural changes, flavor differences, decreased moisture retention (yield) and product appearance. Restructured products made with muscle boning to remove seam and trim fat may require some physical (mechanical) and chemical manipulation to develop texture. A combination of salt systems and manipulation techniques may be required to develop satisfactory comminuted, reformulated or restructured meat products when the sodium content is reduced.

Muscle proteins necessary for binding in processed meats are salt soluble (Trout and Schmidt, 1987) and yield a sticky surface on meat particles that forms a water and fat-holding matrix upon heating (Pepper and Schmidt, 1975; Schmidt, Mawson and Siegel, 1981). Johnson and Henrickson (1970) demonstrated that upon adding sodium chloride to porcine muscle emulsions, more salt soluble proteins were extracted. These solubilized proteins are necessary for development of the structure of the processed product.· According to Sofos (1985), the minimum NaCl concentration needed for adequate binding in meat products also depends on the pH of the meat. The pH of post-rigor muscle is closer to its isoelectric point so that the proteins are less soluble when sodium chloride is added to post-rigor than to pre-rigor meat.

At pH values above 5.7, NaCl levels of at least 2.0% resulted in meat batters with acceptable cooking yields; however, reduction of NaCl to 1.5% or low pH (less than 5.7 as in post-rigor meat) resulted in decreased yields, less desirable product appearance and fat caps, according to Sofos (1983). Sufficiently high pH (6.3), as in pre-rigor meat, with a low NaCl level (1.3%) reportedly has resulted in products with acceptable binding quality. From a protein functionality standpoint, sodium chloride levels in chopped meat products cannot be reduced below 1.5–2.0% (Sofos, 1986c). A salt level of 2.0–2.5% gives optimal bind in processed products, and a range of 2.0–2.5% is necessary under commercial conditions in the absence of any other additives that may supplement the effects of NaCl. Schnell, Vadehra and Baker (1970) reported 2.0% NaCl to be optimum for enhancing bind and decreasing cookout of bound muscle chunks. The degree of solubilization needed and the surface area will depend on the final product (restructured, comminuted, emulsified) requirements. Temperature and manipulation are also critical ingredients of the formulation.

14.3.2 Salt substitution

Attempts to reduce the sodium content of meat products have been directed toward reduction of sodium chloride or its replacement by other forms of NaCl (flaked versus granular) and by use of other chloride salts, mainly potassium chloride. Solubility of the salt form may affect emulsion stability, which could be a critical factor in some products. Flake-type NaCl has been shown to be more functional in model emulsion systems (Campbell, 1979), but not in commercial frankfurters (Sofos, 1983). According to Murphy (1995), flake salt is less coarse and ‘almost two dimensional’ and thus offers better solubility and dispersibility in processed meats.

(a) Various salt substitutes

Several in-depth reviews of the effects of ingredient replacement on the character of various processed meat products have been published (Olson and Terrell, 1981; Terrell and Olson, 1981; Whiting and Jenkins, 1981; Hand, Terrell and Smith, 1982a; Maurer, 1983; Terrell, 1983; Lin, Mittal and Barbut, 1991). Functional properties such as pH, juiciness, yield and sensory attributes are affected when sodium is partially replaced with the chloride salts of potassium, magnesium and lithium (Hand, Terrell and Smith, 1983b, c; Lin, Mittal and Barbut, 1991). According to Hand, Terrell and Smith (1982c), a 100 or 50% replacement of NaCl with KCl produced the most off-flavor of any substitution. Although $LiCl_2$ was preferred to KCl in the products tested, KCl is the only salt with generally recognized as safe (GRAS) status as a replacement for NaCl in processed meats. In fact, $LiCl_2$ is generally considered to be toxic to humans.

Olson and Terrell (1981) and Terrell and Olson (1981) reported that KCl could be substituted for NaCl without loss of functionality. However, at substitution levels greater than 50% of the NaCl, bitter or metallic off-flavors reportedly become objectionable. Several researchers have reported that replacement of NaCl with KCl at 35–50% levels can be accomplished without any loss of functionality (Hand, Terrell and Smith, 1982b; Keeton, 1984). Whiting and Jenkins (1981) reported that replacement of 50% NaCl with KCl in frankfurters provided acceptable flavor, and Hand, Terrell and Smith (1982a) observed similar results by replacing 35% of the NaCl with KCl.

Wirth (1991) contended that removing the common salt in dry sausages would alter the ripening pattern or limit fermentation so that the product would be microbially unstable. Marcy *et al.* (1985) indicated that sausage fermentation was faster with 1.65% compared with 2.48 or 3.3% NaCl in the formulation. Ibanez *et al.* (1994) reported that 50% replacement of NaCl with KCl in a mixture containing 1.37% NaCl and 0.92% KCl produced a dry fermented sausage with acceptable functional properties.

In a study of sectioned and formed hams, Kuo (1986) reported that products processed using 70% NaCl–30% KCl or 70% NaCl–30% $MgCl_2$ were not different in flavor, odor, tenderness or overall acceptability from those made with 100% NaCl; however, formulations using 70% NaCl–30% $CaCl_2$ had lower pH, moisture content, yield and residual nitrite. Jones *et al.* (1988) reported that juiciness ratings decreased in restructured lamb roasts when 50% NaCl was replaced with KCl. Overall acceptability of control samples was greater than in products with a 50:50 mix of NaCl–KCl, but no consistent differences among treatments were reported for juiciness, flavor desirability, off-flavor or overall acceptability. Sodium chloride serves many important functions in processed meats. Not only are textural and functional characteristics altered by reducing or substituting other ingredients for sodium chloride, but overall flavor and acceptability of sodium reduced products are also impacted.

Barbut and Findlay (1989) report that only a partial substitution of NaCl with KCl is possible because of the adverse effects on flavor. Bitter, off-flavor development has been reported in low sodium meat products in which KCl is used as a substitute for NaCl (Maurer, 1983; Barbut, Maurer and Lindsay, 1988). Baccus-Taylor, Garcia and Maurer (1991), however, reported no difference in flavor of turkey braunschweiger made with 50% replacement of NaCl with KCl. They suggested that the spices used and apparent liver flavor of the product may have masked any off-flavors in the products.

(b) Lipid oxidation

The effect of lipid oxidation has also been discussed with regard to sodium chloride and other ingredients in processed meat products. Lipid oxidation limits the flavor quality of processed meat products. A variety of antioxidants have some demonstrated effectiveness for preventing off-flavor and off-color development in processed products containing added sodium salts (Chastain *et al.*, 1982). Chapter 13 discusses lipid oxidation and should be used as a reference for further study.

14.3.3 Product manipulation and temperature changes in reduced sodium products

According to Saffle and Galbreath (1964) and Johnson and Henrickson (1970), the use of pre-rigor beef and pork (normal pH), respectively, results in a roughly 70% increase in extractable salt-soluble protein. Salt extractable proteins were significantly lower (7%) in low pH pre- and post-rigor pork. To obtain the greatest emulsifying ability from pre- and post-rigor low pH meat, NaCl should not be added to the meat before emulsion preparation. According to Bernthal, Booren and Gray (1991), in pre-rigor beef blends the phosphate portion of the ingredient mix should not be added until post-rigor conditions (pH) exist.

(a) Tumbling effects

Mittal, Lin and Barbut (1990) reported on the combined effects of tumbling parameters on reduced sodium restructured hams. Tumbling effectively promotes cohesion during processing and affects textural outcomes of the products. Frye *et al.* (1986) reported that the tumbling process was critical to maximal texture development in ham, with maximum protein extraction being established at 6 h (3 h tumbling, 2 h rest, 1 h tumbling). Inadequate mixing, particularly with reduced sodium products, can result in a rubbery texture and tough skin.

According to Kraus *et al.* (1978), long-term intermittent cycles facilitate the movement of salt soluble protein to the surface of muscle and resulted in higher meat moisture retention. Lin, Mittal and Barbut (1991) reported that 6 h of intermittent tumbling (25 min h^{-1}) was adequate to retain the moisture lost during cooking. In that study, tumbling time (15.6 h) and speed (12 rpm) and NaCl replacement (18.3%) with KCl were found to be optimal to minimize shrinkage and maximize cooked yield and water-holding capacity in a low sodium restructured ham. Parameters to maximize sensory quality of low sodium restructured ham were 12.4 h, 17 rpm and 18.3%, respectively.

(b) Effects of chopping temperature
According to Whiting (1984), chopping temperature in comminuted products greatly affects emulsion stability. There is evidence that proteins have a role in determining the temperature stability of meat emulsions and would be expected to respond to changes in NaCl level. According to Jones and Mandigo (1982), a change in emulsion stability between 16 and 22°C is caused by water loss. The increasing stability from a thickening protein film around the fat globule as the chopping temperature increases is opposed by the film becoming too rigid and a loss of integrity of the protein–gel matrix. In reduced sodium meat batters (1.5%), proteins are capable of binding fat until temperatures reach over 20°C. According to Behablavy and Mandigo (1974), mixing time also affects the textural quality of flaked, formed and sectioned meat.

14.3.4 Use of other ingredients

Binding agents in processed meats can enhance the binding of adjacent pieces of meat and/or increase the water-binding capacity of the finished product. Ingredient substitution for NaCl in processed products requires additional functional components for adequate bind. Phosphates are often used as functional additives. In processed meat, phosphates improve texture by both enhanced binding of meat and increased water-binding capacity of the product matrix. Phosphates are approved for use in some meat products at levels up to 0.5% of total phosphate in the final product (Federal Register, 1982). Phosphates are generally added prior to manipulation of the meat. Uncontrolled use of phosphates in meat products has been reported to lead to soapy flavors and a rubbery texture.

(a) Phosphates and polyphosphates
The functional benefits of phosphates overcome some of those eliminated when sodium is reduced in processed products. For example, phosphates function to bind water and improve texture through changes in protein coagulation and fat emulsification. Combinations of NaCl and phosphates have been reported to act synergistically to improve water binding, meat particle binding, cooking yield and overall texture (Hamm, 1960; Hellen-

doorn, 1962; Maesso *et al.*, 1970; Trout and Schmidt, 1984). Bernthal, Booren and Gray (1991) reported that the utilization of condensed phosphates lowers the concentration of NaCl required to obtain functional benefits.

Polyphosphates may improve protein functionality in reduced NaCl comminuted meats (Seman, Olson and Mandigo, 1980; Puolanne and Terrell, 1983). According to Barbut *et al.* (1988), turkey frankfurters prepared with reduced salt levels (below 2%) without other corrections in formulations resulted in products that exhibited both physical and sensory properties less desirable than those of traditional products. Incorporation of phosphates (sodium tripolyphosphate, sodium hexametaphosphate or sodium acid pyrophosphate) at levels of 0.4% were effective with 20 and 40% salt formulations and improved emulsion stability and sensory properties of reduced sodium poultry frankfurters. Sodium acid pyrophosphate (SAPP) was reported to be the most effective in the reduced salt formulations with SAPP increasing the saltiness and firmness in turkey sausages (Schantz and Bowers, 1993). Overall, tetrasodium pyrophosphate is generally considered superior to sodium tripolyphosphate, tetrapolyphosphate and hexametaphosphate in enhancing meat functional characteristics (Trout and Schmidt, 1984). Increasing levels of polyphosphates are reported to increase the solubility of proteins; increased levels of NaCl do not have the same effect on protein solubilization.

According to Knipe and co-workers (Knipe, Olson and Rust, 1985a; Knipe, Rust and Olson, 1990), of four phosphate treatments studied, addition of tetrasodium pyrophosphate (NaPP) and tetrapotassium pyrophosphate (KPP) produced the most stable emulsions in comminuted meat batters containing 0.75 and 1.5% NaCl after cooking. The relative effect of WHC in raw batter compared similarly to the emulsion stability (gel-liquid released) after cooking for the phosphates tested. In a related study, Knipe, Olson and Rust (1985b) found that NaPP was more effective than added sodium hydroxide (NaOH) on emulsion stability, although the combination of the two was greater. NaOH when added to various phosphates resulted in improved emulsion stability over the control batter, although cooked yield was negatively influenced by the addition of NaOH.

(b) Effects of phosphates on color, flavor and emulsion stability

Although phosphates have been shown to increase emulsion stability, Knipe, Olson and Rust (1988) reported decreased cured color development when phosphates were used. They reported that prepared emulsions would have to be held at least 60 min prior to cooking for cured color development similar to control products. Delaying addition of phosphates was useful for enhancing cured color development, but reduced emulsion stability. Steinhauer (1983) reported that the use of phosphates could both retard oxidation and stabilize color in meats. Sodium phosphates reportedly contribute changed and undesirable flavors when added to meat products (Terrell *et al.*, 1982; Puolanne and Terrell, 1983; Keeton, Foegeding and Patana-Anake, 1984; Schantz and Bowers, 1993). Sodium acid pyrophosphate and sodium tripolyphosphate influenced saltiness in reduced sodium turkey sausages, according to Schantz and Bowers (1993), while reduced NaCl levels in the presence of sodium tripolyphosphate produced a perceived soapiness in turkey sausages.

(c) Other ingredients

A number of other ingredients, both protein and nonprotein, are frequently added to processed meat products for the purposes of increasing bind, often at a reduced cost from muscle tissue. Several of these alternative proteins include a variety of plant sources (Chen and Trout, 1991), soy protein sources (Matulis *et al.*, 1995), nonmeat binders (Tinney *et al.*, 1995) and milk proteins (Ozimek and Poznaski, 1982). The gel matrix formed with these alternative products and proteins differs from that when meat is mixed with sodium chloride, as they provide bind through a combination of protein coagulation and gel formation rather than direct interaction with the muscle proteins. Furthermore, several nonmeat ingredients reportedly affect flavor and other sensory qualities adversely.

In a study designed to assess quality of various low fat, no sodium added restructured beef steaks, Demos *et al.* (1994) reported reduced product quality scores for juiciness, off-flavor and overall desirability. These researchers concluded that 3.0% modified potato starch, 1.0% microcrystalline cellulose, 1.5% surimi, 0.25% iota-carrageenan or 0.25% kappa-carrageenan do not contribute significantly to adhesion/cohesion of meat products in salt-free restructured beef steaks. According to Matulis *et al.* (1995) and Hamann and MacDonald (1992), added carrageenan and surimi gels, respectively, may improve binding in meat emulsions with reduced salt formulations.

14.4 Research needs

Abundant literature describes the breadth and depth of research conducted in the area of reduced sodium processed meats. Scientific exploration has, to date, largely centered around attempts to reconstruct reduced sodium products which mimic their full sodium counterparts. Ample research data provide direction for sodium substitution, replacement/partial replacement and technological capabilities for production of those products. Substantial limitations to the levels of sodium chloride substitution possible exist. Flavor and texture are changed when NaCl is replaced or reduced in meat products. Further, the product benefit is questioned when consumers do not see reduced sodium products as 'tasty' when the justification for the research efforts is to enhance consumer demand for the products. Clearly, sodium reduction in processed meats is still on the agenda of meat processing companies. Reduced sodium processed meats do not generally meet the standards of customers.

The Food and Drug Administration is reviewing the potential for modifying standard meat products with modifications in order to satisfy nutrient claims. For example, the ability of a company to market a 'low sodium bologna' without incumbent language regarding what the 'low sodium' means would be a benefit to the development and marketing of new products. Should the industry encourage the development of such modified standardized processed meats, there would be a marketing advantage to bolster research efforts further in the area of sodium reduction. The fact that the products do not taste like their full sodium counterparts suggests a critical research need if we are to attempt to encourage consumers to reduce sodium in their diets while continuing to eat processed meats. Color and off-color in processed products can further limit their usefulness and stability – both are researchable issues of importance to the industry.

Given the body of research data available, it might be useful to view the outcomes from a clear marketing perspective. What characteristics of processed meats would be possible in products that did not taste like or look like processed meats but were made of the same wholesome components? What else might consumers accept that the industry could provide through existing technology? The industry is well able to produce high quality ready-to-eat meats which have shelf-life characteristics far improved over those of historically cured meats.

How can processed meats be combined with other products or ingredients to produce new foods for the grocery shelves? How can otherwise unused animal products be included in processed meats and products in order to increase the value to the industry? What is the role of preservation technologies such as irradiation and ohmic processing in the future development of processed meats?

14.5 Summary

Processed meats provide significant variety to the diet. One of the characteristics of processed products is the relatively higher sodium content necessitated by processing technologies utilized. Flavor, color stability and texture are altered markedly when 1.5–3.0% NaCl is added to meat formulations during production of processed meats. The addition of sodium chloride to meat results in products with a decidedly different taste.

However, flavor is only one component in food choice. The human taste for salt is acquired and Americans choose less salt added to foods today than they did even a decade ago.

Sodium chloride performs many critical functions in processed meat products. Synergistic effects of sodium chloride with other processed meat ingredients including proteins, salts, nitrites, phosphates and water have been documented here and in other chapters. Any alteration in sodium chloride content of processed meat formulations requires ingredient formulation or manipulation adjustments.

Temperature of production, the thermal process and storage environment significantly affect the character and subsequent functionality of processed meat products. Differentiation among technologies provides for a wide variation in the type of processed meat products available. For example, comminuted meat is a processed meat product with character different from those of restructured processed meat products. Research in the area of sodium reduction continues in the meat industry. Efforts to replace or substitute sodium in processed products have resulted in some products that are successful in the marketplace, but others have been less so.

The nutritional contribution that meats and processed meats make to the average American diet is considerable. The nutritional consequences of increased dietary sodium have long been debated. Health associated agencies have made dietary recommendations based on concerns about presumed health consequences. However, it is critical to recognize that not all individuals require sodium restriction or elimination, and processed meats provide a high level of required nutrients to the diet. For example, the dietary contribution of zinc, iron, B vitamins and some minerals may be significant from processed meats.

It is prudent to recognize the wide variety of meat and meat products available in the marketplace today. Several companies have made a number of reduced sodium products available – consumers have a real choice about the relative level of sodium consumed from processed meats. Processed meats provide variety, taste and nutritional value that should not be discounted for most individuals because of concerns about sodium content of the product.

References

Baccus-Taylor, G., Garcia, H.S. and Maurer, A.J. (1991) Research note: preparation of a low-fat reduced sodium turkey braunschweiger. *Poult. Sci.*, **70**, 1998.

Barbut, S. and Findlay, J.C. (1989) Sodium reduction in poultry products: a review. *CRC Crit. Rev. Poult. Biol.*, **2**, 59.

Barbut, S., Maurer, A.J. and Lindsay, R.C. (1988) Effects of reduced sodium chloride and added phosphates on physical and sensory properties of turkey frankfurters. *J. Food Sci.*, **53**, 62.

Behablavy, R. and Mandigo, R. (1974) The influence of mixing time on chemical and physical properties of flaked, formed and sectioned meat. *J. Anim. Sci.*, **39**, 166.

Bernthal, P.H., Booren, A.M. and Gray, J.I. (1991) Effect of reduced sodium chloride concentration and tetrasodium pyrophosphate on pH, water-holding capacity and extractable protein of pre-rigor and post-rigor ground beef. *Meat Sci.*, **29**, 69.

Breidenstein, B.C. and Williams, J.C. (1985) *Contribution of Red Meat to the U.S. Diet*, National Live Stock and Meat Board, Chicago, IL.

Buncic, S., Fitzgerald, C.M., Bell, R.G. and Hudson, J.A. (1995) Individual and combined listericidal effects of sodium lactate, potassium sorbate, nisin and curing salts at refrigerated temperatures. *J. Food Safety.*, **15**, 247.

Campbell, J.F. (1979) Binding properties of meat blends, effects of salt (sodium chloride) type, blending time and post-blending storage. *PhD Thesis*, Michigan State University, E. Lansing, MI.

Chang, S.F., Huang, T.C. and Pearson, A.M. (1996) Control of the dehydration process in production of intermediate moisture (IM) meat products: a review. *Adv. Food Nutr. Res.*, **39**, 73–152.

Chastain, M.F., Huffman, D.L., Hsieh, W.H. and Cordray, J.C. (1982) Antioxidants in restructured beef/pork steaks. *J. Food Sci.*, **47**, 1779.

Chen, C.M. and Trout, G.R. (1991) Sensory, instrumental texture profile and cooking properties of restructured beef steaks made with various binders. *J. Food Sci.*, **56**, 1457.

Demos, B.P., Forrest, J.C., Grant, A.L. *et al.* (1994) Low-fat, no added salt in restructured beef steaks with various binders. *J. Muscle Foods.*, **5**, 407.

Dietary Guidelines Advisory Committee (1995) *Report of the Dietary Guidelines Advisory Committee on the Dietary Guidelines for Americans*, US Department of Agriculture and US Department of Health and Human Services, Washington, DC.

Federal Register (1982) Meat and poultry products; phosphates and sodium hydroxide. *Fed. Regist.*, **47**(49), 10779.

Federal Register (1991) Food labeling: health claims and label statements; sodium/hypertension. *Fed. Regist.*, **56**, 60825.

Frye, C.B., Hand, L.W., Calkins, C.R. and Mandigo, R.W. (1986) A research note: reduction or replacement of sodium chloride in a tumbled ham product. *J. Food Sci.*, **51**, 836.

Hamann, D.D. and MacDonald, G.A. (1992) Rheology and texture properties of surimi and surimi-based foods, in *Surimi Technology* (eds T.C. Lanier and C.M. Lee), Marcel Dekker, New York, pp. 429–501.

Hamm, R. (1960) Biochemistry of meat hydration. *Adv. Food Res.*, **10**, 355.

Hand, L.W., Terrell, R.N. and Smith, G.C. (1982a) Effects of chloride salts on physical, chemical and sensory properties of frankfurters. *J. Food Sci.*, **47**, 1800.

Hand, L.W., Terrell, R.N. and Smith, G.C. (1982b) Effects of complete or partial replacement of sodium chloride on processing and sensory properties of hams. *J. Food Sci.*, **47**, 1776.

Hand, L.W., Terrell, R.N. and Smith, G.C. (1982c) Effects of chloride salt, method of manufacturing and frozen storage on sensory properties of restructured pork roasts. *J. Food Sci.*, **47**, 1771.

Hellendoorn, B.W. (1962) Water-binding capacity of meat as affected by phosphates. I. Influence of sodium chloride and phosphates on the water retention of comminuted meat at various pH values. *Food Technol.*, **16**, 119.

Ibanez, C., Quintanilla, L., Irigoyen, A. *et al.* (1994) Partial replacement of sodium chloride with potassium chloride in dry fermented sausages: influence on carbohydrate fermentation and the nitrosation process. *Meat Sci.*, **40**, 45.

Johnson, R.G. and Henrickson, R.G. (1970) Effect of treatment of pre- and post-rigor porcine muscles with low sodium chloride concentrations on the subsequent extractability of proteins. *J. Food Sci.*, **35**, 268.

Jones, K.W. and Mandigo, R.W. (1982) Effects of chopping temperature on the microstructure of meat emulsions. *J. Food Sci.*, **47**, 1930.

Jones, D.K., Leu, R., Keeton, J.T. *et al.* (1988) Consumer evaluation of sodium reduced, restructured lamb roasts. *J. Food Qual.*, **11**, 235.

Judge, M.D., Aberle, E.D., Forrest, J.C. *et al.* (1989) Principles of meat processing, in *Principles of Meat Science*, 2nd edn, Kendall/Hunt, Dubuque, IA.

Juven, B.J. and Gertshovki, J. (1976) The effect of salting on the microbiology of poultry meat. *J. Milk Food Technol.*, **39**, 13.
Keeton, J.T. (1984) Effects of potassium chloride on properties of country-style hams. *J. Food Sci.*, **49**, 146.
Keeton, J.T., Foegeding, E.A. and Patana-Anake, C. (1984) A comparison of nonmeat proteins, sodium tripolyphosphate and processing temperature effects on physical and sensory properties of frankfurters. *J. Food Sci.*, **49**, 1462.
Knipe, C.L., Olson, D.G. and Rust, R.E. (1985a) Effects of selected inorganic phosphates, phosphate levels and reduced sodium chloride levels on protein solubility, stability and pH of meat emulsions. *J. Food Sci.*, **50**, 1010.
Knipe, C.L., Olson, D.G. and Rust, R.E. (1985b) Effects of sodium hydroxide and selected inorganic phosphates on the characteristics of reduced sodium meat emulsions. *J. Food Sci.*, **50**, 1017.
Knipe, C.L., Olson, D.G. and Rust, R.E. (1988) Effects of inorganic phosphates and sodium hydroxide on the cooked cured color, pH and emulsion stability of reduced-sodium and conventional meat emulsions, *J. Food Sci.*, **53**, 1305.
Knipe, C.L., Rust, R.E. and Olson, D.G. (1990) Some physical parameters involved in the addition of inorganic phosphates to reduced-sodium meat emulsions. *J. Food Sci.*, **55**, 23.
Kraus, R.J., Ockerman, H.W., Krol, B. *et al.* (1978) Influence of tumbling, tumbling time, trim and sodium tripolyphosphate on quality and yield of cured hams. *J. Food Sci.*, **43**, 853.
Kuo, J.C. (1986) Effect of low sodium salts on the quality of sectioned and formed hams. *Food Sci. China*, **13**, 10.
Lin, G.C., Mittal, G.S. and Barbut, S. (1991) Optimization of tumbling and KCl substitution in low sodium restructured hams. *J. Muscle Foods*, **2**, 71.
Madril, M.T. and Sofos, J.N. (1985) Antimicrobial and functional effects of six polyphosphates in reduced NaCl comminuted meat products. *Abstr. 45th Ann. Mtg. Inst. Food Technol.*, Atlanta, GA.
Maesso, E.R., Baker, R.C., Bourne, M.C. and Vadehra, D.V. (1970) Effect of some physical and chemical treatments on the binding quality of poultry loaves. *J. Food Sci.*, **35**, 323.
Marcy, J.A., Kraft, A.A., Olson, D.G. *et al.* (1985) Fate of *Staphylococcus aureus* in reduced sodium fermented sausage. *J. Food Sci.*, **29**, 316.
Matulis, R.J., McKeith, F.K., Sutherland, J.W. and Brewer, M.S. (1995) Sensory characteristics of frankfurters as affected by salt, fat, soy protein and carrageenan. *J. Food Sci.*, **60**, 48.
Maurer, A.J. (1983) Reduced sodium usage in poultry muscle foods. *Food Technol.*, **37**, 60.
Midgley, J.P., Matthew, A.G., Greenwood, C.M.T. and Logan, A.G. (1996) Effect of reduced dietary sodium on blood pressure: a meta-analysis of randomized controlled trials. *J. Am. Med. Assoc.*, **275**, 1590.
Miller, R.K. and Acuff, G.R. (1994) Sodium lactate affects pathogens in cooked beef. *J. Food Sci.*, **59**, 15.
Mittal, G.S., Lin, C.W. and Barbut, S. (1990) The role of various processes in manufacturing a low sodium restructured ham, *Proc. 35th Int. Congr. Meat Sci. Technol.*, **3**, 730.
Murphy, D. (1995) With a grain of salt. *Natl. Provisioner*, **209**, 58.
Nielsen, H.S. and Zuethen, P. (1985) Sodium chloride and pathogenic bacteria in a vacuum-packaged minced meat product. *J. Food Protect.*, **48**, 150.
NRC/NAS (1989) National Research Council/National Academy of Sciences, *Recommended Dietary Allowances*, 10th edn, National Academy Press, Washington, DC.
Olson, D.G. and Terrell, R.N. (1981) *Proc. Meat Ind. Res. Conf.*, American Meat Institute, Arlington, VA, p. 59.
Ozimek, G. and Poznanski, S. (1982) Influence of an addition of textured milk proteins upon physicochemical properties of meat mixtures. *J. Food Sci.*, **47**, 234.
Pearson, A.M. and Tauber, F.W. (1984) *Processed Meats*, 2nd edn, Van Nostrand Reinhold, New York.
Pepper, F.H. and Schmidt, G.R. (1975) Effect of blending time, salt, phosphate and hot-boned beef on binding strength and cook yield of beef rolls. *J. Food Sci.*, **40**, 227.
Puolanne, E.J. and Terrell, R.N. (1983) Effects of rigor state, levels of salt and sodium tripolyphosphate on physical, chemical and sensory properties of frankfurter-type sausages. *J. Food Sci.*, **48**, 1036.

Robinson, A., Gibson, A.M. and Roberts, T.A. (1982) Factors controlling the growth of *Clostridium botulinum* Types A and B in pasteurized, cured meats. V. Prediction of toxin production: non-linear effects of storage temperature and salt concentration. *J. Food Technol.*, **17**, 727.

Rust, R.E. (1988) Processing options for improving the nutritional value of animal products, in *Designing Foods: Animal Product Options in the Marketplace*, National Research Council/National Academy of Sciences, National Academy Press, Washington, DC.

Saffle, R.L. and Galbreath, J.W. (1964) Quantitative determination of salt-soluble protein in various types of meat. *Food Technol.*, **19**, 1943.

Schantz, R. and Bowers, J. (1993) Response surfaces of sensory characteristics for reduced sodium chloride and phosphate salts in emulsified turkey sausages. *J. Sensory Stud.*, **8**, 283.

Schmidt, G.R., Mawson, R.F. and Siegel, D.G. (1981) Functionality of a protein matrix in comminuted meat products. *Food Technol.*, **35**, 235.

Schnell, P.G., Vadehra, D.V. and Baker, R.C. (1970) Mechanism of binding chunks of meat. 1. Effect of physical and chemical treatments. *J. Inst. Can. Technol. Aliment.*, **3**, 44.

Seman, D.L., Olson, D.G. and Mandigo, R.W. (1980) Effect of reduction and partial replacement of sodium on bologna characteristics and acceptability. *J. Food Sci.*, **45**, 1116.

Sofos, J.N. (1983) Effects of reduced salt (NaCl) levels on the stability of frankfurters. *J. Food Sci.*, **48**, 1684.

Sofos, J.N. (1985) Influence of sodium tripolyphosphate on the binding and antimicrobial properties of reduced NaCl comminuted meat products. *J. Food Sci.*, **50**, 1379.

Sofos, J.N. (1986a) Quality and shelf-life of low sodium chloride meat products, in *The Shelf Life of Foods and Beverages. Proceedings of the 4th International Flavor Conference, Rhodes, Greece*, (ed. G. Charalambous), Elsevier, Amsterdam, pp. 463–482.

Sofos, J.N. (1986b) Antimicrobial activity and functionality of reduced sodium chloride and potassium sorbate in uncured poultry products. *J. Food Sci.*, **51**, 16.

Sofos, J.N. (1986c) Use of phosphates in low-sodium meat products. *Food Technol.*, **40**, 52.

Steinhauer, J.E. (1983) Food phosphates for use in the meat, poultry and seafood industry. *Dairy Food Sanit.*, **3**, 244.

Svetkey, L.P., Shusterman, T. and Lin, P. (1996) *Eating Well and Living Well with Hypertension*, Viking Press, New York.

Terrell, R.N. (1983) Reducing the sodium content of processed meats. *Food Technol.*, **37**, 66.

Terrell, R.N. and Olson, D.G. (1981) Chloride salts and processed meats: properties, sources, mechanisms of action and labeling, in *Proceedings of Meat Industry Research Conference*, American Meat Institute, Arlington, VA, p. 67.

Terrell, R.N., Childers, A.B., Kayfus, T.J. *et al.* (1982) Effect of chloride salts and nitrite on survival of trichina larvae and other properties of pork sausages. *J. Food Protect.*, **45**, 281.

Terrell, R.N., Quintanilla, M., Vanderzant, C. and Gardner, F.A. (1984) Effects of reduction or replacement of sodium chloride on growth of *Micrococcus, Moraxella* and *Lactobacillus* in inoculated ground pork. *J. Food Sci.*, **48**, 122.

Tinney, K.S., Miller, M.F., Ramsey, C.B. and Braden, E.J. (1995) Texture modifying agent and binder effects on the yield, sensory and textural properties of boneless hams. *J. Food Qual.*, **18**, 119.

Trout, G.R. and Schmidt, G.R. (1984) Effect of phosphate type and concentration, salt level and method of preparation on binding in restructured beef rounds. *J. Food Sci.*, **49**, 687.

Trout, G.R. and Schmidt, G.R. (1987) Nonprotein additives. *Adv. Meat Res.*, **3**, 309.

USDA (1990) *Composition of Foods: Beef Products; Raw, Processed, Prepared. Agriculture Handbook 8–13 (revised)*, US Department of Agriculture, Human Nutrition Information Service, Washington, DC.

USDA (1992) *Composition of Foods: Pork Products; Raw, Processed, Prepared. Agriculture Handbook 8–13 (revised)*, US Department of Agriculture, Human Nutrition Information Service, Washington, DC.

USDA/USDHHS (1995) *Nutrition and Your Health: Dietary Guidelines for Americans*, 4th edn, *Home and Garden Bulletin*, No. 232, US Department of Agriculture and US Department of Health and Human Services, Washington, DC.

USDHHS (1988) *The Surgeon General's Report on Nutrition and Health*, US Department of Health and Human Services (PHS) 88-50210, US Government Printing Office, Washington, DC.

Whiting, R.C. (1984) Stability and gel strength of frankfurter batters made with reduced NaCl. *J. Food Sci.*, **49**, 1350.

Whiting, R.C. and Jenkins, R.K. (1981) Partial substitution of sodium chloride by potassium chloride in frankfurter formulations. *J. Food Qual.*, **4**, 259.

Whiting, R.C., Benedict, R.C., Kunsch, C.A. and Woychik, J.H. (1984) Effect of sodium chloride levels in frankfurters on the growth of *Clostridium sporogenes* and *Staphylococcus aureus*. *J. Food Sci.*, **49**, 351.

Whiting, R.C., Benedict, R.C., Kunsch, C.A. and Blalock, D. (1985) Growth of *Clostridium sporogenes* and *Staphylococcus aureus* at different temperatures in cooked corned beef made with reduced levels of sodium chloride. *J. Food Sci.*, **50**, 304.

Wirth, F. (1991) Replacing the fat and sodium content of meat products: what possibilities are there? *Fleischwirtschaft*, **71**, 294.

Ziporyn, T. (1985). The Food and Drug Administration: how 'those regulations' came to be. *J. Am. Med. Assoc.*, **254**, 2037.

15 Low fat and low salt poultry products

D.M. SMITH

15.1 Introduction

Poultry meat is naturally low in fat, which is one of the primary reasons for the large increase in consumption of poultry meat in the last decade. Even further-processed poultry products, such as frankfurters and bologna, have traditionally been 30–50% lower in fat than their red meat counterparts. Recently, however, the poultry industry has been developing products with even lower fat contents to meet consumer demands and remain competitive with the red meat industry. The Nutritional Labeling of Meat and Poultry Products regulations issued in 1993 (USDA, 1994) have made it easier for processors to advertise the low fat attributes of their products and has accelerated the development and marketing of low fat and fat free products.

Although poultry muscle is low in sodium, further-processed poultry products tend to contain high concentrations of sodium. For technological reasons, processors have been unable to lower the sodium content of further-processed products to the point that they can be consumed by hypertensive individuals on low sodium diets. Consequently, consumer interest in low sodium further-processed poultry products has decreased (Murphy, 1995). At the same time, the nutritional labeling regulations (USDA, 1994) have made it more difficult to make claims regarding sodium reduction in processed poultry products. These changes have forced poultry processors to re-evaluate their commitment to aggressive marketing of low sodium products. As a result, fewer low sodium poultry products are available in the marketplace. However, several strategies were developed and used successfully in the 1970s and 1980s to reduce the sodium content of processed products.

Greenwald (1991) suggested that there are two general approaches for altering the nutrient composition of food products: (1) modification of the raw materials and (2) process modifications, including changes in formulations and processing parameters. This chapter discusses how raw material modifications and process modifications have been used to reduce the fat content of poultry meat, and also briefly discusses process modifications currently used to reduce the sodium content of poultry.

15.2 Influence of production practices on fat content

15.2.1 Diet versus genetics

Raw material modifications are based on broiler and turkey production practices that change the composition of the meat. Genetic selection for increased growth rate, body weight and breast meat yield has resulted in concurrent increases in carcass and abdominal fat (Fontana *et al.*, 1993). Havenstein *et al.* (1994) fed 'typical' 1957 and 1991 diets to 1957 random-bred control broilers (these birds have the same genetic makeup as birds grown commercially in 1957) and to broilers of a 1991 Arbor Acres feather-sexable strain. At 43 days of age (typical market age for 1992), total carcass fat and abdominal fat for 1991 birds on 1991 diets averaged 14.3 and 1.6%, respectively. Total carcass fat and abdominal fat for 1957 birds on 1957 diets at 85 days (typical market age for 1957) were lower and averaged 8.9 and 0.7%, respectively. Although carcass fat is greater now than in 1957, genetic selection has been used to decrease carcass fat from that observed about 10–15 years ago when fat pad weights averaged 2.5–3.0% of live body weight.

(a) Energy density of diet

Although genetic makeup has a large influence on the percentage and deposition of carcass fat, nutrition and management practices can also influence the fat content of broilers. The energy density of the ration during different parts of the growth cycle influences growth rate and fat deposition. The fat content is controlled by the calorie to protein ratio of the diet. The fat content of broiler carcasses is usually decreased as dietary protein is increased and dietary energy is decreased (Jackson, Summers and Leeson, 1982). Attempts have been made to reduce growth rate during the early growth period in order to reduce fat accumulation by adjusting the energy density of the diet (Sizemore and Siegel, 1993). Researchers have also reported that the photoperiod can be manipulated to reduce the fat content of broilers (Cartwright, 1991). Other methods to control fat deposition in poultry include the use of sex hormones, growth hormones and repartitioning agents, such as β-adrenergic agonists (Pearson and Dutson, 1991). The response in body composition to most of these treatments is transitory, hence their use must be carefully controlled and timed to minimize fat content at slaughter.

Some poultry producers are using specially formulated low energy diets and a slower growing rate to produce broilers with 25% less fat. This allows the producer to obtain a marketing advantage by labeling the broilers as 'reduced fat – 25% less fat than regular broilers'. This approach will not work for most growers as they are not rewarded for production of lean broilers or turkeys. Producers are paid by weight and need a rapid growth rate and large body size to be profitable. More information describing production practices to reduce fat content of muscle can be found in Chapter 8.

15.3 Fat content of poultry meat

The composition of broiler meat (Table 15.1) and turkey meat (Table 15.2) has been summarized from the USDA (1978) Food Composition Tables. Although there has not been a comprehensive study of chicken and turkey composition since that date, it is likely that some small changes have occurred in composition within the past 18 years due to changes in production practices and poultry genetics. Similarly, the accuracy and precision of many analytical methods have been improved, especially methods to measure cholesterol.

15.3.1 Comparison of chicken and turkey meat with and without skin

Traditionally, poultry has been marketed as a whole eviscerated carcass or as skin-on, bone-in cut-up parts. Although sales of whole birds have declined rapidly since 1981, the popularity of cut-up parts remains high. Broiler meat with skin contains an average of 15.1% fat and 215 kcal per 100 g. In general, light meat parts contain less fat and fewer calories than dark meat, although breast meat with skin contains 9.3% fat whereas drumstick meat with skin contains 8.7% fat. Thighs with skin contain the highest total amount of fat (15.3%). Older chickens, such as roasters and spent hens, contain higher quantities of both fat and cholesterol (USDA, 1978). The composition of turkey meat is shown in Table 15.2. On average, turkey contains less fat and fewer calories than broiler meat. Light turkey meat with skin contains 7.0% fat, whereas dark meat and skin contains 7.9% fat.

(a) Removal of skin and fat pad

The easiest way to reduce the fat content of poultry is to remove the skin, abdominal fat (fat pad) and subcutaneous fat. Poultry muscle has very little marbling or intramuscular fat. Abdominal fat accounts for about 30% of the major fat depots in broilers (Cartwright, 1991), so its removal can decrease carcass fat significantly. Removing the skin can also reduce the fat content of poultry. Broiler skin is composed of 32.4% fat, whereas turkey skin contains 36.9% fat. The fat content of poultry parts can be reduced even more by careful trimming of subcutaneous fat.

The fat content of broiler breast muscle is reduced from 9.3 to 1.2% when the skin is removed. The calorie content of 100 g of breast meat is reduced from 172 to 110 kcal by removing the skin. About 10% of calories are from fat in skinless breast meat. The fat content of broiler thighs

Table 15.1 Composition of broiler meat (per 100 g)[a]

Product	Calories (kcal)	Protein (g)	Total fat (g)	Saturated fat (g)	Monounsaturated fat (g)	Polyunsaturated fat (g)	Cholesterol (mg)	Sodium (mg)
				Raw products				
Flesh and skin	215	18.6	15.1	4.3	6.2	3.2	75	70
Flesh	119	21.4	3.1	0.8	0.9	0.8	70	77
Breast meat and skin	172	20.9	9.3	2.7	3.8	2.0	64	63
Breast meat	110	23.1	1.2	0.3	0.3	0.3	58	65
Thigh meat and skin	211	17.3	15.3	4.3	6.3	3.3	84	76
Thigh meat	119	19.7	3.9	1.0	1.2	1.0	83	86
Drumstick meat and skin	161	19.3	8.7	2.4	3.4	1.9	81	83
Drumstick meat	119	20.6	3.4	0.9	1.1	0.9	77	88
				Cooked products				
Breast meat and skin, roasted	197	29.8	7.8	2.2	3.0	1.7	84	71
Breast meat and skin, fried	260	24.8	13.2	3.5	5.5	3.1	85	275
Breast meat, roasted	165	31.0	3.6	1.0	1.2	0.8	85	74
Breast meat, fried	187	33.4	4.7	1.3	1.7	1.1	91	79

[a]Source: USDA (1978).

Table 15.2 Composition of young tom turkey meat (per 100 g)[a]

Product	Calories (kcal)	Protein (g)	Total fat (g)	Saturated fat (g)	Monounsaturated fat (g)	Polyunsaturated fat (g)	Cholesterol (mg)	Sodium (mg)
	Raw products							
Flesh and skin	154	20.5	7.4	2.1	2.7	1.8	72	68
Flesh	117	21.7	2.7	0.9	0.6	0.8	68	73
Light meat and skin	156	21.6	7.0	1.9	2.7	1.7	67	63
Light meat	114	23.4	1.6	0.5	0.3	0.4	62	67
Dark meat and skin	152	19.1	7.9	2.3	2.7	2.1	77	75
Dark meat	123	20.0	4.1	1.4	0.9	1.2	75	80
	Cooked products							
Light meat and skin, roasted	191	28.5	7.7	2.2	2.7	1.9	75	67
Light meat, roasted	154	29.9	2.9	0.9	0.5	0.8	69	68
Dark meat and skin, roasted	216	27.6	10.9	3.3	3.4	2.9	91	80
Dark meat, roasted	185	28.7	7.0	2.3	1.6	2.1	88	82

[a]Source: USDA (1978).

is reduced from 15.3 to 3.9% if the skin is removed (USDA, 1978). Drumsticks without skin contain 3.4% fat. About 29.5 and 25.7% of calories of thigh and drumstick, respectively, are from fat. Skinless broiler meat meets the Dietary Guidelines (USDA, 1995) recommendations that suggest no more than 30% of total calories should come from fat.

Skinless turkey meat contains less fat than broilers (Table 15.2). Skinless turkey meat contains 2.7% fat and 117 kcal per 100 g. About 20.8% of calories are from fat. Light meat contains less fat and cholesterol than dark meat. Young tom turkey light meat without skin contains 1.6% fat, whereas dark meat without skin contains 4.1% fat. Only 12.6% of breast meat calories are from fat, thus falling well below recommendations of the Dietary Guidelines (USDA, 1995).

(b) Cut-up poultry parts

Cut-up poultry parts are being increasingly marketed without skin owing to consumer demand. Skinless parts, such as boneless chicken breasts and boneless chicken thigh cutlets, are commanding premium prices by capitalizing on their low fat contents. Although less popular than broiler parts, turkey parts are being marketed in several forms to take advantage of their low fat content. Breast slices, breast tenderloin and seasoned turkey breast chops containing 1% fat are among the low fat options available commercially (Table 15.3). These items are prepared from highly trimmed parts and contain no visible fat.

15.4 Fatty acid composition of poultry meat

15.4.1 Proportions of saturated, monounsaturated and polyunsaturated fatty acids

Broiler meat lipid is comprised of about 32% saturated, 37% monounsaturated and 31% polyunsaturated fatty acids (USDA, 1978). Light broiler meat contains a higher proportion of saturated fatty acids than dark meat, although the proportion of polyunsaturated fatty acids is similar. According to the USDA Food Composition Tables (USDA, 1978), turkey meat contains a higher proportion of saturated fatty acids and is comprised of about 40% saturated, 26% monounsaturated and 35% polyunsaturated fatty acids. However, Wong, Sampugna and Dickey (1993) reported that light turkey meat lipids contained 32.9% saturated, 38.3% monounsaturated and 28.9% polyunsaturated fatty acids. Dark turkey meat contained slightly higher amounts of saturated and polyunsaturated fatty acids. Differences in fatty acid composition between the USDA Food Composition Tables (USDA, 1978) and more recent studies may be due to changes in poultry diets in the last 15 years. The principal

Table 15.3 Nutrient composition of selected turkey parts and ground turkey determined from label declarations (brand: The Turkey Store)[a]

Product	Claim	Serving size (g)	Calories (kcal)	Total fat (g)	Saturated fat (g)	Cholesterol (mg)	Carbohydrate (g)	Protein (g)	Sodium (mg)
Breast slices	99% fat free	84	90	1.0	NA	40	0	21	65
Ground breast	99% fat free (extra lean)	112	120	1.5	0.5	55	0	27	75
Seasoned turkey breast chops	99% fat free	113	110	1.0	0	45	3	23	690
Burger patties	7% fat (lean)	112	160	8.0	2	70	1	22	80

[a]Nutrient content is based on serving size.

fatty acids in chicken and turkey are C18:1 (oleic acid), C18:2 (linoleic acid) and C16:0 (palmitic acid).

15.4.2 *Altering fatty acid composition by diet*

Recent research has focused on the use of dietary manipulation to alter the fatty acid composition of poultry meat. Along with a recommendation to decrease the total fat in our diets, the Dietary Guidelines for Americans (USDA, 1995) recommended a diet low in fat, saturated fat and cholesterol. The Guidelines recommend that only 10% of calories should come from saturated fat. Similarly, based on epidemiological and clinical studies suggesting a relationship between fat consumption and cardiovascular disease, some nutritionists are recommending an increase in the consumption of ω-3 (n-3) fatty acids and/or specific monounsaturated fatty acids (Kinsella, 1986; Van Elswyk, 1993). Canadian guidelines recommend that the intake of n-3 fatty acids be increased to provide a minimum of 0.55 g per 1000 kcal or 0.5% of energy (Health and Welfare Canada, 1990).

The fatty acid composition of poultry meat is easily altered by changing the type and amount of dietary fat (Hargis and Van Elswyk, 1993). Thus, poultry could serve as a vehicle to supplement the consumption of beneficial fatty acids.

(a) Effects of feed

Feeding of fish meal, fish oil, flax seed, flax oil or canola seed has been shown to increase the n-3 fatty acid content of poultry. Van Elswyk (1993) reviewed the use of fish oil and fish meal for increasing the n-3 fatty acid composition of poultry. Although the n-3 fatty acid content was increased, off-flavor problems were noted. More recently, terrestrial sources of n-3 fatty acid have been used in hen diets to try and eliminate the off-flavor problems (Hargis and Van Elswyk, 1993). Flaxseed oil or meal incorporated into the diets of broilers (Ajuyah *et al.*, 1991) and spent hens (Ajuyah *et al.*, 1992) increased the n-3 fatty acid composition of the meat. In spent hens, the ratio of n-6 to n-3 fatty acids decreased from 8 in controls to about 2 in hens fed the flaxseed diet. The meat was enriched in linolenic acid (C18:3n3), eicosapentaenoic acid (C20:5n3), decosapentaenoic acid (C22:5n3) and decosahexaenoic acid (C22:6n3), although the increase in the 20 and 22 carbon fatty acids was less than that observed with fish meals.

15.5 Cholesterol content of poultry meat

15.5.1 *Species effects*

The cholesterol content of broiler and turkey meat is similar (Tables 15.1 and 15.2). Light meat contains less cholesterol than dark meat. For example, turkey light meat with skin contains 67 mg cholesterol per 100 g, whereas dark meat with skin contains 77 mg cholesterol per 100 g. Broiler and turkey skin contain about 109 and 95 mg cholesterol per 100 g, respectively, hence removal of skin lowers the cholesterol content of poultry meat. When the skin is removed, turkey light meat contains about 62 mg cholesterol per 100 g, whereas dark meat contains about 75 mg cholesterol per 100 g (USDA, 1978). Label claims indicate that the cholesterol content of commercially available highly trimmed turkey breast meat products is 40–55 mg per 100 g (Table 15.3). The concentration of cholesterol in poultry is similar to that in red meat products. Beef chuck and round (Select grade) contain 68 and 60 mg cholesterol per 100 mg, respectively (USDA, 1990).

Few studies have evaluated specific production practices to lower the cholesterol content of poultry meat. Most research has been designed to lower the total fat content, which concomitantly lowers the cholesterol content (Hargis and Van Elswyk, 1993).

15.5.2 Supercritical fluid extraction

Processing approaches have been investigated to reduce the cholesterol content of poultry meat, but these have not been adapted for large-scale use. Supercritical fluid extraction using carbon dioxide is one process still under consideration for decreasing the cholesterol content of poultry meat. Use of supercritical fluid extraction is considered in greater detail in Chapter 6.

Froning *et al.* (1994) used supercritical fluid extraction with carbon dioxide to reduce the cholesterol and lipid content of dehydrated chunked and powdered chicken meat. Two combinations of pressure and temperature were investigated to provide a fluid density of 0.9 g cm^{-3}. At a pressure of 381 atm and 55°C, about 93% of the lipid and 82% of the cholesterol were removed from chunked products, whereas 89% of the lipid and 90% of the cholesterol were removed from powdered products. Although not in great demand, dehydrated meat is used in soups mixes, gravy mixes, bouillon and casseroles.

15.6 Mechanically deboned poultry

15.6.1 Composition

The fat content of most mechanically deboned products is higher than that of whole muscle (Table 15.4). The lowest fat content is found in mechanically deboned meat prepared from turkey frames (16% fat) and broiler backs and necks without skin (15.5% fat). Mechanically deboned meat from spent hens has a cholesterol content of 143 mg per 100 g, which

Table 15.4 Composition of mechanically deboned chicken and turkey (per 100 g)[a]

Product	Calories (kcal)	Protein (g)	Total fat (g)	Saturated fat (g)	Monounsaturated fat (g)	Polyunsaturated fat (g)	Cholesterol (mg)	Sodium (mg)
Broiler backs and necks with skin	272	11.4	24.7	7.5	10.4	5.0	NA	40
Broiler backs and necks without skin	199	13.8	15.5	4.7	7.4	2.3	NA	51
Mature hens	243	14.7	20.0	4.7	9.3	4.6	143	NA
Turkey frames	201	13.3	16.0	5.3	5.0	4.5	NA	NA

[a]Source: USDA (1978).

is about 2–3-fold higher than that in whole muscle. Mechanically deboned meat is a common ingredient in full fat cured poultry products, but is not used in low fat or fat free products owing to its high fat content. Closely trimmed dark or white poultry meat is used in place of mechanically deboned meat in most low fat products.

(a) Centrifugation effects
Centrifugation has been used to alter the fat content of mechanically deboned poultry (Froning and Johnson, 1973; Maurer and Goble, 1981). Froning and Johnson (1973) found that the fat content of mechanically deboned fowl was reduced from 18.3 to 6.0% by centrifugation. The protein content was increased from 13.9 to 23.2% by the same process.

15.7 Ground poultry products

15.7.1 *Variability in composition*

Ground turkey meat and ground turkey patties are being marketed as a low fat substitute for ground beef. Ground turkey is prepared from surplus parts and thus the composition is variable. It is usually a mixture of dark and light meat, including some skin and visible fat. This product contains an average fat content of 8.5% with a cholesterol content of 81 mg per 100 g (Wong, Sampugna and Dickey, 1993). Highly trimmed ground poultry meat may have a fat content as low as 7% (Table 15.3). The average fat content of extra lean ground beef is 17.1% (USDA, 1990), thus ground turkey contains about 10% less fat.

(a) Cholesterol content
The cholesterol content is similar to that reported in ground beef (USDA, 1990). Ground beef contains about 69–85 mg cholesterol per 100 g, depending on the fat content. Both the fat and cholesterol content of ground turkey products can be reduced by using less skin and more white meat in the formulation. Recently, ground turkey breast meat has been introduced as another low fat alternative for consumers. Label claims indicate this product is 99% fat free and contains 120 cal per 112 g serving (Table 15.3). Close trimming of fat reduces the cholesterol content to 48 mg per 100 g muscle.

15.8 Pre-cooked poultry products

For many years, the most common pre-cooked poultry product was frozen breaded, fried chicken parts. To reduce the fat content, frozen, fried chicken is now being offered without skin. Label claims on this product indicate a 25% reduction in fat.

15.8.1 *Effects of cooking*

Frying causes a large increase in the fat content of poultry products. Other cooking methods are being used to reduce the fat content of pre-cooked products, such as roasting, broiling, grilling and baking. Roasted broiler breast meat contains about 23–41% less fat than the comparable fried product (Table 15.1). Removal of skin prior to roasting also reduces the fat and calorie content of poultry meat (Tables 15.1 and 15.2). Some marinades used on roasted chicken may contain oils that increase the fat content such that it approaches that of fried chicken (Welland, 1994).

Pre-cooked poultry nuggets and patties are also being marketed with lower fat contents. This is accomplished by removing the skin and/or increasing the amount of white meat in the formulation. Eliminating the breading is another technique used to reduce the fat content, as the breading absorbs considerable fat during frying. Pre-cooked unbreaded nuggets are most commonly used in food service applications, such as school lunch programs.

15.9 Composition of other avian species

The meat from many other avian species, such as squab, quail and pheasant, is also naturally low in fat (USDA, 1978). Ostrich meat is being aggressively marketed as a low fat alternative to red meat. Ostrich meat is similar in color and taste to beef, but has a much lower fat content. When cooked to 63°C, ostrich meat contained 2.1–3.8% fat, 23.3–26.3% protein and 76.1–94.7 mg cholesterol per 100 g meat (Harris *et al.*, 1994a). In a consumer acceptability test, there was no difference in overall acceptance when meat from four different ostrich muscles was compared with beef top loin steak (Harris *et al.*, 1994b).

15.10 Further processed products

15.10.1 *Effects of dilution*

In further processed products, dilution is the solution for fat reduction. Low fat raw materials should be used as the starting materials, such as highly trimmed breast meat. Water is added to dilute the fat to the desired concentration. Products containing between 4 and 8% fat can be labeled as 92–96% fat free. Products containing less than 0.5 g of fat per serving can be labeled as fat free. Other labeling claims are discussed in Chapter

3. As with red meat products, poultry processors must follow the '40% rule'. By regulation, fat and added water must not exceed 40% and fat alone must not exceed 30% of the finished product.

15.10.2 Effects of serving size

Serving sizes are regulated under the new labeling regulations and are based on the amount of food customarily eaten. The standard serving size for meat products is 3 oz or 85 g. Processors can also select a serving size based on item size, such as slices or units. For example, common serving sizes for frankfurters are 45 and 50 g and are based on a single frank. Serving sizes for deli breast products can be based on slice size. Serving size may vary with slice diameter and thickness and number of slices selected, although 1 oz or 28 g is a common size for sliced luncheon meats. The variability is illustrated in Table 15.5, where the serving sizes of the three turkey breast products listed range from 28 to 63 g.

15.10.3 Functionality of fat

Addition of water and removal of fat causes several problems in further-processed products. Dilution results in a very wet product with low yields and high purge. As discussed in Chapter 4, removal of fat leads to other problems, such as loss of juiciness, an undesirable texture, changes in mouth-feel, darkening in color and a bland flavor. Low fat products are often more rubbery or chewy than their full fat counterparts. Fat is also a source of essential fatty acids and a carrier for fat soluble vitamins. Other ingredients must be added to substitute for the multiple functions of fat and to bind the excess water. Often combinations of ingredients are used to achieve the desired quality attributes in low fat and fat free products. The composition and function of protein and carbohydrate ingredients for low fat meat and poultry products are presented in Chapter 10. This section will list some ingredients commonly used in further processed poultry product formulations and highlight some unique functionalities of these additives in poultry products.

15.10.4 Deli Breast Products

Turkey and chicken deli breast products are some of the most popular low fat poultry products. They command premium prices as breast meat is in the greatest demand and thus is the most valuable part of the bird. Many varieties of reduced fat (92–98% fat free) and fat free deli breast products are available. The nutrient composition of several representative products is shown in Table 15.5. In general, these products contain 0.5 g of fat or less per serving. Reduced fat deli breasts contain about 107 kcal

Table 15.5 Nutrient composition of selected uncured poultry products determined from label declarations[a]

Brand	Product	Claim	Serving size (g)	Calories (kcal)	Total fat (g)	Saturated fat (g)	Cholesterol (mg)	Carbohydrate (g)	Protein (g)	Sodium (mg)
Tyson	Oven-roasted chicken breast	Fat free	63	50	0	0	25	1	11	680
Butterball	Smoked turkey breast	Fat free	54	50	0	0	25	1	11	400
Louis Rich	Oven-roasted turkey breast	98% fat free	28	30	0.5	0	10	1	5	310
Mr Turkey	Smoked white turkey	98% fat free	28	30	0.5	0	15	2	5	310

[a]Nutrient content is based on serving size.

per 100 g and 15% of calories from fat. Fat free counterparts contain about 74–87% fewer calories than reduced fat brands.

(a) Low fat and reduced fat deli products

Low fat and reduced fat deli breast products are prepared using relatively few ingredients, as the fat content of the muscle is low to start with so very little dilution is required. These products are prepared from closely trimmed turkey or chicken breasts. Water is added to reduce the fat to the desired concentration. Binders are then used to tie up the water, prevent purge during processing and storage and improve sliceability. Common binders in these products include modified food starch and/or carrageenans. Bater, Descamps and Maurer (1992) found that 0.5% kappa-carrageenan increased yields, appearance, sliceability and textural rigidity and decreased expressible moisture of oven-roasted turkey breasts made with 70% added brine as compared with controls. In the same experiment, 1% starch improved product yield but did not increase rigidity.

(b) Use of additives

Use of white colored additives, such as whey protein concentrates, may reduce pinking, a common problem in uncured poultry products. Combinations of salt and phosphate are added to solubilize the muscle proteins for water binding and texture development. Turkey broth or flavorings are often added to enhance the meaty flavor as low fat turkey breast products can be very bland. Many breast meat products are marinated in seasonings and spices to improve flavor and provide consumers with a more varied selection. Sodium lactate may be used as a preservative.

15.10.5 Cured poultry products

(a) Composition

The nutrient composition of representative further-processed cured poultry products is shown in Table 15.6. Further-processed poultry products have typically contained 30–50% of the fat of their red meat counterparts. For example, turkey frankfurters and bologna contain about 17–20% fat, which is about one third less fat than that commonly found in beef franks and bologna. Breakfast sausage links made from turkey meat contain about 50% less fat than pork sausage products. Similarly, bacon, Canadian bacon, sausages, salami and pastrami made from turkey all contain less fat than comparable red meat products. Turkey ham from highly trimmed thigh meat contains 5% fat and can be labeled as extra lean, 95% fat free.

Several fat free and 97% fat free bologna and frankfurters are available. Total fat, total calories, calories from fat and cholesterol levels in low fat

Table 15.6 Nutrient composition of selected cured turkey products determined from label declarations[a]

Brand	Product	Claim	Serving size (g)	Calories (kcal)	Total fat (g)	Saturated fat (g)	Cholesterol (mg)	Carbohydrate (g)	Protein (g)	Sodium (mg)
Mr Turkey	Bologna (turkey)	$33\frac{1}{3}$% less fat than beef bologna (extra lean)	28	70	5	1.5	25	1	3	370
Butterball	Bologna (turkey)	Fat free	28	25	0	0	10	2	5	310
Healthy Choice	Bologna (turkey, pork and beef)	Low-fat (97% fat free)	28	30	1	0.5	15	2	4	240
Mr Turkey	Franks (turkey)	$33\frac{1}{3}$% less fat than red meat franks	45	110	9	2.5	40	2	5	540
Oscar Mayer	Hot dogs (turkey and beef)	Fat free	50	40	0	0	15	2	7	490
Eckrich	Franks (turkey, pork and beef)	Fat free	50	50	0	0	15	5	6	590
Healthy Choice	Franks (turkey, pork and beef)	Low fat (97% fat free)	45	50	1.5	0.5	25	5	5	380
Mr Turkey	Turkey ham	95% fat free (extra lean)	28	35	1.5	0.5	20	0	5	320
Mr Turkey	Turkey cotto salami	40% less fat than beef salami	28	50	3	1	20	1	4	240

[a]Nutrient content is based on serving size.

and fat free formulations are lower than that in full fat formulations. For example, fat free frankfurters and bologna contain 60–67% fewer calories and cholestrol than corresponding full fat counterparts. The percentage of calories from fat decreases from 64% in full fat turkey bologna to less than 30% in 97% fat free products.

Originally, further-processed cured poultry products were designed to use surplus thigh meat, mechanically deboned meat and skin. With the emphasis on low fat products, highly trimmed whole muscle thigh and breast meat is being used to achieve raw materials of the required leanness. The meat is diluted with water or broth to reduce the fat content to the desired concentration. Modified food starch and hydrolyzed milk proteins are the most common binders currently used in low fat and fat free products.

(b) Use of additives

Starch is often used in poultry as it has excellent water holding ability and has a bland flavor and white color. Modified corn and potato starches are also commonly used. Starch binds the added water to prevent purge during processing and improves firmness, although often a protein additive is used in conjunction with starch to achieve the desired texture. Although not allowed in the USA, cellulose derivatives and konjac gum are other carbohydrate-based ingredients commonly used in low fat poultry products marketed in other countries.

Protein additives

Protein ingredients, such as gelatin and milk proteins, are often added to improve texture. They are usually used in combination with other binders such as starch, as they do not have the high water holding capacity of carbohydrate-based binders. Hydrolyzed milk protein, hydrolyzed skim milk and other modified milk proteins (such as sodium caseinates) are common ingredients in low fat and fat free poultry frankfurters and bologna. Although derived from milk proteins, these products impart different functional properties to low fat products. Hydrolyzed sodium caseinate acts as a binder to hold water, prevent purge, improve firmness and sliceability, but does not gel. One major advantage of this product is its ability to mimic fat by imparting a smooth mouth-feel to fat free and low fat products. This ingredient also acts as a whitener to reduce the dark color commonly found in low fat cured poultry products. Whey proteins form gels at meat processing temperatures, thus they impart different functionalities to low fat products. To avoid confusion and possible allergic reactions in consumers, USDA has passed regulations requiring that the specific protein source be identified on the label, although not all labels have been changed at this time.

Salt and phosphates

Many different salt and phosphate blends are available specifically for poultry to maximize muscle protein solubility (for water binding) and gelation (for texture formation). These blends include NaCl, KCl and different combinations of sodium acid pyrophosphate, sodium tripolyphosphate and sodium hexametaphosphate. Careful selection of salts and phosphates is essential to avoid the development of bitter flavors in the blander, low fat poultry products. Also, lower salt levels may be necessary to avoid an excessively salty flavor in the blander products. Potassium lactate may also be used as a preservative.

Hydrolyzed protein (such as autolyzed yeast), broth or natural flavorings are added to improve the meaty flavor of low fat and fat free products. Many low fat and fat free poultry frankfurters also contain smaller quantities of beef and pork in the formulations to enhance the meaty flavor. Oleoresin of paprika may be added to enhance cured meat color.

15.10.6 Process modifications

Some process modifications may be necessary to produce low fat poultry products. The viscosity of low fat formulations is often thin and runny, thus additives are used to maintain and produce a stable batter and improve handling prior to cooking. Careful selection of binders is necessary to stabilize raw batters, yet avoid increasing the viscosity, such that pumping becomes difficult. Longer cooking times may be necessary to maximize yields and optimize texture. Air must be eliminated from the product by vacuum chopping to prevent air pocket formation during thermal processing.

The order in which ingredients are added during the chopping process is important. Dexter, Sofos and Schmidt (1993) found that cooking losses of low fat turkey bologna were decreased if salt and phosphate were allowed to react with the meat protein prior to addition of the other additives. Thus, pre-blending is important so that adequate time is given for myofibrillar proteins to swell and become hydrated. Similarly, adequate mixing time or pre-blending is necessary to hydrate some protein or carbohydrate additives fully.

Dilution can also result in high surface moisture or ‘sweating’, which may prevent even application of smoke or ‘two toning’. Moisture droplets on the surface of a cured product during thermal processing prevent the deposition of smoke, leading to the formation of a mottled appearance.

15.11 Low salt poultry products

There has been much less emphasis on producing reduced salt poultry products. The current USDA (1995) Dietary Guidelines recommend a diet moderate in salt and sodium. Low sodium was a marketing tool in the 1970s and early 1980s, but many companies have discontinued their low sodium product lines owing to a lack of consumer demand and more restrictive changes in labeling regulations for sodium.

15.11.1 Species and cut effects

Poultry muscle is naturally low in sodium. Broiler meat contains about 77 mg sodium per 100 g, whereas turkey meat contains about 73 mg sodium per 100 g (USDA, 1978). A food product must contain less than 140 mg per serving to be labeled as 'low sodium' and contain less than 35 mg per serving to be labeled as 'very low sodium'. Thus, all whole muscle poultry products meet the 'low sodium' labeling requirements. See Chapter 3 for more details on labeling of low sodium products.

Turkey breast products generally contain about 1.2–1.5% salt, turkey ham about 2.0% salt and turkey bologna about 3.0% salt. Chicken patties and nuggets may contain about 0.5–1.0% salt. Salt is not the only source of sodium in these products. Sodium is added to further-processed poultry products in the form of NaCl, sodium phosphates and sodium lactate.

The sodium content in further-processed low fat poultry products ranges from 240 to 680 mg per serving. Most full fat, reduced fat and fat free products available today have about the same or higher sodium contents as the full fat products available 20 years ago. Sodium must be reduced by 75% in further-processed meat products to be labeled 'reduced sodium'. To date, such a large reduction cannot be met in processed products.

15.11.2 Problems encountered in reducing salt content

Salt cannot be easily replaced or eliminated from processed poultry products because of its multiple functions. Salt functions as a preservative, provides flavor and solubilizes the myofibrillar proteins for water binding and gel formation, which improves yields (reduces purge) and texture formation. The inability of researchers to find alternative means to solubilize the myofibrillar proteins without salt has interfered with the rapid development of low salt further-processed poultry products.

Sodium reduction of further-processed products was a major research focus in the 1970s and 1980s. Several excellent review articles describing this research are available (Maurer, 1983; Terrell, 1983). Current approaches used to reduce sodium in further processed poultry products include the use of salt and phosphate blends, phosphate and carrageenan blends and lite salt (a blend of NaCl and KCl). ONC Company claims to have a carrageenan and phosphate blend for low fat poultry products that maintains yields yet reduces the sodium content by 50% (Duxbury, 1991). None of these approaches results in a 75% reduction in sodium content for labeling purposes.

Natura (1993) has developed a non-salt substitute, called eXalt, composed of lysine and succinic acid. This compound is approved for use in Europe and is being tested by a European meat company. The company is seeking GRAS status in the USA. This compound has a salty flavor, and also some antimicrobial and antioxidative properties, and may substitute for some of the functions of salt in meat products. In some low fat and fat free poultry products, sodium lactate is added for its antimicrobial properties. Some processors are using KCl in combination with NaCl to reduce the overall sodium content in products containing sodium lactate. Potassium lactate is less commonly used, as it is less soluble and becomes highly viscous when chilled, which causes handling problems. Also, blends of KCl and NaCl may cause a bitter flavor in poultry products. Additional flavors and seasonings may be added to mask the bitter flavor. Sometimes it is necessary to reduce the salt content of low fat and fat free products owing to the intensity of salty flavor.

15.12 Research needs

15.12.1 Need for mechanistic studies

Very little research has been published on the development of low fat or fat free poultry products. Much of the development work has been done by ingredient suppliers or poultry processors and is proprietary. An extensive literature review revealed only about 10 original scientific papers on low fat poultry products. None of these papers addressed fat free product technology in poultry, although several trade journals discussed the technology. The development of new low fat or reduced fat products is largely a matter of trial and error, as there is a large gap between practical application and fundamental understanding of the basic ingredient interactions in these systems. Fundamental studies are needed in order to understand the science behind the technology and to allow the more efficient development of new and improved fat free products. For example, many trade journal articles state that certain additives are good emulsifiers in fat free products. Obviously the term emulsifier is a misnomer for these products as no fat is present to form an emulsion. More studies are needed to understand the interactions between muscle proteins and carbohydrate and protein additives during gelation and subsequent texture formation and water binding during thermal processing of low fat and fat free products.

15.12.2 Understanding flavor

One of the biggest problems caused by the removal of fat is the loss of flavor. Many low fat, cured poultry products are prepared with red meat to enhance the meaty flavor. It is important that processors understand how flavor compounds interact with fat and fat replacers to improve the flavor quality of fat free and low fat poultry products. Many flavor compounds, such as aldehydes, are derived from the lipids. When the lipids are removed, the flavor is primarily derived from the remaining water soluble components. More research is needed to understand the contributions of water and fat soluble flavors.

Research is also needed to understand the differences in flavor release from fat and water soluble matrices. Flavor interactions are different between carbohydrate and protein fat replacers (Schirle-Keller, Reineccius and Hatchwell, 1994). Fat gradually dissolves in the mouth and releases flavor components slowly. Flavors bound to protein may be released instantaneously or be tightly bound and unavailable. When released instantaneously, there is an initial flavor burst, but on continued chewing the product tastes bland.

15.12.3 Effects of water binders

Similarly, more research is needed to understand moisture release from protein and carbohydrate binders owing to its effect on product juiciness. Starch and protein bind and release water differently. Even though they are high in moisture, many low fat and fat free products are perceived as dry.

15.12.4 Fatty acid profiles and functionality

Finally, more research is needed on fatty acid profile manipulation of muscle prior to commercial application. More studies are needed to understand how changes in fatty acid profile affect muscle functionality, stability and flavor.

15.13 Summary

The poultry industry is making tremendous advances in low fat technology. The quality of low fat and fat free products has improved greatly over the past 5 years. The quality and variety of low fat and fat free products are increasing throughout the food industry, not just in meat products. Obesity is increasing in the USA even with all these new high quality, good tasting, low fat food products. About 21% of US adolescents are overweight by 15–20 lb and one third of the adult population is overweight (FASEB, 1995). The meat industry has responded to consumer demands and has prepared a plethora of low fat products. Now it is time for consumers to take responsibility for their food choices, to eat a healthy diet and maintain ideal body weights.

References

Ajuyah, A.O., Lee, K.H., Hardin, R.T. and Sim, J.S. (1991) Changes in the yield and in the fatty acid composition of whole carcass and selected meat portions of broiler chickens fed full-fat oil seeds. *Poult. Sci.*, **70**, 2304.

Ajuyah, A.O., Hardin, R.T., Cheung, K. and Sim, J.S. (1992) Yield, lipid, cholesterol and fatty acid composition of spent hens fed full-fat oil seeds and fish meat diets. *J. Food Sci.*, **57**, 338.

Bater, B., Descamps, O. and Maurer, A. (1992) Quality characteristics of hydrocolloid-added oven roasted turkey breasts. *J. Food Sci.*, **57**, 1068.

Cartwright, A.L. (1991) Adipose cellularity in *Gallus domesticus*: investigations to control body composition in growing chickens. *J. Nutr.*, **121**, 1486.

Dexter, D.R., Sofos, J.N. and Schmidt, G.R. (1993) Quality characteristics of turkey bologna formulated with carrageenan, starch, milk and soy protein. *J. Muscle Foods*, **4**, 207.

Duxbury, D.D. (1991) Carrageenan/phosphate blends developed for low-sodium, low-fat poultry products. *Food Process.*, June, 40.

FASEB (1995) *Third Report on Nutrition Monitoring in the United States: Executive Summary*. Federation of American Societies for Experimental Biology, US Government Printing Office, Washington, DC.

Fontana, E.A., Weaver, W.D., Denbow, D.M. and Watkins, B.A. (1993) Early feed restriction of broilers: effects on abdominal fat pad, liver, and gizzard weights, fat deposition, and carcass composition. *Poult. Sci.*, **72**, 243.

Froning, G.W. and Johnson, F. (1973) Improving the quality of mechanically deboned fowl meat by centrifugation. *J. Food Sci.*, **38**, 279.

Froning, G.W., Fieman, F., Wehling, R.L., Cuppett, S.L. and Niemann, L. (1994) Supercritical carbon dioxide extraction of lipids and cholesterol from dehydrated chicken meat. *Poult. Sci.*, **73**, 571.

Greenwald, C.G. (1991) Overview of fat and cholesterol reduction technologies, in *Fat and Cholesterol Reduced Foods* (eds C. Haberstroh and C.E. Morris), Advances in Applied Biotechnology, Vol. 12. Gulf Publishing, Houston, TX, Chapter 3, p. 21.

Hargis, P.S. and Van Elswyk, M.E. (1993) Manipulating the fatty acid composition of poultry meat and eggs for the health conscious consumer. *World's Poult. Sci. J.*, **49**, 251.

Harris, S.D., Miller, R.K., Hale, D.S. *et al.* (1994a) Nutritional analysis of cooked ostrich meat. *Poult. Sci.*, **73** (Suppl. 1), 138 (Abstr.).

Harris, S.D., Miller, R.K., Morris, C.A. *et al.* (1994b) Comparison of consumer acceptability of ostrich meat versus beef top loin steak. *Poult. Sci.*, **73** (Suppl. 1), 138 (Abstr.).

Havenstein, G.B., Ferket, P.R., Scheidler, S.E. and Rives, D.V. (1994) Carcass composition and yield of 1991 vs 1957 broilers when fed "typical" 1957 and 1991 broiler diets. *Poult. Sci.*, **73**, 1795.

Health and Welfare Canada (1990) *Summary of Recommended Nutrient Intakes*, Canada Department of Supplies and Services, Ottawa, Ontario, p. 199.

Jackson, S., Summers, J.D. and Leeson, S. (1982) Effect of dietary protein and energy on broiler carcass composition and efficiency of nutrient utilization. *Poult. Sci.*, **61**, 2224.

Kinsella, J.E. (1986) Dietary fish oils. *Nutr. Today.*, Nov./Dec., 7.

Maurer, A.J. (1983) Reduced sodium usage in poultry muscle foods. *Food Technol.* **37**(7), 60.

Maurer, A.J. and Goble, J.W. (1981) Partial dewatering of mechanically deboned poultry meat. *Poult. Sci.*, **60**, 750.

Murphy, D. (1995) With a grain of salt. *Natl. Provisioner*, **209**(1), 58.

Natura (1993) Metal free and low metal salt substitutes containing lysine. *US Pat.*, 5 229 161.

Pearson, A.M. and Dutson, T.R. (eds) (1991) *Growth Regulation in Farm Animals*, Advances in Meat Research, Vol. 7, Elsevier Applied Science, London.

Schirle-Keller, J.P., Reineccius, G.A. and Hatchwell, L.C. (1994) Flavor interactions with fat replacers: effect of oil level. *J. Food Sci.*, **59**, 813.

Sizemore, F.G. and Siegel, H.S. (1993) Growth, feed conversion, and carcass composition in females of four broiler crosses fed starter diets with different energy levels and energy to protein ratios. *Poult. Sci.*, **72**, 2216.

Terrell, R.N. (1983) Reducing the sodium content of processed meats. *Food Technol.*, **37**(7), 66.

USDA (1978) *Composition of Foods: Poultry Products, Agriculture Handbook No. 8-5,* US Department of Agriculture, Washington, DC.

USDA (1990) *Composition of Foods: Beef Products, Agriculture Handbook No. 8-13,* US Department of Agriculture, Washington, DC.

USDA (1994) *Nutrition Labeling of Meat and Poultry Products,* Feed and Nutrition Information Center, Food Safety and Inspection Service, US Department of Agriculture, Washington, DC.

USDA (1995) *Dietary Guidelines for Americans,* 4th edn, *Home and Garden Bulletin,* No. 232, US Department of Agriculture, Washington, DC.

Van Elswyk, M.E. (1993) Designer foods: manipulating the fatty acid composition of meat and eggs for the health conscious consumer. *Nutr. Today,* March/April, 21.

Welland, D. (1994) Claims of a better bird don't always fly. *Environ. Nutr.*, **17**(4), 5.

Wong, M.K., Sampugna, J. and Dickey L.E. (1993) Moisture, total lipid, fatty acids and cholesterol in raw ground turkey. *J. Agric. Food Chem.*, **41**, 1229.

16 Low fat and reduced fat fish products

M.T. MORRISSEY

16.1 Introduction

Seafood is an important food source for a number of developed and developing countries. Japan, Iceland, Norway and Portugal are countries where fish constitute a main source of animal protein in the diet. Several developing countries, especially in the Asian region, also have high seafood consumption rates. Coastal communities in West Africa, Indonesia and the Philippines are very dependent on fish for protein, calories and micronutrients. Although fish as 'brain' food has been a nutritional lore for centuries, it is only recently that scientific research has focused on the relationship between fish and health. There have been copious articles, books and symposia over the last decade on the beneficial effects of fish on health. Much of this work focuses on the role of ω-3 fatty acids and their role in inhibiting cardiovascular disease (section 16.4). Consequently, fish is considered a healthy food by many consumers. The reasons for this are the low caloric intake for many types of fish products consumed and the consumption of fish for its potential nutritional/physiological effects.

The world fishing industry involves more than 200 nations fishing in a variety of waters and environmental conditions and under widely different technological expertise. More than 4000 species of finfish, shellfish and molluscs are harvested throughout the world for local consumption and international trade (FAO, 1993). Fisheries production has maintained a level of approximately 100 million metric tons (t) since 1990. Aquaculture contributes 15–20 million t of current fisheries production and has shown a slow but steady growth. The USA ranks fifth in world production, harvesting 4.7 million t in 1993, while annual consumption of seafood products in the USA is about 6.8 kg (15 lb) per person. The main products consumed in the USA (canned tuna, shrimp, Alaska pollock) emphasize the diversity of the industry. About two thirds of the US products are consumed as either frozen or fresh seafood. US consumers depend on imports for approximately 55% of their total supply of fishery products.

There is considerable effort in the seafood industry to develop value-added products. These are based primarily on developing new markets for seafood and meeting the changing demands of consumers. There have been attempts at generic marketing of seafood focusing on its health aspects, but these efforts were limited and with mixed results. The seafood industry is conscious of the reduced fat trends for products in the US marketplace, although they have not exploited this line of marketing as vigorously as other animal protein sectors. The following sections of this chapter describe fish as food from a caloric standpoint, focusing on fat content, different processing techniques that help reduce fat in fish products, a brief review of the health benefits of seafood and the characterization of some of the newer processes in the seafood industry.

16.2 Nutrient content of fish

16.2.1 Categories of seafood

Seafood can be divided into three broad categories: finfish, crustaceans and molluscs. Finfish are consumed primarily for their muscle meat in the form of fillets. Crustaceans are also consumed for their muscle meat, which is usually in the tail section. However, it is not uncommon among several ethnic groups to consume other body parts of the animal such as whole small shrimp, or the 'tamalley' of the lobster (liver and pancreas). Molluscs, especially the bivalves, are often consumed whole. There are, naturally, exceptions such as scallops, where only the muscle is consumed in Northern European cultures and the USA, and with the gonads in Latin America. Squid and octopus (which are molluscs) are consumed mainly for their outer tube and tentacles, although in some cultures they are consumed whole.

The 20 most important species consumed in the USA and corresponding nutritional information are given in Table 16.1. These are species purchased in the fresh and frozen state and do not include canned tuna, which has the highest actual consumption. Protein content ranges from 12 to 29% and fat content from 1 to 12% among the species. Because of the variety of species consumed worldwide, it is difficult to present an accurate picture regarding the consumption of seafood. Nonetheless, some generalizations will be necessary to give an overall picture of the composition of seafood and consumption patterns. The following sections focus on the fat content of different species of seafood.

16.2.2 Fat content of finfish

The fat content of finfish varies depending on the species. Most researchers categorize finfish into three nutritional composition categories: high fat, medium fat and low fat. Table 16.2 lists some commercial species and their fat content. However, the fat content will vary with the individual, season, sex, physiological status and geographical location of the fish (Love, 1988). For example, many of the small and intermediate sized pelagic fish (sardine, herring, mackerel) will have wide seasonal variation

Table 16.1 Nutritional information on seafoods[a,b]

Fish	Calories	Protein (g)	Carbohydrate (g)	Fat (g)	Saturated (g)	Cholesterol (mg)	Sodium (mg)
Shrimp	110	22	0	2	0	160	155
Cod	90	19	0	1	0	50	60
Pollock	100	21	0	1	0	80	90
Catfish	120	19	0	5	1	60	65
Scallops	150	29	2	1	0	60	275
Salmon, coho and Atlantic	150	22	0	7	1	50	50
Flounder	100	20	0	1	0	50	85
Sole	100	21	0	1	0	60	90
Oysters	120	12	7	4	1	90	190
Orange roughy	70	16	0	1	0	20	70
Mackerel	190	21	0	12	3	60	95
Ocean perch	100	20	0	2	0	50	80
Rockfish	100	20	0	2	0	40	65
Whiting	100	19	0	1	0	70	75
Clams	130	22	4	2	0	60	95
Haddock	90	20	0	1	0	60	70
Crab, blue	90	19	0	1	0	80	310
Trout, rainbow	130	22	0	4	1	60	30
Halibut	120	22	0	2	0	30	60
Lobster	100	20	1	1	0	100	320

[a]Based on 3 oz edible portion, cooked without fat or seasoning.
[b]Source: National Seafood Educators (1992).

in composition, depending on feeding and spawning cycles. The Atlantic mackerel (*Scomber scombrus*) muscle will be low in fat during the spring but high in fat in the autumn (Ackman, 1995). The dark muscle will have a continuously high (9–18%) lipid content throughout the year, whereas in the white muscle it will vary from 2.2% in the spring to up to 10% later in the year. In the European sardine (*Sardina pilchardus*) whole body fat can vary between 1.6 and 22.4% during the year (Nunes, Batista and Marao de Campos, 1992).

Pacific salmon is another case in point. There are five different species of Pacific salmon (chinook, chum, coho, pink and sockeye), all with varying amounts of fat depending on the season and location of harvest (Crapo and Kramer, 1993). Pink salmon (*Oncorhynchus gorbuscha*) is a low to medium fat fish that is primarily used in canned salmon products. Because of their lower fat content, low cost and abundance, pink salmon are also being developed into frozen food products such as breaded nuggets. King (or chinook) salmon (*Oncorhynchus tschawytscha*) is a highly prized salmon that is sought for its distinctive flavor and high fat content. When chinook salmon are caught in the ocean, before they enter their home river, they may have a fat content as high as 19% (Sidwell *et al.*, 1974). Once salmon begin their journey upstream they stop feeding

Table 16.2 Fat levels in marine fish commonly found in the marketplace[a]

Low (<5% fat)	Medium (5–10% fat)	High (>10% fat)
Anchovy	Albacore	Dogfish (spiny)
Argentine	Arctic char	Greenland turbot
Bass, hybrid striped	Butterfish	Halibut (black, European)
Black seabass	Capelin	Herring
Bluefish (Boston)	Orange roughy	Mackerel (various)
Cod (Atlantic or Pacific)	Pompano	Sablefish
Croaker	Salmon (various)[b]	Salmon (various)
Cusk	Salmon (canned)[c]	Shad
Drum	Sea bass[c]	Sprat
Flounder (various)	Sea trout[c]	
Grenadier	Shark (mako)	
Grouper	Spot	
Haddock	Sturgeon (white)[c]	
Hake (various, also whiting)	Swordfish[c]	
Halibut (Atlantic or Pacific)		
Hoki		
Kingclip		
Lemon sole		
Ling		
Lingcod		
Mahimahi		
Monkfish		
Mullet[b]		
Ocean perch		
Oreo (smooth or black)		
Pacific rockfish		
Plaice		
Pollock (Alaska)		
Pollock (Atlantic)		
Porgy		
Pout (ocean)		
Redfish		
Red snapper		
Saithe		
Scup		
Shark (most)[c]		
Skate		
Snapper		
Sole		
Striped bass		
Surimi products		
Tilefish		
Tuna (canned in water)		
Turbot (European)		
Whiting (Pacific)		
Wolffish[c]		
Yellowtail[c]		

[a]Adapted from Ackman (1995).
[b]Medium to high fat fish.
[c]Low to medium fat fish.

and use their fat reserves as the fish move toward their spawning ground. In general, the closer salmon are to their spawning ground the less fat is found in the muscle tissue. Spawned chinook salmon can have a fat content as low as 2.2% and a mealy texture. Spring chinook salmon that enter the Columbia river in March do not spawn until autumn (unlike other salmon that usually enter their home rivers close to their spawning period). They will contain relatively high levels of fat in the river through the spring and summer seasons and will only decrease rapidly in fat content in the last month before spawning. Conversely, fall chinook salmon, in the Columbia river system, return directly to their spawning ground and have been shown to have a medium range fat content (5–7%) before they enter the Columbia (Wander and Patton, 1991). Even low fat fish, such as Pacific whiting (*Merluccius productus*), will have a significant difference in fat during the season when they are harvested, varying from below 1% in spring to close to 3% in fall (Morrissey, 1995).

16.2.3 Fat content of crustaceans and molluscs

The fat content of crustaceans and molluscs is given in Table 16.3. The fat content represents the edible portion of the seafood, which for many of the bivalves is the whole animal. Both types of shellfish usually contain less than 2% fat. Seafood is noted for its high concentration of polyunsaturated fatty acids. Eicosapentaenoic (EPA) and docosahexaenoic (DHA) are ω-3 fatty acids having 20 carbon atoms and 5 double bonds (20:5ω3) and 22 carbon atoms and 6 double bonds (22:6ω3), respectively. They are found in seafoods in large amounts and have been shown to have a number of beneficial health effects (section 16.4). EPA and DHA are synthesized by phytoplankton and are consumed by crustaceans, mol-

Table 16.3 Fatty acid content (g per 100 g) of selected molluscs (raw) and crustacea[a]

Type	Species	Total fat	Saturates	Monounsaturates	Polyunsaturates	ω-3 Fatty acids (EPA and DHA)
Molluscs	Eastern oysters	2.47	0.63	0.25	0.74	0.44
	Blue mussels	2.24	0.42	0.51	0.61	0.44
	Clams (mixed species)	0.97	0.09	0.08	0.28	0.14
	Scallops (mixed species)	0.76	0.08	0.04	0.26	0.20
	Pacific oysters	2.30	0.51	0.36	0.89	0.69
	Squid (mixed species)	1.38	0.36	0.11	0.52	0.49
Crustacea	Tropical shrimp (5 species mixed)[b]	0.90	0.14	0.09	0.24	0.17
	Blue crab	1.08	0.22	0.19	0.39	0.32
	Dungeness crab	0.97	0.13	0.17	0.32	0.29
	American lobster	0.90	0.11	0.16	0.09	0.08

[a]Exler (1987).
[b]Krzynowek and Panunzio (1989).

luscs and finfish. Linolenic acid (LNA) is also an important ω-3 fatty acid (commonly found in plants) that is present in seafoods, but usually at lower levels than EPA and DHA.

16.2.4 Cholesterol content of seafoods

The cholesterol content in seafoods is of little concern, with the exception of squid. Most seafoods have less than 100 mg per 100 g of cholesterol in raw muscle meat (Table 16.1). Early reports did not distinguish among the different steroids and often cholesterol was reported in the total steroids. There are conflicting data in the literature, especially for shrimp samples, that are related to both the testing methodology and the species tested. Shrimp cholesterol will vary between a low of 60 to a high of 160 mg per 100 g of sample, depending on the species. There is some concern about cholesterol in squid as the majority of squid species have very high levels of cholesterol ($>200\,mg\,g^{-1}$ sample). Pacific oysters (*Crassostrea gigas*) showed seasonal variation with a low cholesterol content of 37 mg per 100 g of sample for February and a high of 60 mg per 100 g of sample in October (Gordon, 1982).

16.2.5 Unusual fatty acids in fish and seafoods

Seafoods may contain several unusual fatty acids not found in terrestrial animals. These may be odd chain, branched and furanoid fatty acids that are poorly metabolized, but are in sufficiently low enough concentrations not to cause any physiological effects (Ackman, 1989). Wax esters are common in some commercial species, such as orange roughy, and care must be taken when processing the fish to remove the majority of fat (Sargent, 1989). Some unusual monoenoic fatty acids such as gadoleic (20:1ω9) and cetoleic acid (22:1ω9) are found in liver oils of cod, sithe and monkfish as well as the flesh of mackerel, capelin and herring (McGill and Moffat, 1992). At certain times of the year, cetoleic acid may reach as high as 20% of the fish oil in herring and capelin.

16.2.6 Effect of aquaculture on fatty acids

Aquaculture affects the lipid content of several species of finfish. By 1990, the volume of aquaculture in the world's fisheries production reached 15 million mt (Hempel, 1993). If current trends continue, it is estimated that it will be approximately 30 million mt by the year 2010. Salmon and catfish aquaculture has grown significantly over the past 10 years and has had an important impact on the consumption patterns in the US seafood diet. Because of the dependence on artificial diets in finfish aquaculture, there are differences in the general makeup of the fat content when compared with similar wild-caught species. The fat content of cultivated catfish and coho salmon fillets can be double that of wild caught species, according to Nettleton and Exler (1992). Their findings showed that increased fat content was at the expense of water content and did not significantly alter the cholesterol content. Other researchers have shown similar trends, with fatty acid profiles in wild-caught fish and shrimp having higher levels of ω-3 fatty acids than those raised by aquaculture (Chanmugan *et al.*, 1986; Jahncke *et al.*, 1988). The fatty acid profiles in the muscle meat often reflect the diets of the cultivated animals. Cultivated oysters are often not different in fat or other nutrients from their wild-caught counterparts as their diet depends largely on phytoplankton.

16.2.7 Distribution of fat in muscle tissue

Although there are numerous exceptions, the majority of seafood is consumed as muscle meat. Fish muscle contains three main tissues: white muscle, dark muscle and connective tissue (Love, 1988). Demersal fish, such as the cod and snapper, have a predominance of white muscle. Pelagic fish, such as tuna and mackerel, have a higher percentage of dark muscle and are considered oily fish. The principal part of connective tissue is in the skin. Fish skin has the highest concentration of fat (40–50%) of any organ, and in some species may represent 40% of the fat of the whole fish (Ackman, 1980). The skin is either removed or readily converted to gelatin during cooking or heat processing, thus causing few textural or sensory problems.

The dark muscle runs along the lateral line of the fish and has a higher concentration of lipids (Figure 16.1). The lipid concentration of the dark

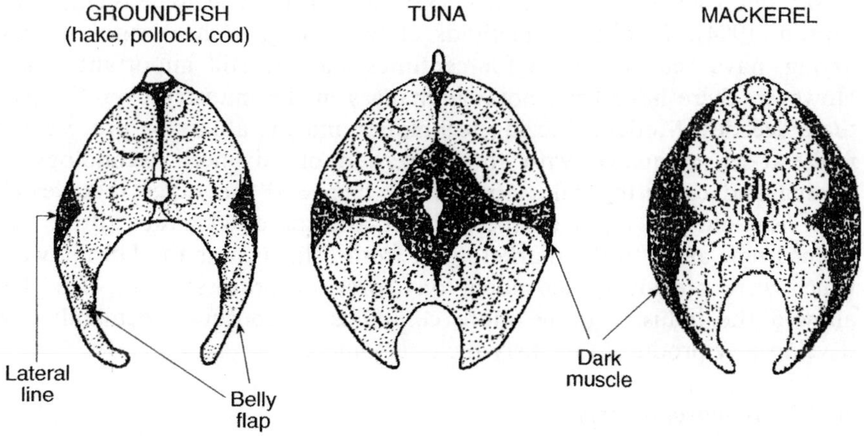

Figure 16.1 Cross-section of groundfish, tuna and mackerel muscle (adapted from Spinelli and Dassow, 1982).

Table 16.4 Lipid and lipid-class composition of dark muscle, white muscle and myocommata of mackerel and cod (% total lipids)[a]

Substance	Atlantic mackerel		Atlantic cod		Myocommata
	White muscle	Dark muscle	White muscle	Dark muscle	
Total lipids (% wet tissue)	2.3	13.1	0.7	1.8	0.8
Phospholipids	24.0	19.0	87.7	87.1	59.5
Cholesterol	4.6	2.0	9.7	8.4	31.9
Triglycerides	70.9	76.8	1.6	1.5	2.3

[a]Love (1988).

muscle and white muscle from two different species is given in Table 16.4. The Atlantic mackerel is considered a fatty fish and has substantially more red muscle than the Atlantic cod. The mackerel has a concentration of 2.3% lipid in the white muscle while the dark muscle contains 13.1% lipid, although this will vary depending on the season. The belly flaps of fish are muscle in the lower ventral areas that surround the visceral cavity. This muscle will have a higher fat content than the dorsal muscle area, especially for the medium to high fat fish.

16.3 Seafood processing

16.3.1 General

There are several methods for preserving and processing the catch, which have been described in several publications (Martin and Flick, 1990; Pigott and Tucker, 1990; Sikorski, 1990; Regenstein and Regenstein, 1991; Martin, 1994). Traditional methods of processing, such as smoking and drying, have been practised for centuries and are still important today. However, there have been notable changes in the industry over the past three decades. Modern freezing techniques and the ability to process and freeze-at-sea combined with the utilization of advanced technology of finding and capturing wild stocks have increased the range that vessels now fish and the types of species that are harvested. Aquaculture has opened new areas to the marketing of fresh fish, such as the US midwest, which were previously dependent on frozen or processed fish. This has allowed the industry to be more conscious of consumer demands and develop new products to satisfy their customers.

16.3.2 Removal of lipids

The removal of lipid from fillets or fish mince is often done not for nutritional/caloric reasons, but to lessen the effects of lipid oxidation and to obtain certain sensory characteristics. The dark meat of tuna, rich in lipids, is usually removed by hand before the final pack and canning operation. Fish lipids are high in polyunsaturated fatty acids and will readily undergo nonenzymatic lipid oxidation (Hultin, McDonald and Kelleher, 1982; Stansby, 1990). Lipid oxidation will lead to a number of off-odors and off-flavors, and make the product less acceptable as food. Lipid oxidation can be a critical problem in fatty fish, such as mackerel, and for non-fatty fish that are minced during frozen storage. Toughening of the flesh is also, in part, due to lipid oxidation and the reaction of intermediate compounds with proteins (Shenouda, 1980). However, it should be noted that high fat fish, despite their tendency for lipid oxidation, may be preferred for a number of processes. A high fish oil content in the flesh contributes to the 'moist' sensation in smoked salmon or lox. Mackerel that is high in fat is preferred to low fat mackerel for the canning process as the flesh does not dry out and maintains the proper texture.

16.3.3 Filleting of fish

There are a number of processing steps that can be undertaken partially to remove the lipids and reduce the caloric intake of the product. Although a number of processors depend on hand filleters for several species, the use of automatic filleting machines is becoming more practical. This is especially true in large-volume single-species fisheries, such as Alaska pollock. It is common in these fisheries for a single tow to capture 100,000–250,000 lb of the same species and same age class. Automatic fillleters are capable of filleting 2–3 fish per second. The resultant fillets are skin-on and usually are sent through an automatic skinning machine to produce skinless fillets (Bykowski, 1990). Removal of the fat layer under the skin and along the lateral line can be done by several methods. The skinning machine can be set at a deeper cut to remove all of the fat under the skin except that which remains along the lateral line. The skinned fillet can also undergo a deep V-cut to remove the fat along the lateral line as well as intramuscular bones that are found in certain species. A deep-skinned fillet, with a V-cut to remove fat along the lateral line and pin-bones, will have a lipid content well below 1.0%. The yields of such a product are often less than 25% of the whole fish, but the extra work required often translates into significantly higher prices. The belly flaps of medium to fatty fish, which are higher in lipid content than the dorsal region of fish musculature, may be removed to lessen the problems of oxidation during frozen storage.

16.3.4 Value-added products

The seafood industry has moved toward the production of value-added products of a number of species. These processes include smoking of dif-

Table 16.5 Nutritional content of some Norwegian smoked seafood[a]

Seafood	Energy (kcal)	Protein (g)	Fat (g)
Smoked mackerel	180	21.5	10.6
Smoked salmon fillet	160	21.4	8.4
Kippered herring	205	21.1	13.4
Brisling sardines in oil[b]	340	18.6	29.6
Brisling sardines in tomato sauce[b]	195	16.8	14.3
Sild sardines in oil[b]	355	18.3	29.2
Sild sardines in tomato sauce[b]	180	17.1	12.2
Greenland halibut[b]	135	13.4	8.8

[a]Adapted from Nettleton (1985). Values are for 100 g portion
[b]Canned.

ferent fish species, shaping of different forms from mince and dried seafood products. There are also a number of products that are manufactured for the convenience industry, such as microwavable entrees, sous-vide style products and pre-cut portions for single service. The fat content mainly depends on the pre-processing undertaken before final processing and packaging. Traditionally, fish that is medium–high in fat content is preferred for smoked fish products. The higher fat content prevents drying out of the product while retaining low enough water activity (in combination with salt) to prevent microbial growth. Salmon and mackerel are examples of finfish that lend themselves to superior smoked products. Table 16.5 shows the nutritional content of some smoked Norwegian products.

Mince blocks are manufactured from several finfish species, such as pollock and whiting. These are low fat fish and the mince reflects the fat content of the fillet (<1%). Mince can also be made from underutilized species, industrial fish and waste materials, such as fish-frame muscle tissue that remains after filleting (Regenstein, 1986). Mincing releases a number of enzymes and incorporates air into the block itself. Because mince is considered a lower grade product by most of the seafood industry, often care is not taken with its handling and storage. Lipid oxidation of minced blocks can be a problem, causing off-flavors and texture defects. Most of the mince is used for frozen breaded products. Because of the negative publicity surrounding fried foods and batter and breading, the production of breaded products (fish sticks, fish portions and breaded shrimp) has decreased in production in the USA from a high of 537 million lb in 1986 to 335 million lb by 1994 (NMFS, 1995). Minced fish from low fat species is an excellent source of raw material for fish sausages and other products. Several sausage-style products made from fish mince or fish-frame waste have shown initial success in the marketplace but have not been long-lasting (Regenstein and Regenstein, 1991).

With the advent of convenience foods, several companies are producing specific portion-size cuts of fish muscle toward the single- or two-person household. These include marinated seafood steaks, pre-cut portions ready for grilling or barbecuing, seafood burgers, hams and other deli-style items. The fat content can be controlled by the choice of species and type of cut that is made from the product. These are often advertised as low fat or rich in ω-3 fatty acids. A new process has recently been patented to decrease the fat content in catfish (a medium fat species) by 40% (Anon., 1995). The process uses a vacuum tumbler and standard food processing ingredients and is sold as a 'reduced fat' product.

16.3.5 Drying of fish

Dried foods are common in a number of countries where preservation methods such as freezing are expensive or unavailable or in regions (Scandinavia) where dried fish products are traditional. Drying is often done with the addition of salt to aid the moisture removal in the fish product (Doe and Olley, 1990). Fat will slow the drying process and high fat areas of fish (belly flaps) may be removed and dried separately. Dried fishery products often attain a water activity (a_w) between 0.75 and 0.85. Salted and marinated fish are also common in several countries where this method of preservation has contributed an array of traditional products (Shenderyuk and Bykowski, 1990). The nutritional composition of the final products depends on the raw material used. The addition of salt for preservation dramatically increases the sodium content. The final salt content will depend on the degree of brining, the fat content of the fish and its stage of freshness. The salt content for certain salted products can be reduced by soaking.

16.3.6 Canned fish

The most consumed seafood product in the USA is canned tuna (NMFS, 1995). Canned tuna in water is considered a low calorie, high protein meal and is incorporated into a number of dietary meals. The total fat content will depend on the species and the canning medium. Yellowfin tuna (*Thunnus albacares*) will have a fat content of 1.8 g per 100 g when canned in water and 6.8 g per 100 g when canned in vegetable oil (Nettleton, 1985). Canned salmon varies from a low of 6.4 g fat per 100 g for pink salmon to a high of 14.8 g per 100 g for chinook.

16.4 Advantages of fish muscle

16.4.1 Beneficial effects of lipids from seafood

The health advantages of eating fish and fishery products have been well documented in the literature. For those interested in more extensive reviews on this topic, the publications by Simopoulos, Kifer and Martin (1986), Kinsella (1987), Harris (1989), Stansby (1990), Nestel (1993) and Nettleton (1995) are recommended. There is general agreement among researchers that the lipid component of seafood, which is high in ω-3 fatty acids, provides numerous beneficial health effects.

The earliest recorded study of the health effects of fish oil was more than two centuries ago and described the effects of cod-liver oil on arthritis (Percival, 1783). Over 40 years ago a beneficial effect on heart disease was noted. Bronte-Stewart *et al.* (1956) found that fish oils lowered blood cholesterol more effectively than vegetable oils. Nelson (1972) undertook clinical trials during the 1950s and 1960s with patients who had suffered heart attacks. He found a significant increase in the survival of patients consuming a high fish diet. Kromhout, Bosschieter and Coulander (1985) reported similar results. In a 20 year study in The Netherlands of middle-aged men who had not suffered a heart attack, they found that the consumption of as little as 20 g of fish per day was predictive of decreased heart disease.

16.4.2 Eicosapentaenoic acid (EPA) and docosahexaenoic acid (DHA)

Research has focused on the roles of EPA and DHA, which are present in relatively high levels in fish oils, to explain the above effect on heart disease. Although it is beyond the scope of this chapter to describe the biochemical and physiological effects of EPA and DHA in detail, they will be addressed briefly.

Several researchers have shown EPA and DHA to reduce blood cholesterol, lower blood triglyceride levels, become incorporated into erythrocytes and other cell membranes, be readily taken up by heart muscle and affect the electrical behavior of the heart (Dyerberg, 1982; Swanson and Kinsella, 1986; Harris, 1989; McLennan, Abetwardena and Charnook, 1990). EPA and DHA inhibit platelet formation and ischemic heart disease associated with clotting (Nelson, Schmidt and Corash, 1991; Silvermen *et al.*, 1991). These biochemical studies have been correlated with epidemiological research in human populations. Studies in Greenland (Dyerberg *et al.*, 1978) and Japan (Hirai *et al.*, 1987) have confirmed that diets high in fish oils result in an antithrombic action in populations with high fish consumption.

16.4.3 ω-3 Fatty acids

The ω-3 fatty acids have a dramatic effect on eiconsanoid metabolism and production. Eicosanoids are short-lived hormone-like compounds (including prostaglandins, thrombexanes, prostacyclins and leukotrienes) that serve several physiological functions. They are produced by oxygenation of 20-carbon fatty acids such as arachidonic acid (AA) and EPA. By far the greatest amount of research has explored how the metabolism of AA produces these highly potent compounds. It has become increasingly apparent, however, that the eicosanoids produced from EPA are equally important. Eicosanoids produced from both fatty acids play a role in atherosclerosis, asthma, arthritis, diabetes, psoriasis, tumors, inflammatory conditions and septic shock (Nettleton, 1995). In addition to the unique role that the eicosanoids derived from EPA play, EPA and DHA competitively inhibit cyclooxygenase from interacting with AA (Corey, Shih and Cashman, 1983). For instance, thromboxane A_2 is produced by platelets and is a powerful aggregating agent. Inhibition of its production reduces clotting in the blood (Dyerberg *et al.*, 1978).

Leukotrienes are recognized as important biochemical mediators in vasoregulation, neurotransmission, hormone regulation, inflammation and immune response (Landauer, Davis and Walden, 1990). EPA and DHA also inhibit the lipoxygenase enzyme necessary for leukotriene formation, thereby reducing their synthesis (Lokesh, Black and Kinsella, 1989).

Because of the effects of ω-3 fatty acids on the production of eicosanoids, the consumption of fish oils has been thought to have several other health benefits, especially in the area of inflammation and immune response (Nettleton, 1995). Inflammatory responses are mediated by prostaglandins and leukotrienes produced by neutrophils, macrophages and mast cells. Chronic inflammation is associated with several diseases, such as rheumatoid arthritis, psoriasis and gout. Eicosanoids from EPA tend to be less inflammatory than those from AA. Although there have been numerous studies showing the beneficial effects of diets high in fish oils in alleviating a number of the symptoms of these diseases, there have also been conflicting results that show little effect of EPA and DHA. Future research in this area may help to clarify the role of ω-3 fatty acids in mediating the immune response.

There has been considerable research regarding the effects of ω-3 fatty acids on the early development of retina and brain cells (Neuringer and Connor, 1986; Carlson and Salem, 1991; Innis, 1992). Currently, it is thought that DHA is necessary for these functions in premature infants and may be necessary for full term infants (Carlson *et al.*, 1994). The Japanese dietary allowances for important nutrients now recommend that pregnant women should eat sufficient seafood to obtain 0.5–1.0 g of DHA per day (Linko and Hayakawa, 1996).

16.4.4 Other effects of fish oils

There are several other reported nutritional benefits related to fish oils. These include partial remediation of diseases such as diabetes (Malasanos and Stacpoole, 1991), multiple sclerosis (Holman, Johnson and Kokman, 1989) and various types of cancer (Nettleton, 1995). Because of the broad range of physiological effects of ω-3 fatty acids and the roles of genetics and life-style factors in these diseases, however, it is difficult to formulate definitively a cause–effect relationship.

In contrast to these studies, which show a positive effect of fish oil, there are several studies that have shown either minimal or no effects of a fish diet in preventing heart disease (Vollset, Heugh and Bjelke, 1985; Ascherio *et al.*, 1995). There is also concern that the consumption of highly unsaturated fatty acids may increase *in vivo* lipid peroxidation. This may be particularly important to oxidation of low-density lipoproteins (ox-LDL) (Thomas *et al.*, 1994; Wander *et al.*, 1995). An ox-LDL particle is thought to be more atherogenic than the unmodified particle (Steinberg *et al.*, 1989). If the LDL particle is enriched with EPA and DHA, as would occur after consuming fish, they may be prone to oxidize more readily. However, current data suggest a protective effect is obtained in EPA/DHA-enriched LDL. The wide range of effects that ω-3 fatty acids have in human physiology and health shows the need for continued research in this area.

16.5 Surimi

Surimi is washed fish mince that is stabilized in long-term frozen storage by the addition of cryoprotectants. It is an intermediate product from which a number of surimi seafood products are made. One may think of the surimi process as a method of concentrating the myofibrillar proteins in fish muscle. Fish myofibrillar proteins have unique gelling characteristics that allow them to be used for a number of surimi-based seafood products, such as crab analog. These proteins form heat-set gels at moderate temperatures (80–90°C) that have both high gel strength and high elastic properties.

16.5.1 Production of surimi and surimi-based seafood

Production of surimi is an ancient process first invented by the Japanese more than eight centuries ago (Lee, 1984). In early times, fish captured near the shore were minced and washed and the remaining insoluble proteins were squeezed dry, mixed with salt and other ingredients and steam cooked. It was not until the early 1960s that the modern surimi industry began to grow significantly. Discovery of the use of common low cost compounds as cryoprotectants allowed the Japanese fishing industry to harvest and process fish in distant waters for surimi production (Park, 1994). The majority of surimi produced on processing ships and at shoreside plants uses sucrose and sorbitol as the main cryoprotectants in frozen surimi blocks.

The industry today produces a total of 530,000 mt of surimi (Ishikawa, 1996). The USA is the leading producer, followed by Japan, Thailand,

Korea and South America. The production of surimi-based seafood products is over one million mt annually, with Japan being the largest producer and consumer nation. Low fat fish of the gadoid family (1–2% fat), such as pollock and several whitings, constitute ca 60–70% of all surimi produced.

The majority of surimi in the USA is utilized for the production of imitation crab, while smaller amounts are used for shrimp, scallop and lobster analogs (Freeman, 1993). Seafood salads with shredded crab analog are a common sight in delicatessens and grocery stores. New products such as oriental potstickers and tempura-type burgers also are being marketed. In Asia there are over 300 products made from surimi. These include traditional products, such as kamaboko, chikuwa, hampen and satsumaage, that are important in the Japanese diet (Wu, 1992).

16.5.2 Processing of surimi

Although the surimi operation will vary from plant to plant, surimi operations have similar processing steps as shown in Figure 16.2. Washing

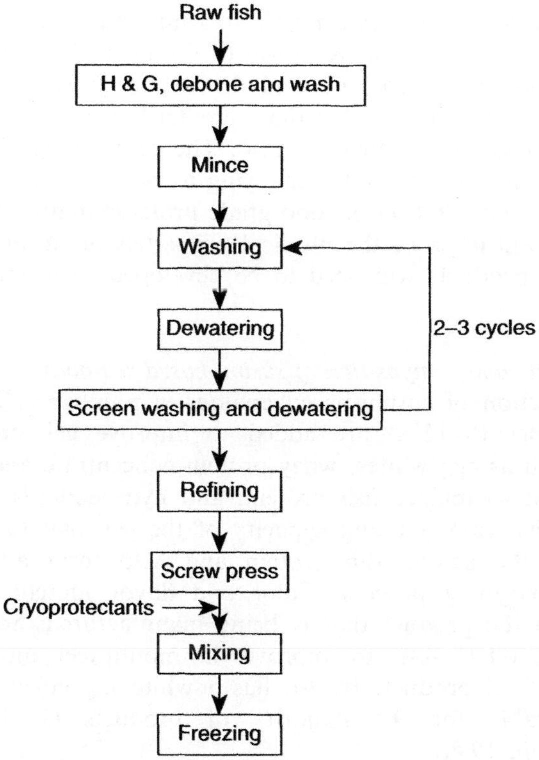

Figure 16.2 Flow diagram of surimi production.

is a crucial step in the process. It removes blood components, skin pieces, soluble proteins, fat, minerals and other water soluble components. As the mince undergoes several washing and dewatering cycles, the concentration of myofibrillar protein increases and the mince becomes whiter. Fresh washed mince, after the final dewatering step at the screw press and before the addition of cryoprotectants and other additives, has a proximate composition of 80–81% moisture, 18–19% protein, 0.7–0.9% fat and 0.4–0.5% ash (Lin, Park and Morrissey, 1995). There are a number of food ingredients that may be added to surimi at the blender step depending on the manufacturer's and customer's specifications. For surimi made from Alaska pollock these include sucrose (4%) and sorbitol (4–5%) as cryoprotectants and polyphosphates (0.3%) for improving water holding capacity. Because of residual protease activity in Pacific whiting washed mince, beef plasma (0.9–1.4%) is usually added as a protease inhibitor and calcium compounds may be added as gel enhancers (Morrissey *et al.*, 1993).

Production of surimi from medium and high fat fish, such as sardines and mackerel, has been shown to be feasible (Shimuzu, Toyohara and Lanier, 1992; Nishioka, 1993; Oshima, Suzuki and Koizumi, 1993). These fish tend to have a higher red muscle to white muscle ratio and a higher protease activity in the flesh. Removal of the fat is facilitated by washing with 0.2% sodium bicarbonate and 0.1% pyrophosphate. The leached sardine meat is darker in color than standard surimi, but has good gel strength and a low fat content (<1%). The gel-forming ability of surimi from these species is usually lower owing to protease activity and can be improved with the addition of food-grade protease inhibitors. Technological advances will improve the physical properties of surimi from high-fat fish, but new products will need to be developed to increase the market potential.

(a) Ingredients and composition of surimi-based seafoods

In the production of surimi-based analogs, a number of ingredients are added. Starches (8–12%) are added to improve gel strength; protein additives, such as egg whites, whey protein concentrate and soy proteins, may be added to replace fish protein; and hydrocolloids may be added to improve the water binding capacity of the gel. Salt (1–2%) is added to solubilize the actomyosin protein and help form a uniform paste during the chopping process. Color and flavor ingredients are added depending on the product that is being manufactured. Some companies add vegetable oil (1–4%) to improve the mouth-feel and textural attributes of the final product. It also has a whitening effect, which is considered desirable for the majority of products (J. Park, personal communication, 1996).

Surimi-based seafoods are considered a low cost alternative to shellfish.

Table 16.6 Comparison of the nutrient content of pollock surimi analog with baked pollock, boiled shrimp and raw Alaska king crab legs[a]

Nutrient	Surimi analog	Baked pollock	Boiled shrimp	Raw Alaska king crab legs
Calories (kcal per 100 g)	96	106	102	80
Protein (g per 100 g)	15.2	22.4	21.1	15.4
Fat (g per 100 g)	0.9	1.0	1.2	1.6
Carbohydrate (g per 100 g)	6.8	<1	–	1.0
Cholesterol (mg per 100 g)	30	95	226	42
Sodium (mg per 100 g)	143	108	299	836
Potassium (mg per 100 g)	78	374	209	204

[a]Adapted from Nettleton (1985).

Table 16.6 describes the different nutrients in a number of seafood products including surimi. The surimi analog is the lowest in fat and cholesterol among all the products listed. Since there is not a standard of identity for surimi-based seafoods, there is a wide range of nutrients that may be found in the final product. Higher quality analogs may have 45–50% surimi as the main ingredient. Lower cost surimi analogs may have 30–35% surimi and a corresponding lower protein content. The other nutrients may be adjusted high or low depending on the formulation. Although the majority of surimi is used to produce shellfish analogs, surimi has been promoted as a potential resource for several foods and also nutritional supplements and specialty products (Pigott and Tucker, 1990). These include low calorie snack foods and frozen desserts, high protein, low fat health food specialty items, low fat emulsions for sauces, cheese analogs and binders.

(b) Criticisms of surimi-based seafoods

Surimi seafood is not without its critics from a nutritional standpoint. This criticism centers on three areas: (1) a number of ingredients that make seafood a healthy food (ω-3 fatty acids, minerals, vitamins, etc.) are removed in the washing process, (2) large amounts of carbohydrates in the form of mono- and disaccharides are added into the products and (3) a high sodium content. More than 90% of its calories are derived from protein and carbohydrate. The sole source of carbohydrate would be the cryoprotectants sucrose and sorbitol. Because of the necessity for salt for the solubilization of myofibrillar protein, the sodium content is high (600 mg per 100 g).

(c) Improving the nutritional value of surimi-based seafoods

Several researchers have investigated methods to improve the nutritional content of surimi. Work is under way to replace some of the fish lipids lost in the washing process. The main constraint in adding back fish lipids is the high degree of polyunsaturated fatty acids and the potential development of rancidity, especially during frozen storage. Minerals and vitamins can be added if the consumer demands this of the product.

Sucrose is added as a low cost cryoprotectant and contributes carbohydrate calories to the final product. There have been a number of research efforts into replacing sucrose as a cryoprotectant. These include work with compounds that would reduce the caloric load, such as polydextrose and lactitol (Sych *et al.*, 1990; MacDonald and Lanier, 1991). Polydextrose is a zero calorie additive that awaits approval in meat compounds as well as in surimi. The European Community has assigned sorbitol a reduced calorie value ($2.5\,cal\,g^{-1}$) compared with other simple sugars. A 100% sorbitol formulation for cryoprotection would significantly reduce the caloric value for surimi. However, with least cost formulation being a driving force in surimi seafood production, it is unlikely that this will occur at current sorbitol prices. Sodium substitutes, such as KCl, are currently being used by a number of surimi seafood companies, a 40% replacement of NaCl by KCl having been found to be effective without causing a bitter off-flavor (J. Park, personal communication, 1996).

16.6 Other processes

Several processes that remove fish oils are currently being used for making animal feeds. There has been considerable discussion about upgrading these processes for the production of high protein, low fat products for human consumption. These processes include fish protein concentrate (FPC, Figure 16.3b), marine beef (MB, Figure 16.3b), and fish protein hydrolysate (FPH, Figure 16.4).

16.6.1 Fish protein concentrate (FPC)

FPC was investigated world-wide in the 1950s as a low cost protein additive for developing countries (Tannenbaum, Stillings and Scrimshaw, 1974; Regenstein and Regenstein, 1991). The main resource was high fat, small pelagic fish, such as sardines or anchovies, although underutilized species, such as whole Pacific Whiting, have also been used. The process is similar to fish meal production except that the fish oils are removed with organic solvents. The final product (Type A FPC) is high in protein (>67.5) and low in fat ($<0.75\%$). There has been little acceptance of FPC for human foods. This is mainly due to the lack of functionality of the proteins that are recovered after the oils are removed (Barzana and Garcia-Garibay, 1994). These include poor

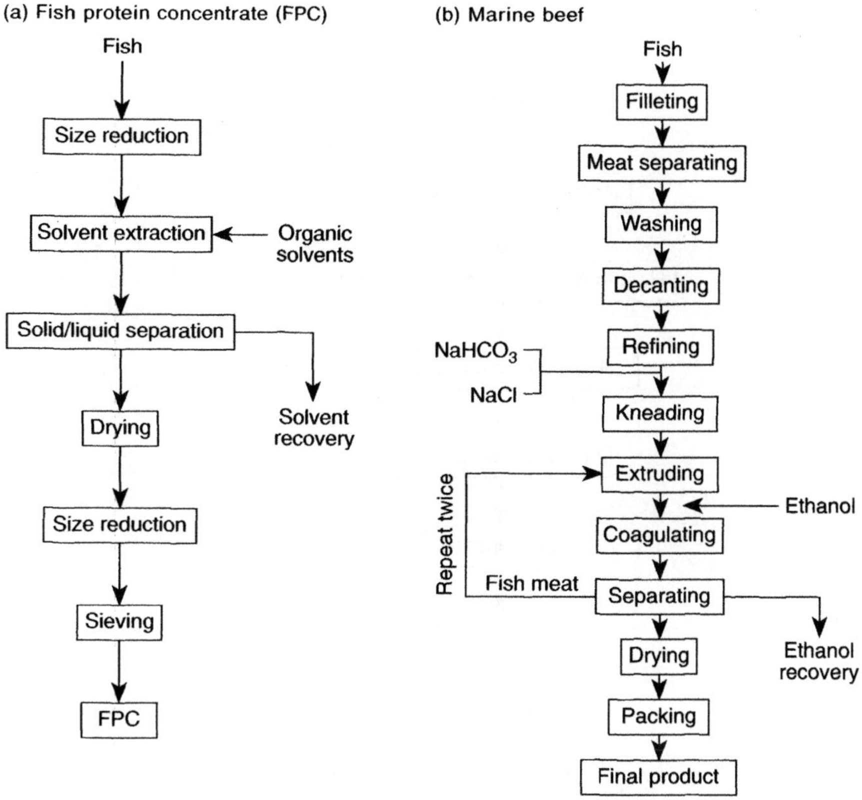

Figure 16.3 Flow diagrams of fish protein concentrate and marine beef production.

emulsifying properties and low water-holding capacity and solubility. Although considerable research effort has been made to improve the utility of FPC, a number of technical, regulatory and economic problems could not be overcome to achieve its acceptance in the marketplace (Finch, 1977).

16.6.2 Marine beef (MB)

MB was developed by the Japanese in a similar process to concentrate fish proteins from small pelagics (Suzuki, 1981). Whereas the FPC process often uses the whole fish, MB uses minced meat, excluding the head, skin and viscera. The oils are removed by ethanol extraction and the remaining protein has high functionality. Although a processing plant for making MB from underutilized species was constructed in Peru, little product was produced owing to the high cost of ethanol and the lack of acceptance of the product by the local population.

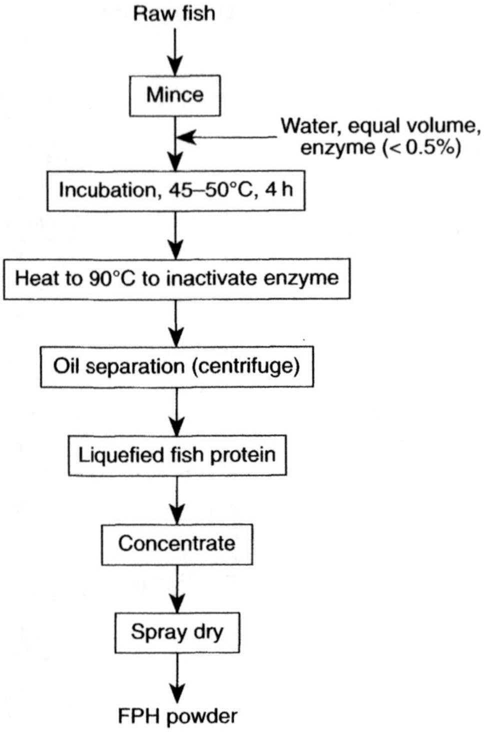

Figure 16.4 Flow diagram of fish protein hydrolyzate production (Venugopal, 1994).

16.6.3 Fish protein hydrolysate (FPH)

Protein hydrolysis has been shown to improve markedly the functionality of proteins recovered in the process (Barzana and Garcia-Garibay, 1994). FPH has also proved useful in developing a serviceable fish paste from the high amounts of waste generated from most finfish processing operations. The FPH process uses endogenous proteases or added proteolytic enzymes to break down the proteins resulting in a uniform paste (Venugopal, 1994). After the proteases have been inactivated by heat, the oils are separated by centrifugation. The remaining liquified fish protein is reduced in water content by evaporators and then dried to a powder by spray drying. FPH will differ in its characteristics depending on the source of raw material, the enzymes used and the degree of hydrolysis. A low fat fish will yield an FPH of 90% protein and 1.5% fat (Ritchie and Mackie, 1982). Although all FPH currently produced is used for animal feeds and plant food, several researchers feel that it has a role in human nutrition. They have shown the feasibility of producing a human food FPH from Pacific whiting carcass frames as well as headed and gutted fish (Pigott and Tucker, 1990). The functionality of the protein, relatively low cost of production and versatility of the process can provide a high protein food that can be used as a functional ingredient in several foods (Owens and Mendoza, 1985).

16.7 Research needs

Future research needs can be divided into four areas: (1) developing products to satisfy consumer needs, (2) forming management strategies to assure adequate supply and good quality of the raw material, (3) improving the functionality and palatability of products extracted from underutilized species and recovered seafood protein and (4) exploring new opportunities in aquaculture.

The seafood industry is slowly changing from being primarily a resource supplier to becoming a modern industry that is more competitive in the marketplace of protein-based value-added products and convenience foods. Few seafood companies have the luxury of having product development teams and are dependent on proven technology or outside institutions for their research needs. This inevitably has led to a lack of coordination in research and marketing goals in the industry. A long-term plan is needed for industry to meet consumer needs in the twenty-first century. This would include working closely with universities, the National Marine Fisheries Service, National Fisheries Institute and other agencies in setting general objectives and specific goals in the area of quality and product development. In the area of product development, goals need to include more research with consumer groups to understand better their needs and attitudes toward fish products. Understanding the range of processing technologies in the food industry and their potential application to low fat seafood products is critical to the development of wholesome products for the marketplace.

The seafood industry, as a whole, needs to become more involved in the management of fisheries from a quality standpoint. Wild-caught fisheries are still a hunting–capturing process that encompasses a large degree of uncertainty. The allowed harvest is defined by quotas, fishing areas and specific seasons. Fisheries management practices are involved primarily with the biological aspect of the stock assessment based on the estimation of fish populations. There has been little effort to manage the fisheries according to their characteristics as a food source or through harvesting practices that would lead to optimizing quality. There has been some initial effort to incorporate quality aspects into the management scheme. In the Pacific whiting fishery, processors have been successful in asking for a delay of the seasonal opening until fish have reached a higher protein and lower moisture content in the muscle tissue (Sylvia, Larkin and Morrissey, 1996). This should improve both the yield and quality of the final product and also economic benefits to the processors.

There has also been success in changing the Alaska halibut fishery from a derby fishery (where the total quota is caught within 48 h) to an individual transferable quota system (ITQ). The ITQ system has, in effect, extended the season from 2 days to several months, allowed fishermen to be more responsive to market demands and improved the quality of the raw material. Because of the seasonal variation of certain species of fish, one could envision harvesting practices targeting fish at different times of the year for a lower fat content. Involving the seafood industry, with its focus on quality and markets, in fisheries management will, in the long run, decrease some of the uncertainties that now exist in fisheries and allow for capital investment necessary for manufacturing products of high quality which meet consumer needs.

The seafood industry utilizes only 30–50% of the seafood harvested as primary products. The remaining material is made into a variety of products, such as fish meal and fertilizers, or discarded. There is an important research need to develop more products for human consumption from these underutilized products. Methods for protein extraction from surimi wash water by membrane filtration could lead to a high protein cake with good functional properties. Research should continue in the production of food-grade FPH and FPC. Additional research is needed in the areas of fat removal and/or fat stabilization and in increasing the functionality of extracted proteins. Processing techniques that focus on producing a human food-grade product that is wholesome and palatable could provide millions of pounds of high protein products from what was formally considered seafood wastes.

Aquaculture presents several opportunities for the development of new product forms with altered fat levels. Although early studies have reported higher fat levels in aquacultured fish, it appears to be diet related. This would allow researchers to investigate the manipulation of the fat content of fish muscle by diet or culturing practices. The high degree of freshness in aquacultured products also provides processors with increased possibilities of producing product forms with varying degrees of fat content. Research in this area will depend greatly on market demand and consumer preferences for low fat seafood products.

16.8 Summary

Seafood varies in fat content depending on the species, season, geographical location and physiological status. The majority of finfish found in the USA and northern European markets are low in fat and high in protein and in itself is considered a healthy, low fat food by consumers. Fish oils in all finfish species are high in polyunsaturated fatty acids and can have therapeutic effects on a number of human diseases, most notably on cardiovascular diseases. The fatty acids EPA and DHA have been shown to affect the production of eicosanoids which have a number of physiological functions. Consequently, fatty fish, to a certain extent, may be viewed as desirable from a health or nutritional standpoint. Most finfish and shellfish are low to moderate in cholesterol content (<100 mg per 100 g sample), with the exception of squid (>200 mg per 100 g sample). Aquacultured fish vary in fat content depending on their diet.

Reducing fat content in fish is done for organoleptic reasons, to prevent the effects of lipid oxidation and to reduce caloric intake. These methods include removal of the dark muscle, which is higher in fat, deep skinning to remove the lateral line in many of the low-fat species and removal of the belly flaps from moderate fat species. New engineered seafood products may also have reduced fat content. Surimi is a process where the myofibrillar protein is concentrated through extensive washing and removal of soluble proteins and fat. Traditional surimi products such as crab analogs or kamaboko usually have <1% fat. Other seafood products such as formed products from mince and canned products will have fat contents similar to those of the raw materials used. Smoked and salted/dried products will usually increase in fat content as moisture is removed. The production of FPC, MB and FPH reduces the fat to protein ratio through the partial removal of fat by centrifugation and solvent extraction.

Acknowledgement

I would like to thank Rosemary Wander, Department of Foods and Nutrition, Oregon State University, for her assistance with the nutritional information.

References

Ackman, R.G. (1980) Fish lipids. Part I, in *Advances in Fish Science and Technology* (ed. J.J. Connell), Fishing News Books, Farnham, pp. 86–103.

Ackman, R.G. (1989) Nutritional composition of fats in seafood. *Prog. Food Nutr. Sci.*, **13**, 161.

Ackman, R.G. (1995) Composition and nutritive value of fish and shellfish, in *Fish and Fishery Products* (ed. A. Ruiter), CAB International, Wallingford, Oxon, p. 387.

Anon. (1995) The fat in the cat came out. *Seafood Leader*, **15**(6), 24.

Ascherio, A., Rimm, E.B., Stampfer, M.J. *et al.* (1995) Dietary intake of marine $n-3$ fatty acids, fish intake, and the risk of coronary disease among men. *N. Engl. J. Med.*, **332**, 977.

Barzana, E. and Garcia-Garibay, M. (1994) Production of fish protein concentrates, in *Fisheries Processing: Biotechnological Applications* (ed. A.M. Martin), Chapman & Hall, London, pp. 206–222.

Bronte-Stewart, B.A., Antonis, A., Eales, L. and Brock, J.F. (1956) Effects of feeding different fats on serum cholesterol levels. *Lancet*, **i**, 521.

Bykowski, P.J. (1990) The preparation of catch for preservation and marketing, in *Seafood: Resources, Nutritional, Composition and Preservation* (ed. Z.E. Sikorski), CRC Press, Boca Raton, FL, pp. 77–92.

Carlson, S.E. and Salem, N. (1991) Essentiality of omega -3 fatty acids in growth and development of infants, in *Health Effects of Omega -3 Polyunsaturated Fatty Acids in Seafoods* (eds R.R. Kifer, R.E. Martin and S.M. Barlow), Karger, New York, pp. 74–86.

Carlson, S.E., Werkman, S.E., Peeples, J.M. and Wilson, W.M., III (1994) Growth and development of premature infants in relation to ω3 and ω6 fatty acid status, in *World Review of Nutrition and Dietetics*, (ed C. Galli, A.P. Simopoulos and E. Tremoli), Vol. 75, Karger, New York, pp. 63–69.

Chanmugan, P., Boudreau, M. and Hwang, D.H. (1986) Differences in omega−3 fatty acid content in pond-reared and wild fish and shellfish. *J. Food Sci.*, **51**, 1556.

Corey, E.J., Shih, C. and Cashman, J.R. (1983) Docosahexaenoic acid is a strong inhibitor of prostaglandin but not leukotriene biosynthesis. *Proc. Natl. Acad. Sci. USA*, **80**, 3581.

Crapo, C. and Kramer, D.E. (1993) *Salmon Quality Almanac: a Manual for Tender Boat Operators*, Alaska Seafood Marketing Institute, Juneau, AK.

Doe, P. and Olley, J. (1990) Drying and dried fish products, in *Seafood: Resources, Nutritional, Composition and Preservation* (ed. Z.E. Sikorski), CRC Press, Boca Raton, FL, pp. 125–145.

Dyerberg, J. (1982) Observations on populations in Greenland and Denmark, in *Nutritional Evaluation of Long-chain Fatty Acids* (eds S.M. Barlow and M.E. Stansby), Academic Press, New York, pp. 245–261.

Dyerberg, J., Bang, H.O., Stoffersen, E. *et al.* (1978) Eicosapentaenoic acid and prevention of thrombosis and atherosclerosis? *Lancet*, **ii**, 117.

Exler, J. (1987) *Composition of Foods: Finfish and Shellfish Products. Aquaculture Handbook No. 8-15*, US Department of Agriculture, Washington, DC.

FAO (1993) *Yearbook of Fisheries Statistics. Vol. 72: Catches and Landings.* Food and Agriculture Organization of the United Nations, Rome.

Finch, R. (1977) Whatever happened to fish protein concentrate? *Food Technol.*, **31**(5), 44.

Freeman, K. (1993) Surimi update; seafood analogs venture into new frontiers. *Seafood Bus.*, **12**(4), 89.

Gordon, D. (1982) Steroids in mollusks and crustacea of the Pacific Northwest, in *Chemistry and Biochemistry of Marine Food Products* (eds R.E. Martin, G.J. Flick, C.E. Hebard and D.R. Ward), AVI, Westport, CT, pp. 93–104.

Harris, W.S. (1989) Fish oils and plasma lipid and lipoprotein metabolism in humans: a critical review. *J. Lipid Res.*, **30**, 785.

Hempel, E. (1993) Constraints and possibilities for developing aquaculture. *Aquacult. Int.*, **1**, 2.

Hirai, A., Terano, T., Saito, H. *et al.* (1987) Clinical and epidemological studies of eicosapentaenoic acid in Japan, in *Proceedings of the AOAC Short Course on Polyunsaturated Fatty Acids and Eicosanoids* (ed. W.E.M. Lands), American Oil Chemists Society, Champaign, IL, pp. 9–24.

Holman, R.T., Johnson, S.N. and Kokman, E. (1989) Deficiencies of polyunsaturated fatty acids and replacement by nonessential fatty acids in plasma lipids in multiple sclerosis. *Proc. Natl. Acad. Sci. USA*, **86**, 4720.

Hultin, H.O., McDonald, R.E. and Kelleher, S.D. (1982) Lipid oxidation in fish muscle microsomes, in *Chemistry and Biochemistry of Marine Food Products* (eds R.E. Martin, G.J. Flick, C.E. Hebard and D.R. Ward), AVI, Westport, CT, pp. 1–12.

Innis, S.M. (1992) $n-3$ Fatty acid requirements of the newborn. *Lipids*, **27**, 879.

Ishikawa, Y. (1996) World surimi market outlook. *INFOFISH Int.*, No. 1/96, 16.

Jahncke, M., Hale, M.B., Gooch, J.A. and Hopkins, J.S. (1988) Comparison of pond-raised and wild red drum (*Sciaenops ocellatus*) with respect to proximate composition, fatty acid profiles, and sensory evaluations. *J. Food Sci.*, **53**, 286.

Kinsella, J.E. (1987) *Seafoods and Fish Oils in Human Health and Disease*, Marcel Dekker, New York.

Kromhout, D., Bosschieter, E.B. and Coulander, C.L. (1985) The inverse relation between fish consumption and 20-year mortality from coronary heart disease. *N. Engl. J. Med.*, **312**, 1206.

Krzynowek, J. and Panunzio, L.J. (1989) Cholesterol and fatty acids in several species of shrimp. *J. Food Sci.*, **54**, 237.

Landauer, M.R., Davis, H.D. and Walden, T.L. (1990) Behavioral and physiological effects of leukotriene, C_4. *Prostaglandins, Leukotrienes Essent. Fatty Acids* **39**, 247.

Lee, C. (1984) Surimi process technology. *Food Technol.*, **38**(11), 69.

Lin, T., Park, J. and Morrissey, M.T. (1995) Recovered protein and reconditioned water from surimi processing waste. *J. Food Sci.*, **60**, 4.
Linko, Y.-Y. and Hayakawa, K. (1996) Docosahexaenoic acid: a valuable nutraceutical? *Trends Food Sci. Technol.*, **7**, 59.
Lokesh, B.R., Black, J.M. and Kinsella, J.E. (1989) The suppression of eicosanoid synthesis by peritoneal macrophages is influenced by the ratio of dietary docosahexaenoic acid to linoleic acid. *Lipids*, **24**, 589.
Love, M.R. (1988) *The Food Fishes*, Farrand Press, London.
MacDonald, G.A. and Lanier, T. (1991) Carbohydrates as cryoprotectants for meats and surimi. *Food Technol.*, **45**(3), 150.
Malasanos, T.H. and Stacpoole, P.W. (1991) Biological effects of omega−3 fatty acids in diabetes mellitus. *Diabetes Care*, **14**, 1160.
Martin, A.M. (ed.) (1994) *Fisheries Processing: Biotechnological Applications*, Chapman & Hall, London.
Martin, R.E. and Flick, G.J. (eds) (1990) *The Seafood Industry*, Van Nostrand Reinhold, New York.
McLennan, P.L., Abetwardena, M.Y. and Charnock, J.S. (1990) Reversal of the arrhythmogenic effects of long-term saturated fatty acid intake by dietary $n-3$ and $n-6$ polyunsaturated fatty acids. *Am. J. Clin. Nutr.*, **51**, 53.
McGill, A.S. and Moffat, C.F. (1992) A study of the composition of fish liver and body oil triglycerides. *Lipids*, **27**, 360.
Morrissey, M.T. (1995) *Report to Oregon Trawl Commission*. Oregon State University Seafood Laboratory, Astoria, OR.
Morrissey, M.T., Wu, J.W., Lin, D.D. and An, H. (1993) Effect of food grade protease inhibitor on autolysis and gel strength of surimi. *J. Food Sci.*, **58**, 1050.
National Seafood Educators (1992) *Nutritional Breakdown of Top 20 Seafoods*, National Seafood Educators, Richmond Beach, WA.
Nelson, A.M. (1972) Diet therapy in coronary disease – effects of mortality of high protein, high seafood, fat controlled diet. *Geriatrics*, **27**, 103.
Nelson, G.J., Schmidt, P.C. and Corash, L. (1991) The effect of a salmon diet on blood clotting, platelet aggregation and fatty acids in normal adult men. *Lipids*, **26**, 87.
Nestel, P.J. (1993) Effects of $n-3$ fatty acids on lipoproteins and atherosclerosis, in *Omega*−3 Fatty Acids: Metabolism and Biological Effects (eds C.A. Drevon, I. Baksaaa and H.E. Kroran), Birkhauser Verlag, Basle, pp. 201–210.
Nettleton, J. (1985) *Seafood Nutrition: Facts, Issues, and Marketing of Nutrition in Fish and Shellfish*, Osprey Books, Huntington, NY.
Nettleton, J.A. (1995) *Omega−3 Fatty Acids and Health*, Chapman & Hall, New York.
Nettleton, J.A. and Exler, J. (1992) Nutrients in wild and farmed fish and shellfish. *J. Food Sci.*, **57**, 257.
Neuringer, M. and Connor, W.E. (1986) $n-3$ Fatty acids in the brain and retina: evidence for their essentiality. *Nutr. Rev.*, **44**, 285.
Nishioka, F. (1993) Frozen surimi from sardine. *INFOFISH Int.*, No. 1/93, 31.
NMFS (1995) *Fisheries of the United States*, US Department of Commerce, National Oceanic and Atmospheric Administration, National Marine Fisheries Service, Washington, DC.
Nunes, M.L., Batista, I. and Marao de Campos, R. (1992) Physical, chemical and sensory analysis of sardines (*Sardina pilchardus*) stored in ice. *J. Sci. Food Agric.*, **59**, 37.
Ohshima, T., Suzuki, T. and Koizumi, C. (1993) New developments in surimi technology. *Trends Food Sci. Technol.*, **4**, 157.
Owens, J.D. and Mendoza, L.S. (1985) Enzymatically hydrolyzed and bacterially fermented fishery products. *J. Food Technol.*, **20**, 273.
Park, J. (1994) Cryoprotection of muscle proteins by carbohydrates and polyalcohols – a review. *J. Aquat. Food Product Technol.*, **3**(3), 23.
Percival, T. (1783) Observation on the medicinal uses of the oleum jeconis aselli or cod liver oil in chronic rheumatism and other painful disorders. *London Med. J.*, **3**(4), 393.
Pigott, G.M. and Tucker, B.W. (1990) *Seafood: Effects of Technology on Nutrition*, Marcel Dekker, New York.
Regenstein, J.M. (1986) The potential for fish mince. *Food Technol.*, **40**(3), 101.

Regenstein, J.M. and Regenstein, C.E. (1991) *Introduction to Fish Technology*, Osprey Books, Van Nostrand, New York.

Ritchie, A.H. and Mackie, I.M. (1982) Preparation of fish protein hydrolysates. *Anim. Feed Sci. Technol.*, **7**, 125.

Sargent, J.R. (1989) Wax esters, long chain monoenoic fatty acids and polyunsaturated fatty acids in marine oils. Resources and nutritional implications, in *Fish, Fats and Your Health. Proceedings International Conference on Fish Lipids and Their Influence on Human Health*, Svanoy Foundation, Svanoybukt, Norway.

Shenderyuk, V.I. and Bykowski, P.J. (1990) Salting and marinating of fish, in *Seafood: Resources, Nutritional, Composition and Preservation* (ed. Z.E. Sikorski), CRC Press, Boca Raton, FL, pp. 147–162.

Shenouda, S. (1980) Theories of protein denaturation during frozen storage of fish flesh. *Adv. Food Res.*, **26**, 275.

Shimizu, Y., Toyohara, H. and Lanier, T.C. (1992) Surimi production from fatty and dark-fleshed species, in *Surimi Technology* (eds T.C. Lanier and C.M. Lee), Marcel Dekker, New York, pp. 181–207.

Sidwell, V.D., Foncannon, P.R., Moore, N.S. and Bonnett, J.C. (1974) Composition of the edible portion of raw (fresh or frozen) crustaceans, finfish and mollusks. I. Protein, fat, moisture, ash, carbohydrate, energy value, and cholesterol. *Mar. Fish. Rev.*, **368**, 21.

Sikorski, Z.E. (ed.) (1990) *Seafood: Resources, Nutritional, Composition and Preservation*, CRC Press, Boca Raton, FL.

Silvermen, D.I., Ware, J.A., Sacks, F.M. and Pasternak, R.C. (1991) Comparison of the absorption and effect on platelet function of a single dose of $n-3$ fatty acids given as fish or fish oil. *Am. J. Clin. Nutr.*, **53**, 1165.

Simopoulos, A.P., Kifer, R.R. and Martin, R.E. (eds) (1986) *Health Effects of Polyunsaturated Fatty Acids in Seafoods*, Academic Press, New York.

Spinelli, J. and Dassow, J.A. (1982) Fish proteins: their modification and potential uses in the food industry, in *Chemistry and Biochemistry of Marine Food Products* (eds R.E. Martin, G.J. Flick, C.E. Hebard and D.R. Ward), AVI, Westport, CT, pp. 13–26.

Stansby, M. (ed.) (1990) *Fish Oils in Nutrition*, Van Nostrand Reinhold, New York.

Steinberg, D., Parthasarathy, S., Khao, J.C. and Witzum, J.L. (1989) Cholesterol modification of low-density lipoprotein that increases its atherogenicity. *N. Engl. J. Med.*, **320**, 3603.

Suzuki, T. (1981) *Fish and Krill Protein Processing Technology*, Applied Science, London.

Swanson, J.E. and Kinsella, J.E. (1986) Dietary $n-3$ PUFA: modification of rat cardiac and fatty acid composition. *J. Nutr.*, **116**, 514.

Sych, J., Lacroix, C., Adambounou, L.T. and Castiaigne, F. (1990) Cryoprotective effects of lactitol, palatinit and polydextrose on cod surimi proteins during frozen storage. *J. Food Sci.*, **55**, 356.

Sylvia, G., Larkin, S.L. and Morrissey, M.T. (1996) Intrinsic quality and fisheries management: bioeconomic analysis of the Pacific whiting fishery. Presented at the 2nd Annual World Fisheries Congress, Brisbane, Australia, July 28–August 3, 1996.

Tannenbaum, S.R., Stillings, B.R. and Scrimshaw, N.S. (eds), (1974). *The Economics, Marketing and Technology of Fish Protein Concentrate*, MIT Press, Cambridge, MA.

Thomas, M.J., Throngburg, T., Manning, J. *et al.* (1994) Fatty acid composition of low-density lipoprotein influences its susceptibility to autooxidation. *Biochemistry*, **33**, 1828.

Venugopal, V. (1994) Production of fish protein hydrolysate by microorganisms, in *Fisheries Processing: Biotechnological Applications* (ed. A.M. Martin), Chapman & Hall, London, pp. 223–243.

Vollset, S.E., Heugh, I. and Bjelke, E. (1985) Letter to the Editor. *N. Engl. J. Med.*, **313**, 820.

Wander, R.C. and Patton, B.D. (1991) Lipids and fatty acids of three species of Northeast Pacific finfish harvested in summer. *J. Food Compos. Anal.*, **4**, 128.

Wander, R.C., Shi-Hua, D., Ketchum, S.O. and Rowe, K.E. (1995) Effects on the interaction of RRR-α-tocopherol acetate and fish oil on low-density-lipoprotein oxidation in post-menopausal women with and without hormone replacement therapy. *Am. J. Clin. Nutr.*, **63**, 184.

Wu, M.-C. (1992). Manufacture of surimi-based products, in *Surimi Technology* (eds T.C. Lanier and C.M. Lee), Marcel Dekker, New York, pp. 245–272.

17 Microbial stability and safety of healthy meat, poultry and fish products

L. LEISTNER

17.1 Introduction

Muscle food products based on meat, poultry or fish behave similarly with regard to their microbial stability and safety since most of them belong in the categories of high moisture foods ($a_w > 0.90$) and of low acid foods ($pH > 4.5$), thus they are easily perishable and in general need refrigeration during storage. Much is known about the microbial stability of and the proper storage conditions for conventional muscle food products; nevertheless, spoilage or even food poisoning happens from time to time. If for nutritional reasons the fat and/or the salt content of muscle food products are reduced, then most likely their microbial stability and safety will be diminished, whereas a reduced cholesterol content probably will not impair microbial stability. Since many different fat replacers and salt (sodium) substitutes or enhancers are under discussion or are already being applied, for which quantitative data about their influence on the microbial stability of muscle food products are lacking, the microbial consequences of their use are generally unknown. Therefore, the effect of these ingredients must be measured and evaluated by taking important stability factors of foods (hurdles), such as a_w, pH, and preservatives, into consideration in addition to the potential pathogenic and spoilage microorganisms inherent to muscle foods. The emerging quantitative data could become the basis for the microbial stability management of low fat and/or low salt muscle food products.

The toxicology, and thus the safety, of additives which are suggested to replace fat or salt (sodium) in foods have been thoroughly investigated in many cases, especially if such substances must be legally approved as additives. However, relatively little is known about the microbiological stability and safety of foods prepared with these additives. Therefore, this chapter will discuss microbial aspects of low fat and/or low salt foods based on meat, poultry and fish, which will be summarized as muscle food products.

17.2 Microbial risks of muscle food products

The microorganisms which cause food poisoning or spoilage of foods derived from meat, poultry or fish are similar because all of these foods are excellent nutrient sources for microorganisms and the preservation methods used for these muscle foods are also similar. Only a few of the microorganisms that occur are especially related to the origin of the muscles used, because they are indigenous to the different raw materials, i.e. meat, poultry or fish.

17.2.1 Pathogenic microorganisms

The most important pathogenic or toxigenic microorganisms belong to the genera *Salmonella*, *Campylobacter*, *Listeria*, *Escherichia*, *Staphylococcus*, *Vibrio*, *Clostridium* and *Bacillus* (Todd, 1989). These food-poisoning bacteria have been extensively studied and much is known about their properties and behavior (Gould and Russell, 1991; Shapton and Shapton, 1991). They may occur in all muscle foods; however, for red meats *Salmonella* spp. in general and *Clostridium botulinum* type A and B as well as *Listeria monocytogenes* and verotoxin-producing *Escherichia coli* (VTEC) attract particular attention, and *Salmonella enteritidis*, *Campylobacter jejuni* and *C. coli* are most frequently found in poultry, whereas *Vibrio parahaemolyticus*, *V. vulnificus*, *V. cholerae* and *Clostridium botulinum* type E are almost exclusively encountered in marine foods. In food-poisoning cases related to muscle food products, the bacteria are the main causes. However, molds, which produce mycotoxins, should also be considered because they might cause chronic diseases in humans. Toxinogenic molds are ubiquitous, and their occurrence depends more on the physical properties (e.g. a_w and/or pH) of the food than on the type of muscle used for the products.

17.2.2 Spoilage microorganisms

Whereas the number of types of microorganisms that can cause food poisoning is relatively small, the number of types that can cause spoilage is very large (Gould and Russell, 1991). Spoiled food is not necessarily injurious for the consumer, but it may be unappetizing and unpleasant. Spoilage of food is not always caused by microbial changes, but might be due to enzymatic changes or even caused by physical and/or chemical changes under the influence of oxygen and/or light. However, microbial food spoilage is by far the most important mechanism (Shapton and Shapton, 1991). Since spoilage of food often leads to severe economic losses and is detrimental to the reputation of the product and its manufacturer, it must be avoided, especially with respect to low fat and/or low salt muscle food products newly introduced into the market. The organisms causing spoilage in muscle foods are bacteria, molds and sometimes yeasts. If the spoilage proceeds in the direction of sourness then lactic acid bacteria of the genera *Lactobacillus*, *Enterococcus* and *Pediococcus* are usually involved, whereas putrefaction of foods at chill temperatures is caused mainly by the genus *Pseudomonas* and also by Enterobacteriacea and the genera *Bacillus* and *Clostridium* at ambient temperatures. Yeasts and molds are major spoilage organisms in foods with low a_w and/or pH values below the level at which bacteria can grow.

Since potential spoilage organisms are often eliminated or inhibited by similar preservation procedures as food-poisoning organisms, for both groups the same preservative factors (hurdles) are important, and therefore they will be discussed together.

17.3 Preservation of muscle food products

The microbial stability and safety of most foods are based on a combination of preservative factors (hurdles) which the microorganisms present cannot overcome. This is true for traditional foods with inherent hurdles and also for novel products for which hurdles have been intelligently selected and are intentionally applied (Leistner, 1995a).

17.3.1 Important preservative factors (hurdles)

The most important hurdles in foods are temperature (high or low), water activity (a_w), acidity (pH), redox potential (E_h), several preservatives and the competitive microorganisms (e.g. lactic acid bacteria). However, more than 50 potential hurdles which influence the preservation and/or quality of foods have been already identified and described (Bøgh-Sørensen, 1994), and the list of possible hurdles used in food preservation is by no means complete. At present, physical methods (hurdles) for non-thermal processes are of interest, since they may be used in combination with other hurdles for minimally processed, fresh-like food products with little induced degradation of their nutritional and sensory properties (Barbosa-Cánovas, Pothakamury and Swanson, 1995). Furthermore, effective preservatives derived from 'nature' (e.g. extracts from spices and herbs, lysozyme from egg white, pectin hydrolysates from grape skin, phytic acid from rice bran, protamines from fish roe, chitosan from crab or shrimp shells, paprika glycoprotein from paprika seeds and hop extract from hop) are especially appealing to the consumer of modern foods, and are currently being researched and promoted in Japan (M. Yajima, personal communication).

17.3.2 Application of combined processes (hurdle technology)

The hurdle concept was first proposed by Leistner (1978), and from an understanding of the so-called hurdle effect, hurdle technology has been devised by Leistner (1985). More recently, from the comprehension of hurdle technology a widened concept for food safety has emerged. The latter is focusing on homeostasis, metabolic exhaustion and stress reactions of microorganisms and the multi-target preservation of foods (Leistner, 1995b). The hurdle technology approach is applicable not only to safety, but also to quality aspects of foods. In order to secure an optimal total quality of a food, the preservation and the quality hurdles must be adjusted to an optimal range (Stanley, 1991; Leistner, 1994a). This approach should also be applicable for low fat and low salt muscle food products.

17.4 Microbial stability of low fat and/or low salt muscle food products

Fat and common salt have fundamental technological, microbiological and sensory functions to perform in meat products, and there are many measures that have to be taken to make up for any reduction in the content of these substances (Wirth, 1991). Much is known about fat replacement in European-type meat products. Although fat reduction mainly relates to texture, smoothness, succulence, colour and flavour of the food products, their microbial stability might be also impaired. As Wirth (1988) pointed out, this may be due to several reasons: (1) low fat meat products often have a higher moisture content caused by the replacement of fat by collagen, blood plasma, lean meat, or water; (2) since low fat products in general taste more salty, the addition of salt has to be reduced by 20–25% of the usual salt level; (3) in Germany and many other European countries, the nitrite is added mixed with salt (due to the use of nitrite curing salt), and therefore with the reduction of the salt content the nitrite addition also is reduced; and (4) vacuum-packaged sliced emulsion-type meats, that are low in salt, may have a short shelf-life, especially if the water binding capacity of the sausages is impaired by the lowered salt addition, which might result in more liquid accumulation in the packages. Jones (1995) pointed out that the removal of fat from a food formulation usually results in an increased moisture content, and thus in decreased microbial safety and stability. Therefore, low fat products might require addition of a preservative, e.g. potassium sorbate, which usually is not needed for the unmodified foods.

By using appropriate technologies, minimum fat contents in the finished products of 10, 15 and 20–30% are feasible in emulsion-type sausage, liver sausage and fermented sausages, respectively. Hence the possible fat reduction, compared with the usual products, is substantial (Wirth, 1988).

About 20% of the daily sodium chloride ingestion of the consumers in Europe may come from meat products. According to Wirth (1989), common salt performs three main functions in meat products: (1) flavour development, (2) water and fat binding and (3) microbial stability. However, these desirable effects require very different concentrations of salt in the various types of meat products. The high tolerance to the taste of common salt often noted in the adult is acquired, but consumers can become accustomed to lower salt levels (Chapter 5). Therefore, the development of salt reduced meat products should be primarily based on the technological and microbiological requirements. From the technological point of view, in liver and blood sausage (Kochwurst) the sodium level of 600–800 mg normally added to Kochwurst, Brühwurst and cooked ham per 100 g of product (which corresponds to about a 2% salt content) could be reduced substantially to 100–120 mg in Kochwurst, because here salt is only essential for the flavour. With emulsion-type sausage (Brühwurst), the limiting factor is water and fat binding, and here the sodium content may be reduced to 500 mg per 100 g. Similar values are satisfactory for cooked ham. The decreased microbial stability of the low salt products mentioned above must be compensated for by strict refrigeration requirements.

However, most critical from the microbiological point of view is the salt reduction in raw meat products, which are generally stored without refrigeration. If for fermented sausage a special ripening regime (temperatures of less than 15°C and a rapid decrease in pH) is applied, the sodium values of otherwise 1400–1800 mg per 100 g may be slightly lowered to 1000–1200 mg by addition to the sausage batter. In raw hams with a necessary salt content of 5–6% in the final product, any worthwhile reduction of the usually high sodium content of 2000–3000 mg per 100 g would involve too great a microbial risk (Wirth, 1989). Bone-in ham, for instance, must be cured for 3 months at low temperatures (<5 °C), in order to prevent growth of relevant food-poisoning bacteria (e.g. non-proteolytic strains of *Clostridium botulinum* type B) and important spoilage bacteria (e.g. psychrotolerant Enterobacteriaceae). However, after the water activity in all parts of the raw ham has been reduced by applying salt to <0.96 (which corresponds to a sodium chloride content of 4.5%), the products are safe and stable. The hams are now ready to be further ripened or smoked at ambient temperature and can be held for weeks or months without bacteriological risk. If the ripening is skilfully controlled, then the sodium chloride content in the final products will not be higher than 5–6% (Leistner, 1990). The longer the products are ripened the less salty they taste because the sodium chloride present in long ripened bone-in hams apparently binds more tightly to the proteins, and then is less mobilized during mastication of the product.

Several substances with low sodium contributions have been investigated as substitutes for common salt in muscle foods, with the aim of reducing the sodium content of these products (Chapter 12). The various salts evaluated include lithium chloride, potassium chloride, calcium chloride, magnesium chloride, potassium sulfate, magnesium sulfate and potassium glutamate. However, most have severe drawbacks since they either have an unpleasant taste, or are less 'salty' than sodium chloride, or from a toxicological point of view are not acceptable. Potassium chloride is the only substitute for sodium chloride in widespread use, but the amount of NaCl that can be replaced by KCl is limited by the sensory properties (bitterness) of potassium chloride (Maurer, 1983; Terrell, 1983). Therefore, 50:50 blends of NaCl and KCl ('half-salt') are often used for lowering the sodium content of foods, including cured meats. However, at this level KCl might still impair the flavour, and NaCl replacement by KCl up to 40% seems to be preferable (Terrell, 1983; Askar *et al.*, 1994). It has been reported that in Pasterma (dried raw beef) and in beef bouillon (made into bouillon cubes) the sodium chloride content without sensory impairment could be substituted by potassium chloride or potassium lactate, if not more than 40% of the sodium chloride is replaced. The sensory properties of the samples substituted with potassium chloride were superior to those of the samples containing 40% potassium lactate. The effects of potassium chloride and potassium lactate on the pH and the microbiological status in the partially substituted products were similar to the normal products made only with sodium chloride (Askar *et al.*, 1994). Similar bacteriological findings were reported by Terrell *et al.* (1983).

The growth of microorganisms depends on the water activity (i.e. the ionic strength) of the foods. Sodium chloride is of primary importance for lowering the a_w of many muscle foods. If NaCl is replaced by other chloride salts (K, Ca, Mg) or polyphosphates, then their contribution (particularly of KCl) to the ionic strength is similar to that of NaCl (Leistner, Rödel and Krispien, 1981), and several commercial blends of potassium salts and polyphosphates are available as NaCl substitutes (Terrell, 1983; Chapter 12). On the other hand, sugars are much less effective with respect to the ionic strength, because about three times the quantity of sugars (sucrose, glucose, lactose) compared with NaCl is needed to achieve the same a_w depression in meat. LiCl depresses the a_w of meat even more strongly than NaCl (Leistner, Rödel and Krispien, 1981), and it comes closest to producing the same flavour as NaCl with little off-taste (Terrell, 1983). Unfortunately, LiCl is not suitable as a substitute for NaCl because of toxicological concerns.

The substitution for the flavour effects of common salt is another challenge and may or may not contribute to the ionic strength impact of a sodium chloride reduction (Chapter 12). Food acids (e.g. lactic, acetic and citric acid) and their potassium salts can add to the 'saltiness' of meats, but they contribute less to the a_w depression in meat than sodium chloride. Furthermore, it has been demonstrated by Hofmann and Marggrander (1989) that a particular collagen hydrolysate enhances saltiness and aroma in some meat products and ready-to-eat meals with meat components, which were produced with a reduced sodium chloride content. However, the collagen hydrolysate will not have much effect in increasing the ionic strength of meats, and thus cannot replace NaCl in this respect.

Replacement of fat and also reduction or substitution of salt (sodium) might influence important preservative factors (hurdles) in foods, and therefore the microbial stability and safety of the products. The water activity (a_w) hurdle is probably most affected, but the acidity (pH) as well as the effects of temperature and preservatives will also be influenced. Furthermore, the redox potential (E_h) and the microstructure might be altered in low fat foods. All these aspects need further studies, and will be only briefly discussed here.

17.4.1 Water activity (a_w)

Reduction of salt or replacement of fat in high moisture muscle foods can often result in an increase of water activity and thus diminish the microbial stability and safety of the products. The effects of salt and sugar on the water activity are obvious, since these solutes reduce a_w by increasing the ionic strength in the fluid phase of the foods. In low salt emulsion-type sausage (Brühwurst), compared with the normal products, the a_w only increases slightly (i.e. from 0.980 to 0.985), but in low salt liver and blood sausage (Kochwurst) the increase in a_w is much greater (from 0.970 to above 0.985). The latter is due to the above-explained larger reduction of sodium chloride possible in Kochwurst. High yield cooked hams (shoulders and loins) with a technologically feasible reduction of the salt content have a very high water activity close to 0.990, and thus need strict refrigeration (<5 °C). Fermented sausages should soon achieve an $a_w < 0.95$ during ripening in order to inhibit growth of *Salmonella* spp. However, if the salt addition is reduced from the normal 2.8 to 2.4% or even lower, then the products remain above this critical a_w level much longer, and thus are at risk in relation to *Salmonella* spp. and also from spoilage bacteria (Wirth, 1989). As mentioned above, raw hams should have a salt content of 5–6% in the finished products, and in general should not be sold with an $a_w > 0.90$.

Fat influences the water activity of foods also, but only indirectly. If a food is composed of muscle (contains about 75% moisture) and fat (may contain only 5–15% moisture), then foods high in muscle and low in fat will have a higher water content (and a_w) than foods with a low muscle and high fat content. Also other dry ingredients (such as soy protein or milk powder) introduce little water into the food system and thus reduce the water activity. However, if fat replacers (e.g. starches, gums, pectins, alginates, inulin, hemicellulose, plant or animal proteins, surimi made from fish, red meat or poultry muscle) are added to bind water in a hydrated food by increasing the water holding capacity (WHC), they will increase the water content of the product, and thus, the water activity. On the other hand, certain fat replacers, such as polydextrose, may act as humectants and reduce the a_w of a food.

Therefore, the effects of the many known fat replacers (so far more than 100) on the water activity of foods are complex. It is safe to assume, however, that in most cases the reduction of the salt and the fat content of a food will increase the water activity of the products and thus decrease the microbial stability and safety of the modified muscle foods.

17.4.2 Acidity (pH)

Some low salt meat products have a slightly higher pH than the normal products, and this is true for emulsion-type sausage (Brühwurst) and liver and blood sausage (Wirth, 1989). Therefore, these salt reduced meats are also less stable with regard to their pH. This must be of particular concern with liver and blood sausages, which already have, even in the normal products, relatively high pH values of about 6.3 and 7.2, respectively, whereas normal Brühwurst products have a pH of about 5.9–6.0.

If fat is replaced by proteins then the pH and/or the buffering capacity of heated muscle foods might increase, and therefore the microbial stability and safety of these products would decrease. The antimicrobial effect of weak organic acids is also greatly influenced by the pH of the food, because their action results from the undissociated molecules rather than the anion, and they are more undissociated at lower pH. Hence less preservation due to organic acids and their salts (e.g. acetate, sorbate) is to be expected at higher pH values of foods.

On the other hand, in foods which undergo a desirable fermentation by microorganisms (e.g. salami), a high buffering capacity of the food might delay the fermentation process. More important, in such foods the replacement of fat by fermentable carbohydrates (e.g. starches) will lead to a very strong and thus undesirable acidification of the product. Therefore, the pH and changes in acidity during processing and storage of low fat muscle foods need particular attention.

17.4.3 Temperature (heating and chilling)

Diminished microbial stability and safety in low fat or low salt muscle foods require more careful chilling during storage of the products. In the case of temperature abuse, e.g. during retail display or in the home of the consumer, however, such foods might become hazardous. Therefore, in addition to the temperature hurdle some additional hurdles should be incorporated into chilled products, and this precaution is sometimes called 'invisible technology' (Leistner, 1995a).

As far as the heat treatment is concerned, a higher water activity in low fat or low salt muscle foods might aid in the inactivation of microorganisms by heat (Leistner, 1997). This may cause an increased stability, but in general this effect should be only minor.

17.4.4 Preservatives (additives)

Healthy muscle foods should contain as little additives as possible, since a 'clean label' is expected for such products. However, some additives (e.g. lactate, acetate or sorbate) might become essential for preservation if the water activity of the products increases because of the substitution of other ingredients for fat or salt. If low fat products contain more water owing to the increased water holding capacity caused by fat replacers, then water soluble preservatives (e.g. nitrite) might become diluted, and thus become less effective. In addition, nitrite is less effective at higher pH levels, and thus the nitrite hurdle in modified muscle foods will often be reduced. This aspect should be taken into consideration. In general, the addition of preservatives should be calculated by taking the water phase of the food into account.

17.4.5 Redox potential (E_h)

The influence of fat replacers and salt (sodium) substitutes on the redox potential of modified muscle foods has not been investigated. Proteinaceous fat replacers might increase the E_h buffering capacity of a food, whereas foods with a high water content should have a low E_h capacity. The exclusion of oxygen (by application of vacuum packaging or modified atmospheres) should have a synergistic effect on other hurdles in the food, especially with respect to a_w and pH (Gould, 1995). Therefore, anaerobic conditions could increase the microbial stability and safety of modified muscle products.

17.4.6 Microstructure (submerged colonies)

The microstructure of some foods is significant from the standpoint of their microbial stability and safety (Wimpenny *et al.*, 1995). This is already well known for emulsions (e.g. mayonnaise) and fermented foods (e.g. salami). The microstructure of foods could change owing to the addition of some fat replacers; however, this effect has not yet been investigated, but should be kept in mind.

17.5 Stability management of low fat and/or low salt muscle food products

It was pointed out above that in many instances low fat and/or low salt products based on meat, poultry or fish will have a diminished microbial stability and safety, which might be compensated for by better refrigeration. However, it seems risky to rely solely on refrigeration because temperature abuse could happen, and would be a constant threat. Therefore, the reduced microbial stability of such foods should be better defined and possibly compensated for by the application of hurdle technology.

17.5.1 Measurement and evaluation of hurdles

Since knowledge about which respect fat replacers and salt (sodium) substitutes will influence the microbial stability and safety of muscle foods is still scarce and the effects are uncertain, the important hurdles in modified products have to be measured and evaluated. The water activity (a_w) and the acidity (pH) in the final products before shipment and during storage especially are significant. The instruments employed for measuring the a_w or the pH must give reliable data. With respect to the a_w measurement it has to be remembered that some volatile food components (e.g. glycol, ethanol) might lead to erroneous data. If the water activity is measured by determination of the freezing point of the sample with the so-called a_w kryometer developed by Rödel, Scheuer and Wagner (1990), then volatile components are not a problem. If other a_w instruments are used, employment of suitable filters which protect the sensors from volatile substances might be necessary.

The data obtained must be evaluated with respect to the microbial stability and safety of the modified product. This should be done in close cooperation between technologists and microbiologists. The technologist must determine which processes or additives are suitable for the enhancement of hurdles in a food, by taking the legal, sensory, nutritional, and technological limitations into account. The microbiologist must determine which types and intensity of hurdles in a particular food are needed for the desired safety and stability of the product. Since the stability hurdles in foods are to a certain degree interchangeable, the deficient hurdles (e.g. a_w) in a modified muscle food could be compensated for by other hurdles (e.g. pH, E_h, temperature). However, which hurdles are applicable always depends on the total quality of the product.

17.5.2 Product design and process control

Hurdle technology as a concept has proven to be successful in the optimization of traditional foods and also in the development of novel products.

However, it should be combined, if possible, with HACCP (hazard analysis critical control point) and, if feasible, with predictive microbiology (Leistner, 1994b). The hurdle technology is primarily used for food design, the HACCP concept for process control and predictive microbiology for process refinement. Considering these three concepts, an overall strategy for securing stable, safe and tasty foods could be accomplished.

With this aim in mind, a 'user guide' for product development has been outlined by Leistner (1994c) which includes all three concepts. In this guide 10 steps have been suggested which proved to be appropriate in solving real product development tasks in the food industry. Therefore, this user guide could be applicable for the optimization of microbial stability and sensory quality of low fat and/or low salt foods. The backbone of the guide is hurdle technology, but in two steps of the guide predictive microbiology (McClure *et al.*, 1994) has been incorporated as an important tool for food design, because it can narrow considerably the range over which challenge tests with relevant microorganisms needs to be performed. Although predictive microbiology will not render challenge testing obsolete, it may greatly reduce both time and costs spent in product development. As the last step in the suggested guide, after the critical control points and their monitoring have been established, it is recommended that the manufacturing process of the designed low fat and/or low salt products should be controlled by HACCP.

17.6 Research needs

In the design and production of low fat and/or low salt muscle food products, the nutritional aspects have been emphasized much more than the microbial aspects. Since such foods often have a diminished microbial stability and safety, however, the latter should not be neglected, because these foods will only be continuously accepted by the consumer if they cause no food poisoning and do not spoil easily.

Therefore, microbiologists together with nutritionists, technologists and marketing experts should take part in the design, processing and marketing of low fat and/or low salt muscle foods. The microbiologists should investigate and define quantitatively the effects the many potential ingredients for low fat and/or low salt (sodium) foods will have on the major stability hurdles, especially water activity, acidity and preservatives of these products. Since different fat replacers are often used in purposely chosen combinations of ingredients which result in more optimal fat mimicking properties, the effects of such combinations of fat replacers, especially on the a_w and pH of modified foods, must also be studied. Furthermore, safe temperature regimes for the processing and storage of low fat and/or low salt foods must be defined. Since novel ingredients for modified muscle foods might also influence the redox potential or the microstructure of such products, even if little is known about the consequences, research needs to exist in these respects too.

Based on the data obtained, safe processing procedures for low fat and/or low salt muscle food products should be defined by establishing appropriate HACCP plans. Furthermore, guidelines should be proposed for the proper handling and storage of these foods during distribution and use.

17.7 Summary

Healthy foods, containing less fat and/or salt, in general will have diminished microbial stability and safety, i.e. they are more prone to spoil or even to cause food poisoning than the usual products. Therefore, the microbial aspects of healthy foods are certainly of relevance and need attention.

In this chapter, the microbial risks of modified muscle foods related to pathogenic and spoilage microorganisms have been discussed, and the preservative factors (hurdles) significant for the inhibition or elimination of these organisms have been outlined. Of major importance for the microbial stability of low fat and/or low salt muscle foods is apparently the increased water activity (a_w) of the modified products. However, the acidity (pH) might also be unfavourable, as added preservatives might be diluted and thus rendered less effective. Furthermore, the redox potential (E_h) and the microstructure might change in modified muscle foods, even though these aspects have not yet been investigated. To keep low fat and/or low salt muscle foods microbiologically stable and safe, the refrigeration during storage of these products must be perfect. Since this cannot be always guaranteed, the weak hurdles in modified muscle foods should be compensated for by other preservative factors and by applying the concept of hurdle technology.

Therefore, microbiologists should take an active part in the design, processing and marketing of low fat and/or low salt muscle foods. Quantitative data on the influences of the many different replacers for fat and the substitutes for salt (sodium) on the important preservative factors (hurdles) of modified muscle foods are needed. The design of these foods should be based on the application of intentional hurdle technology, including predictive microbiology, and the process control should be based on HACCP.

References

Askar, A., El-Samahy, S.K., Shehata, H.A. and Tawfik, M. (1994) Pasterma and beef bouillon. The effect of substituting KCl and K-lactate for sodium chloride. *Fleischwirtsch. Int.*, (1), 50.

Barbosa-Cánovas, G.V., Pothakamury, U.R. and Swanson, B.G. (1995) State of the art technologies for the stabilization of foods by non-thermal processes: physical methods, in *Food Preservation by Moisture Control, Fundamentals and Applications* (eds G.V. Barbosa-Cánovas and J. Welti-Chanes), Technomic Publishing, Lancaster, PA, pp. 493–532.

Bøgh-Sørensen, L. (1994) Description of hurdles, in *Food Preservation by Combined Processes* (eds L. Leistner and L.G.M. Gorris), FLAIR Final Report, EUR 15776 EN, European Commission, Directorate-General XII, Brussels, pp. 7–24.

Gould, G.W. (1995) Homeostatic mechanisms during preservation by combined methods, in *Food Preservation by Moisture Control Fundamentals and Applications* (eds G.V. Barbosa-Cánovas and J. Welt-Chanes), Technomic Publishing, Lancaster, PA, pp. 397–410.

Gould, G.W. and Russell, N.J. (1991) Major food-poisoning and food-spoilage microorganisms, in *Food Preservatives* (eds N.J. Russell and G.W. Gould), Blackie, Glasgow, pp. 1–12.

Hofmann, K. and Marggrander, K. (1989) Reducing the common salt content of meat products by using collagen hydrolysates. *Fleischwirtsch. Int.*, (3), 34.

Jones, S.A. (1995) Fat replacers, the broad perspective. Part 2. *World of Ingredients*, Sept.–Oct., 12.

Leistner, L. (1978) Hurdle effect and energy saving, in *Food Quality and Nutrition*. (ed. W.K. Downey), Appl. Science, London, pp. 553–557.

Leistner, L. (1985). Hurdle technology applied to meat products of the shelf stable product and intermediate moisture food types, in *Properties of Water in Foods in Relation to Quality and Stability* (eds D. Simatos and J.L. Multon), Martinus Nijhoff, Dordrecht, pp. 309–329.

Leistner, L. (1990). Fermented and intermediate moisture products, in *Proceedings of 36th ICoMST, August 27–September 1, 1990, Havana, Cuba*, Vol. III, pp. 842–855.

Leistner, L. (1994a). Further developments in the utilization of hurdle technology for food preservation. *J. Food Eng.*, **22**, 421.

Leistner, L. (1994b). *Food Design by Hurdle Technology and HACCP*. Adalbert Raps Foundation, Kulmbach.

Leistner, L. (1994c). User guide to food design, in *Food Preservation by Combined Processes* (eds L. Leistner and L.G.M. Gorris), FLAIR Final Report, EUR 15776 EN, European Commission, Directorate-General XII, Brussels, pp. 25–28.

Leistner, L. (1995a). Principles and applications of hurdle technology, in *New Methods of Food Preservation* (ed. G.W. Gould), Blackie, Glasgow, pp. 1–21.

Leistner, L. (1995b). Emerging concepts for food safety, in *Proceedings of 41st ICoMST, August 20–25, 1995, San Antonio, TX, USA*, Vol. II, pp. 321–322.

Leistner, L. (1997) Water activity and hurdle technology, in *Production of Processed Meats and Convenience Foods* (ed. C. Genigeorgis), European Consortium for Continuing Education in Advanced Meat Science and Technology, in press.

Leistner, L., Rödel, W. and Krispien, K. (1981). Microbiology of meat and meat products in high- and intermediate-moisture ranges, in *Water Activity: Influences on Food Quality* (eds L.B. Rockland and G.F. Stewart), Academic Press, New York, pp. 855–916.

Maurer, A.J. (1983) Reducing sodium usage in poultry muscle foods. *Food Technol.*, **37**(7), 60.

McClure, P.J., Blackburn, C. de W., Cole, M.B. *et al.* (1994) Modelling the growth, survival and death of microorganisms in foods: the UK Food Micromodel approach. *Int. J. Food Microbiol.*, **23**, 265.

Rödel, W., Scheuer, R. and Wagner, H. (1990) A new method of determining water activity in meat products. *Fleischwirtsch. Intn.*, (4), 22.

Shapton, D.A. and Shapton, N.F. (1991) *Principles and Practices for the Safe Processing of Foods*. Butterworth–Heinemann, Oxford, pp. 283–301.

Stanley, D.W. (1991) Biological membrane deterioration and associated quality losses in food tissues. *Crit. Rev. Food Sci. Nutr.*, **30**, 487.

Terrell, R.N. (1983) Reducing the sodium content of processed meats. *Food Technol.*, **37**(7), 66.

Terrell, R.N., Quintanilla, M., Van der Zant, C. and Gardner, F.A. (1983) Effects of reduction or replacement of sodium chloride on growth of *Micrococcus*, *Moraxella* and *Lactobacillus* inoculated ground pork. *J. Food Sci.*, **48**, 122.

Todd, E.C.D. (1989) Preliminary estimates of costs of foodborne disease in the United States. *J. Food Protect.*, **52**, 595.

Wimpenny, J.W.T., Leistner, L., Thomas, L.V. *et al.* (1995) Submerged bacterial colonies within food and model systems: their growth, distribution and interactions. *Int. J. Food Microbiol.*, **28**, 299.
Wirth, F. (1988) Technologies for making fat-reduced meat products. *Fleischwirtschaft*, **68**, 1153.
Wirth, F. (1989) Reducing the common salt content of meat products, possible methods and their limitations. *Fleischwirtsch. Int.*, (3), 25.
Wirth, F. (1991) Reducing the fat and sodium content of meat products, what possibilities are there? *Fleischwirtsch. Int.*, (2) 3.

Index

9 780751 403909

GPSR Compliance

The European Union's (EU) General Product Safety Regulation (GPSR) is a set of rules that requires consumer products to be safe and our obligations to ensure this.

If you have any concerns about our products, you can contact us on ProductSafety@springernature.com

In case Publisher is established outside the EU, the EU authorized representative is:

Springer Nature Customer Service Center GmbH
Europaplatz 3
69115 Heidelberg, Germany

Batch number: 10755284

Printed by Printforce, the Netherlands